Antarctica

Antarctica

An Encyclopedia from Abbott Ice Shelf to Zooplankton

Edited by Mary Trewby

FIREFLY BOOKS

A FIREFLY BOOK

Published by Firefly Books Ltd. 2002

First published in 2002 by David Bateman Ltd., 30 Tarndale Grove, Albany, Auckland, New Zealand in association with Natural History New Zealand Ltd

First Printing

National Library of Canada Cataloguing in Publication Data

Antarctica : an encyclopedia from Abbott Ice Shelf to zooplankton /
edited by Mary Trewby
Includes bibliographical references and index.
ISBN 1-55297-590-8
1. Antarctica—Encyclopedias. I. Title.
G855.C43 2002   998'.9'003   C2002-900768-2

Publisher Cataloging-in-Publication Data (U.S.)

Antarctica : an encyclopedia from Abbott Ice Shelf to zooplankton /
Mary Trewby. edited by — 1st ed.
[208] p. : col. ill. : photos. ; cm.
Includes bibliographical references and index.
Summary: An alphabetical encyclopedia of Antarctica, including
entries on history, weather, geology, wildlife, scientific research and tourism.
ISBN 1-55297-590-8
1. Antarctica — Encyclopedias. I. Title.
919.8/ 9 21   CIP   G855.C54 2002

Published in Canada in 2002 by
Firefly Books Ltd.
3680 Victoria Park Avenue
Toronto, Ontario
M2H 3K1

Published in the United States in 2002 by
Firefly Books (U.S.) Inc.
P.O. Box 1338, Ellicott Station
Buffalo, New York
14205

Acknowledgements
Natural History New Zealand extends special thanks to:
Antarctica New Zealand for generous access to their photo archives and assistance with research;
Gateway Antarctica, University of Canterbury, for invaluable research support.

Project and editorial manager: Rebecca Tansley MA (Hons), DipBrC
Principal researcher: Kate Sinclair MSc
Researchers: Dillon Shiel Burke MA (Hons); Kate Guthrie BSc (Hons);
Andrew Joel BA (Hons); Fiona Stephenson MA, MPhil; Aileen Wallace BSc; Jim Wilson

Photographic research: Darryl Sycamore
Design: Grace Design
Maps: Standout Maps and Graphics
Printed in China through Colorcraft Ltd., Hong Kong

Page 1: An icebreaker forms a channel through sea ice in McMurdo Sound.

Pages 2–3: At Cape Bird on Ross Island an Adélie penguin (*Pygoscelis adeliae*) makes its way over pack ice that has been driven into shore and piled up by the prevailing wind.

Opposite: A crack in the sea ice in Erebus Bay, Ross Island, in early summer, before the fast ice has broken up. Such gaps prove convenient exit and entry points for Weddell seals (*Leptonychotes weddelli*).

Cover credits: Front cover (main photo) Mike Single/NHNZ; (clockwise from top left) Jeanie Ackley/NHNZ; Alexander Turnbull Library, National Library of New Zealand; Max Quinn/NHNZ; Mike Single/NHNZ; Mike Single/NHNZ; Jeanie Ackley/NHNZ; Jeanie Ackley/NHNZ. Back cover (clockwise from top left) Alexander Turnbull Library, National Library of New Zealand; Mike Single/NHNZ; George Chance; Mike Single/NHNZ.

Contents

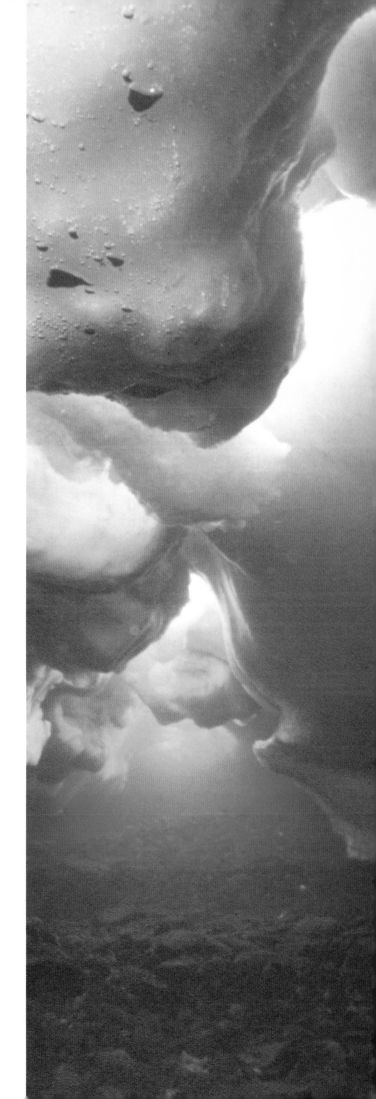

Consulting Editors

Environment and Conservation
Gordon Brailsford BSc
Atmospheric Chemist
National Institute of Water and Atmospheric Research

Cetaceans
Peter W. Carey BSc, PhD
SubAntarctic Foundation for Ecosystems Research

Climate and Atmosphere
Blair Fitzharris BSc (Hons), MA, PhD
Professor of Geography
University of Otago

Birds
Peter Harper BSc (Hons), PhD
Course Coordinator, Antarctic Studies
University of Canterbury

History
David L. Harrowfield BSc
Director, South Latitude Research Ltd
Honorary Fellow in Geography
University of Canterbury

Oceanic Processes
Tim Haskell BSc (Hons), PhD
Sea Ice Researcher
Industrial Research Limited

Optical Phenomena
Gordon Keys MSc
Research Physicist
National Institute of Water and Atmospheric Research

Glaciological Processes
Harry Keys BSc, MSc, PhD
Conservancy Advisory Scientist
New Zealand Department of Conservation

Marine Ecology and Fishes
John A. Macdonald AB, PhD
Senior Lecturer
School of Biological Sciences
University of Auckland

Optical Phenomena, Astronomical Research
Richard L. McKenzie BSc (Hons), MSc, DPhil
Principal Scientist (Radiation); Project Leader, UltraViolet
Radiation Studies
National Institute of Water and Atmospheric Research

History
Baden Nolan Norris QSO
Curator of Antarctic History
Canterbury Museum
Christchurch

Politics
Michelle Rogan-Finnemore BSc (Hons), LLB
Project Manager
Gateway Antarctica
University of Canterbury

Geology
Bryan Storey BA, PhD
Professor of Antarctic Studies
Director, Gateway Antarctica
University of Canterbury

Seals, Cetaceans
Joseph R. Waas BSc (Hons), PhD
Senior Lecturer
Department of Biological Sciences
University of Waikato

Terrestrial Fauna Ecology
David Alan Wharton BSc, PhD, DSc
Senior Lecturer in Zoology
University of Otago

Ozone and Atmosphere
Stephen Wood BSc (Hons), PhD
Research Scientist
National Institute of Water and Atmospheric Research

Transport
Paul Woodgate
Movements Controller
Antarctica New Zealand

Survival
Gillian Wratt BSc (Hons), MBA
Chief Executive
Antarctica New Zealand

The project team also extends special thanks to:

Jeanie Ackley, MSC, Assistant Producer/Researcher, NHNZ, for additional research and fact checking;
Max Quinn, Senior Producer, NHNZ, for research assistance;
Marcus Turner, Research and Information Co-ordinator, NHNZ, for research assistance;
Timothy Corballis, Alexander Turnbull Library, National Library of New Zealand/Te Puna Mātauranga o Aotearoa; for assistance sourcing historic photographs;
David Harrowfield for assistance with historic research;
Peter Wilson PhD, for preliminary scientific assistance on birdlife.

Foreword

Antarctica is a region of great fascination for millions of people around the world; for its ice, its wild seas, its geographic and historic significance, for its mystery as an unknown place to so many people, and as an environment in which there is so much potential for research. The Antarctic continent, the surrounding ocean and subantarctic islands are the last great global wilderness, and their preservation as a place for peace and science is central to the nations of the Antarctic Treaty.

New Zealand, as a Treaty partner, takes its Antarctic responsibilities very seriously and charges the New Zealand Antarctic Institute to carry out its mandate.

The Institute has a vision of "Antarctica: refreshing global eco-systems and the human spirit." This vision is the basis for a commitment to research, environmental stewardship and encouraging public awareness and education. It is in this latter role that the Institute welcomes this anthology of Antarctica—a comprehensive and detailed account of the A to Z of Antarctica and the Southern Ocean.

We have supported Natural History New Zealand for many years in their endeavours to document the wildlife, the beauty and the activities that occur in Antarctica and the Southern Ocean. This partnership has enabled people around the world to get a closer glimpse of the frozen continent via screen technology.

With the publication of this book, Natural History New Zealand has produced a new perspective on the Antarctic, with contributions from noted authors and a selection of superb images, many of which have come from the New Zealand Antarctic Institute's library.

We are proud to endorse this publication and warmly congratulate Natural History New Zealand on its latest initiative.

Gillian Wratt
Chief Executive
New Zealand Antarctic Institute

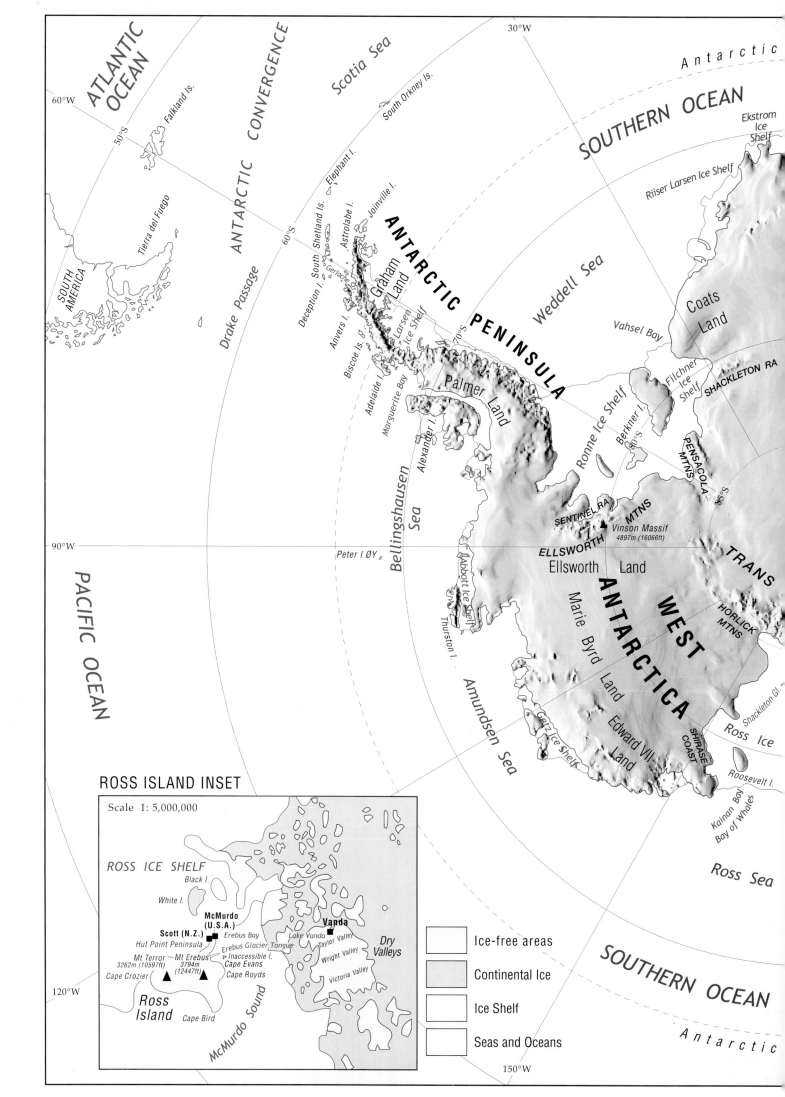

ATLANTIC OCEAN

ANTARCTIC CONVERGENCE

Scotia Sea

SOUTHERN OCEAN

Ekstrom Ice Shelf

60°W

50°S

Falkland Is.

South Orkney Is.

Riiser Larsen Ice Shelf

30°W

Antarctic

Tierra del Fuego

Elephant I.

Joinville I.

Astrolabe I.

ANTARCTIC

60°S

Deception I. South Shetland Is.

Gerlache

Graham Land

Weddell Sea

Coats Land

SOUTH AMERICA

Drake Passage

Anvers I.

Larsen Ice Shelf

70°S

PENINSULA

Vahsel Bay

Filchner Ice Shelf

SHACKLETON RA

Biscoe Is.

Adelaide I.

Marguerite Bay

Palmer Land

Ronne Ice Shelf

Berkner I.

80°S

PENSACOLA MTNS

PACIFIC OCEAN

Bellingshausen Sea

Alexander I.

SENTINEL RA

MTNS

Vinson Massif
4897m (16066ft)

TRANS

90°W

Peter I ØY

Abbott Ice Shelf

ELLSWORTH

Ellsworth Land

WEST

HORLICK MTNS

Thurston I.

Marie Byrd Land

ANTARCTICA

Amundsen Sea

Getz Ice Shelf

Edward VII Land

SHIRASE COAST

Ross Ice

Shackleton Gl.

Roosevelt I.

Kainan Bay
Bay of Whales

Ross Sea

120°W

SOUTHERN OCEAN

Antarctic

150°W

ROSS ISLAND INSET

Scale 1: 5,000,000

ROSS ICE SHELF

Black I.

White I.

McMurdo (U.S.A.)

Scott (N.Z.)

Erebus Bay

Hut Point Peninsula

Erebus Glacier Tongue

Inaccessible I.

Mt Terror
3262m (10597ft)

Mt Erebus
3794m (12447ft)

Cape Evans

Cape Royds

Cape Crozier

Ross Island

Cape Bird

McMurdo Sound

Vanda

Lake Vanda

Taylor Valley

Wright Valley

Dry Valleys

Victoria Valley

	Ice-free areas
	Continental Ice
	Ice Shelf
	Seas and Oceans

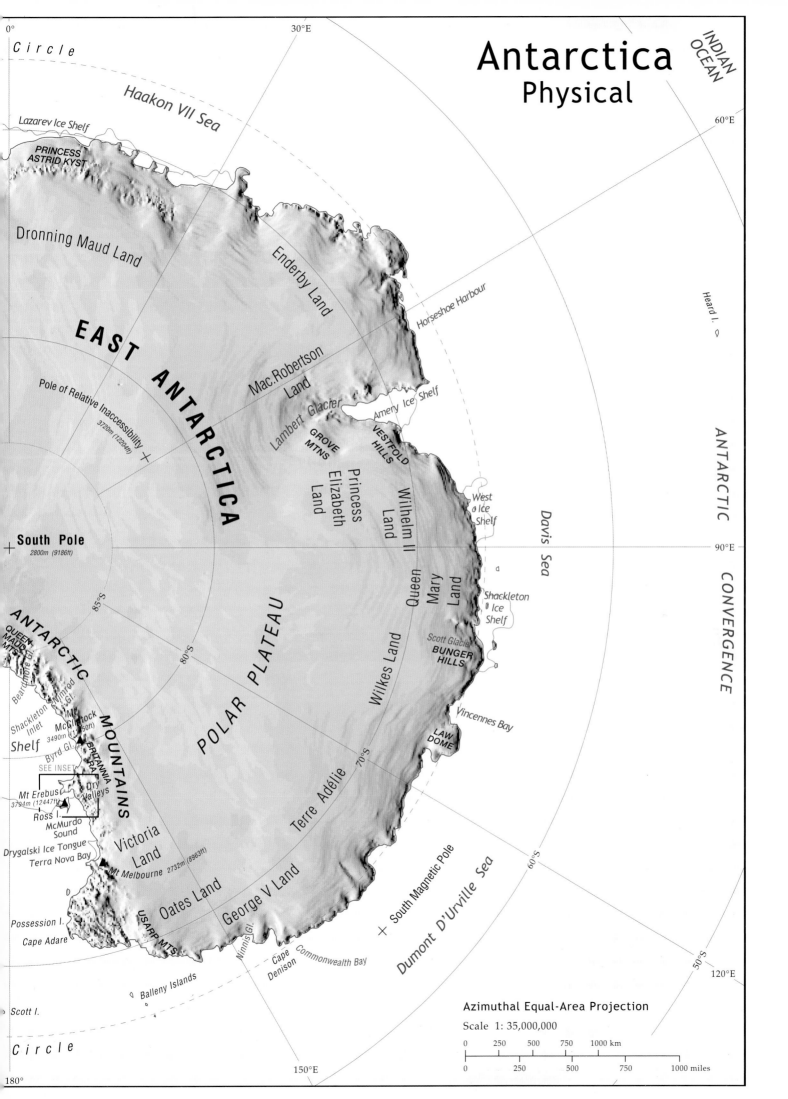

Antarctica
Physical

0°

Circle

30°E

INDIAN OCEAN

60°E

Haakon VII Sea

Lazarev Ice Shelf

**PRINCESS
ASTRID KYST**

Dronning Maud Land

Enderby Land

Horseshoe Harbour

Heard I.

EAST

Mac.Robertson Land

Pole of Relative Inaccessibility

3720m (12204ft) +

Lambert Glacier

GROVE MTNS

Amery Ice Shelf

VESTFOLD HILLS

ANTARCTICA

Princess Elizabeth Land

Wilhelm II Land

West Ice Shelf

ANTARCTIC

+ South Pole

2800m (9186ft)

Davis Sea

90°E

85°S

Queen Mary Land

Shackleton Ice Shelf

CONVERGENCE

80°S

POLAR PLATEAU

Scott Glacier

BUNGER HILLS

ANTARCTIC

QUEEN MAUD MTS.

Beardmore Gl.

Shackleton Inlet

Nimrod Gl.

Mt McClintock

3490m (11450ft)

Byrd Gl.

Shelf

SEE INSET

BRITANNIA RA.

Mt Erebus

3794m (12447ft)

Ross I.

McMurdo Sound

Dry Valleys

Drygalski Ice Tongue

Terra Nova Bay

Victoria Land

Mt Melbourne 2732m (8963ft)

Possession I.

Cape Adare

USARP MTS.

MOUNTAINS

Oates Land

George V Land

Wilkes Land

70°S

Terre Adélie

LAW DOME

Vincennes Bay

+ South Magnetic Pole

60°S

Ninnis Gl.

Cape Denison

Commonwealth Bay

Dumont D'Urville Sea

Balleny Islands

120°E

Scott I.

Circle

180°

150°E

Azimuthal Equal-Area Projection

Scale 1: 35,000,000

0   250   500   750   1000 km

0   250   500   750   1000 miles

50°S

9

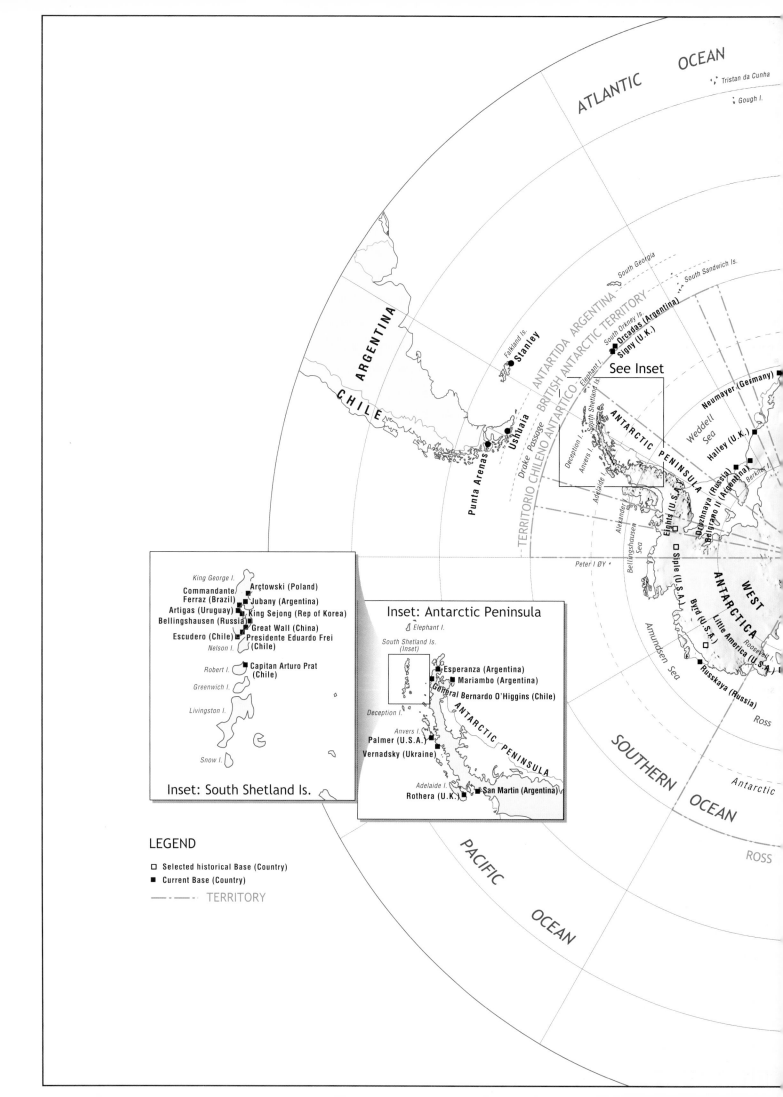

ATLANTIC OCEAN

Tristan da Cunha

Gough I.

ARGENTINA

CHILE

Falkland Is.
● Stanley

Ushuaia ●

Punta Arenas ●

Drake Passage

ANTARTIDA ARGENTINA
BRITISH ANTARCTIC TERRITORY
TERRITORIO CHILENO ANTARTICO

South Georgia

South Sandwich Is.

South Orkney Is.
Orcadas (Argentina) ■
Signy (U.K.) ■

See Inset

Deception I.
Elephant I.
South Shetland Is.
Anvers I.
Adelaide I.
Alexander I.

ANTARCTIC PENINSULA

Weddell Sea

Neumayer (Germany)

Halley (U.K.) ■

Druzhnaya (Russia) ■
Belgrano II (Argentina) ■
Berkner I.

Eights (U.S.A.) □
□ Siple (U.S.A.)

Bellingshausen Sea

Peter I ØY •

Amundsen Sea

WEST ANTARCTICA

Byrd (U.S.A.) □
Little America (U.S.A.) □
Roosevelt I.

Russkaya (Russia) ■

Ross

SOUTHERN OCEAN

Antarctic

PACIFIC OCEAN

ROSS

Inset: South Shetland Is.

King George I.

Commandante Ferraz (Brazil) ■
Arçtowski (Poland) ■
Jubany (Argentina) ■
Artigas (Uruguay) ■
King Sejong (Rep of Korea) ■
Bellingshausen (Russia) ■
Great Wall (China) ■
Escudero (Chile) ■
Presidente Eduardo Frei (Chile) ■

Nelson I.

Robert I.
Capitan Arturo Prat (Chile) ■

Greenwich I.

Livingston I.

Snow I.

Inset: Antarctic Peninsula

Elephant I.

South Shetland Is. (Inset)

Deception I.

Esperanza (Argentina) ■
Mariambo (Argentina) ■
General Bernardo O'Higgins (Chile) ■

ANTARCTIC PENINSULA

Anvers I.
Palmer (U.S.A.) ■
Vernadsky (Ukraine) ■

Adelaide I.
Rothera (U.K.) ■
San Martin (Argentina) ■

LEGEND

□ Selected historical Base (Country)
■ Current Base (Country)
—·—·— TERRITORY

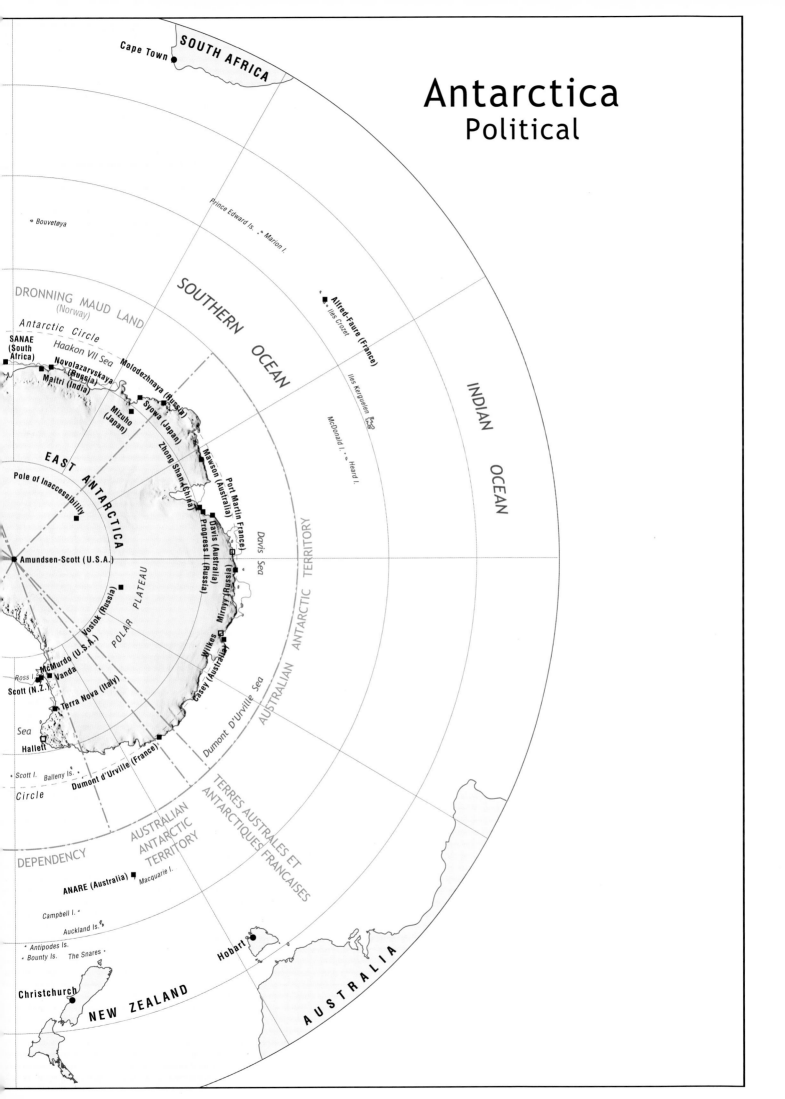

Antarctica
Political

Cape Town ● **SOUTH AFRICA**

Bouvetøya

Prince Edward Is. ● Marion I.

DRONNING MAUD LAND
(Norway)

SOUTHERN OCEAN

Alfred-Faure (France)
Îles Crozet

Antarctic Circle

SANAE
(South Africa)

Haakon VII Sea

Molodezhnaya (Russia)

Novolazarevskaya
(Russia)

Maitri (India)

Mizuho
(Japan)

Syowa (Japan)

Îles Kerguelen

INDIAN OCEAN

McDonald I. ● Heard I.

Zhong Shan (China)

Mawson
(Australia)

EAST ANTARCTICA

Pole of Inaccessibility

Port Martin France

Davis (Australia)

Progress II (Russia)

Davis Sea

Amundsen-Scott (U.S.A.)

POLAR PLATEAU

Vostok (Russia)

Mirnyy (Russia)

AUSTRALIAN ANTARCTIC TERRITORY

McMurdo (U.S.A.)

Ross I. Vanda

Scott (N.Z.)

Terra Nova (Italy)

Wilkes

Casey (Australia)

Dumont D'Urville Sea

Sea

Hallett

Scott I. Balleny Is.

Dumont d'Urville (France)

Circle

TERRES AUSTRALES ET
ANTARCTIQUES FRANÇAISES

AUSTRALIAN
ANTARCTIC
TERRITORY

DEPENDENCY

ANARE (Australia) ● Macquarie I.

Campbell I.

Auckland Is.

Antipodes Is.
Bounty Is. The Snares

Hobart

AUSTRALIA

Christchurch

NEW ZEALAND

11

a

Abbott Ice Shelf First sighted from aircraft in February 1940, by members of the UNITED STATES ANTARCTIC EXPEDITION, the Abbott Ice Shelf is 402 km (250 miles) long and 62 km (38 miles) wide, and borders Eights Coast. It partly or entirely encloses eight COASTAL ISLANDS.

Ablation Loss of material from a GLACIER or ICE SHEET caused by melting at the surface or base, evaporation and calving. Most ablation occurs in Antarctica in the ICE SHELVES and glacier tongues—through melting at their bases, and when ice calves (breaks) off their edges to form ICEBERGS. Some melt also occurs in summer: vertical ice cliffs at the margin of some polar glaciers are caused by ice calving or shearing off their sides and also melting. The melt rates of these steep faces are up to eight times greater than those of horizontal surfaces because of the low angles of SOLAR RADIATION.

Académie des Sciences Created in 1666, the French Académie des Sciences drew up instructions for physical observations for Jules-Sébastien DUMONT D'URVILLE when he was preparing for his southern voyage in 1837 but on the whole were not supportive of the explorer. Six decades later, the Académie was one of the financial supporters of Jean-Baptiste CHARCOT on his 1903–05 FRENCH ANTARCTIC EXPEDITION. The Académie publishes the findings of Antarctic scientists.

Adare, Cape The site of the first confirmed landing on the Antarctic continent and nesting ground

Above: A slab of ice about to break off the steep face of an iceshelf is the result of ablation.

of a huge colony of ADÉLIE PENGUINS. Cape Adare juts out from the western littoral of the ROSS SEA on the northern tip of VICTORIA LAND. It was named by James Clark ROSS in 1841 and is the site of the largest known colony of Adélie penguins.

On 24 January 1895, the members of the ANTARCTIC EXPEDITION made landfall on the Antarctic mainland at Cape Adare. Captain Leonard Kristensen and Carsten BORCHGREVINK made competing claims of being first to land, and another member of the expedition, Alex von Tunzelman, maintained he was first ashore, after jumping from the bow to steady the boat.

Borchgrevink returned to Cape Adare leading the 1898–1900 BRITISH ANTARCTIC EXPEDITION, and he and nine others became the first men to winter on the mainland from March 1899 to January 1900. During this time the expedition's zoologist Nikolai Hanson died and was buried on the top of Cape Adare—the first GRAVE in Antarctica.

In 1911, the Northern Party of Robert SCOTT's fateful final expedition, led by Victor Campbell, wintered at Cape Adare.

Adelaide Island Located off the west coast of the ANTARCTIC PENINSULA, the island was discovered by John BISCOE on 16 February 1832. He described the island as 'imposing and beautiful ... with one high peak shooting up into the clouds, and occasionally appearing both above and below them, and a lower range of mountains extending about four miles, from north to south, having only a thin covering of snow on their summits, but towards their base buried in a field of snow and ice of the most dazzling brightness ...'

ROTHERA STATION, on the eastern coast, is the main air operations centre for the BRITISH ANTARCTIC SURVEY.

Adélie penguin (*Pygoscelis adeliae*) The most recognizable, and most studied, of all PENGUINS, Adélies are the 'dinner-suit' birds—jet black and white, with distinctive white rings around their eyes. Their average height is 70 cm (28 in) and they weigh around 5.5 kg (12 lb). Named after the wife of Jules-Sébastien DUMONT D'URVILLE, they are the most widely distributed and most abundant of all the Antarctic species: there are thought to be about 2.5 million pairs in identified localities, and it is likely there are colonies yet to be discovered.

Adélies spend the winter at sea, in small groups among ICEBERGS and PACK ICE, diving for food, particularly KRILL. They begin coming ashore in early summer to breed in large traditional shoreside colonies; because the inshore ice has not yet broken up, they may have to trek—in single file—up to 100 km (62 miles) over ice. Colonies are situated on the more southerly SUB-ANTARCTIC ISLANDS and all around the Antarctic continent. The largest known colony is at Cape ADARE, where there are hundreds of thousands of breeding pairs. The noise—shrieking, squawking, groaning—from these often enormous colonies has been picked up 50 km (31 miles) away; they are notable, too, for their stench.

Adélies build nests of pebbles, collected from the rocky slopes of their breeding sites or 'stolen' from other nests. Eggs are laid in November, and both parents take turns incubating the eggs for around 35 days. Once the chicks hatch, they are

Below: Place of the first confirmed landing on the Antarctic continent, Cape Adare is also the site of a large Adélie penguin colony.

*Opposite: The Adélie penguin (*Pygoscelis adeliae*) population is thought to comprise around 2.5 million breeding pairs.*

Adélie penguin

Above: Adélie penguins (Pygoscelis adeliae) are popularly dubbed 'dinner-suit birds', for obvious reasons.

brooded for 25 days, after which both parents must forage to keep up with the demands of the growing chick.

While their parents are away, the chicks gather together in crèches (large groups of young birds). This offers some protection from the cold and from predators, particularly LEOPARD SEALS and SKUAS, but many chicks die during this stage. Fledging takes place at 50–54 days old.

The raised beaches the Adélies favour for their colonies are also prime sites for human bases. A joint New Zealand-USA base was established at Cape Hallett in the middle of an Adélie breeding colony in 1957 (the base closed in the 1960s); more controversially, in 1983 FRANCE began constructing an AIRSTRIP, using explosives to level the surface through Adélie colonies near its base on TERRE ADÉLIE.

Admiralty Bay A large bay, 8 km (5 miles) wide at its entrance, on the southern coast of King George Island in the SOUTH SHETLAND ISLANDS. It was named after the British Admiralty in or before 1822 by George Powell, a sealing captain. A visitor to the island's WHALING station, which was established in 1906, described it as 'a sordid habitation. Scores of squalid and dilapidated wooden huts and buildings clustered as near as possible to the foreshore. A curious ozone smell pervaded the whole area—a smell which only a whaling station can produce.'

From 1946 to 1961, the British operated a sci-

Right: A United States Air Force Hercules—affectionately known as 'Hercs' among the Antarctic community—lands at the South Pole.

Above: The 1979 Air New Zealand plane crash on Mount Erebus resulted in more fatalities than any other Antarctic aircraft crash and led to the continent's largest coordinated search and recovery effort.

entific station, Base G, at Admiralty Bay. In 1977 Poland opened the year-round ARĊTOWSKI STATION. It is also the location of BRAZIL's Commandante Feraz research station, and Peru's Machu Picchu Station.

The western shore of Admiralty Bay has been designated a SITE OF SPECIFIC SCIENTIFIC INTEREST.

Agreed Measures for the Conservation of Antarctic Flora and Fauna In 1964 general rules of conduct for scientific expeditions in Antarctica, which had been drawn up by the SCIENTIFIC COMMITTEE ON ANTARCTIC RESEARCH (SCAR) to minimize the human impact on species and the environment, were incorporated into the Agreed Measures for the Conservation of Antarctic Flora and Fauna. The main points of these measures are: prohibition of harming any native mammal (excluding WHALES, which are covered under INTERNATIONAL WHALING COMMISSION conventions); minimizing harmful interference in the environment of mammals and BIRDS; prohibition on the introduction of non-indigenous species,

parasites and diseases; and the creation of SPECIALLY PROTECTED AREAS (SPA).

Air crashes Pilots in Antarctica face the most severe flying conditions. If caught in a WHITEOUT or BLIZZARD, it may be impossible to see landscape features. The full fury of Antarctic blizzards has swept planes away when they have been securely tied down to the ice. Radio communications are also regularly disrupted by SOLAR WINDS in the magnetosphere (see ATMOSPHERE).

There have been numerous fatal crashes in such conditions. The worst loss of life was the crash of an Air New Zealand DC10 on a non-stop tourist flight in November 1979. All 257 people aboard were killed when the plane ploughed into the side of Mount EREBUS near MCMURDO STATION in dense cloud. The accident was attributed to mistakes in the programming of the aircraft's navigation computer, and whiteout is also believed to have been a factor. The crash site has been declared a tomb by ANTARCTIC TREATY nations.

Aircraft The first successful flights over Antarctica

Above: A Ford monoplane used in Richard Byrd's 1928–30 expedition.

were made in 1928–29 by Carl Eielson and Hubert WILKINS in a Lockheed Vega monoplane; they surveyed and took aerial photographs of the ANTARCTIC PENINSULA. But the first aircraft was Douglas MAWSON's air-tractor, which crashed in 1911 during a demonstration flight in Adelaide, Australia, before it had arrived in Antarctica: the strange machine managed several short land journeys at Cape DENISON before being abandoned.

Richard BYRD made a number of aircraft-supported expeditions over uncharted ice fields: on 29 November 1929, it was long believed, he made the first flight over the SOUTH POLE (although it has lately been shown he could not have flown over the Pole), and in 1946–47 he led OPERATION HIGHJUMP, the first large-scale deployment of aircraft in Antarctica.

The aerial crossing of the continent by Lincoln ELLSWORTH and Herbert Hollick-Kenyon in 1935 demonstrated the feasibility of landing aircraft in inland Antarctica. Later, ski-equipped planes, wheeled landing gear and jet-assisted take-offs enabled shorter take-offs and landings on smaller AIRSTRIPS.

The US Navy made the first intercontinental flights to Antarctica in the summer of 1955–56, from Christchurch, New Zealand, to MCMURDO SOUND. In the largest Antarctic air transport operation, equipment and personnel are now flown from Christchurch each summer to Italian, American and New Zealand bases (other national programmes use ship TRANSPORT). Hercules, Globemaster and Starlifter planes land on a SEA ICE airstrip in McMurdo Sound. Until 1995, planes also made resupply airdrop flights to the South Pole in winter. However, winter airdrops over the Pole are now only made in EMERGENCIES.

An Antarctic aircraft is not comfortable. The planes are unlined metal shells that vibrate in time with their engines. Passengers, wearing up to six layers of stifling clothing and earmuffs, perch on webbing seats throughout the trip, which may last up to nine hours. Often bad weather forces flights to turn back up to five hours into the journey.

Smaller ski planes, such as Twin Otters, are used to ferry parties and supplies to deep field camps out of range of HELICOPTERS. Used extensively by the British, Chilean, Argentinian, USA, New Zealand and Italian Antarctic programmes, Twin Otters are common over the Antarctic

Peninsula in summer. In 1988 Dick Smith and Giles Kershaw flew a Twin Otter 14 hours from Hobart, Australia, to CASEY STATION, then on to the South Pole; over five weeks they traversed 41,448 km (25,700 miles) of Antarctica and the SOUTHERN OCEAN.

LAN Chile flew the first tourist flights over Antarctica on 22 December 1956. In 1977 Qantas and Air New Zealand began regular flights over the ROSS SEA region. These ended abruptly in 1979, when an Air New Zealand DC10 crashed on Mount EREBUS, killing all passengers and crew (see AIR CRASHES). Qantas resumed regular trips in 1994. These flights are now in high demand: within the space of a day tourists are able to fly to Antarctica and back, after circling over the continent for several hours. The only Antarctic airline is operated by Adventure Network International (ANI), which supports land-based expeditions, offers trips from its base at Patriot Hills to almost anywhere on the continent, including the South Pole, and flies to an EMPEROR PENGUIN rookery in the WEDDELL SEA.

Above: The controversial hard-rock airstrip France built near Dumont D'Urville Base.

Airstrips Ski-equipped AIRCRAFT can land on snow surfaces, but aircraft equipped with wheels require a solid surface. A network of airstrips support aircraft operations in Antarctica. They are constructed on permanent rock in ice-free areas, or on ICE, in which case they need constant maintenance. MCMURDO STATION has three airstrips: a hard-ice runway, a snow runway, and a temporary strip built on sea ice and only useable from October to December. ARGENTINA, BRITAIN and CHILE have constructed rock airstrips on the ANTARCTIC PENINSULA.

In 1983 FRANCE began building a hard rock airstrip near its DUMONT D'URVILLE BASE close to an ADÉLIE PENGUIN colony. Construction was disrupted because of worldwide concerns about the penguins and damage to the environment. The airstrip was finally abandoned in 1994 after it was damaged by a large wave.

Albatrosses (Diomedeidae) The *Concise Oxford Dictionary* defines an albatross as being 'any long-winged stout-bodied bird of the family Diomedea'; it is also 'a source of guilt or encumbrance.' Sailors believed that albatrosses were the souls of dead seafarers who were destined to wander the oceans forever and, thus, it was bad luck to kill them. Nevertheless, the birds were captured for their meat, and their webbed feet were made into tobacco pouches and their bones, pipe-stems.

Of the 13 species of albatross, most nest in the Southern Hemisphere. They occupy the windiest

Right: The wandering albatross (Diomedea exulans), *found throughout the Southern Ocean, breeds on many subantarctic islands.*

regions and they rely on wind above about 20 knots to keep them comfortably airborne: their distribution is effectively limited in the north by the calms of the Doldrums. Seven have been reported in cold, southern waters: SOUTHERN ROYAL ALBATROSS, WANDERING ALBATROSS, WHITE-CAPPED MOLLYMAWK, BLACK-BROWED MOLLYMAWK, GREY-HEADED MOLLYMAWK, LIGHT-MANTLED ALBATROSS and SOOTY ALBATROSS. Albatrosses belong to the Procellariiformes, a large order of pelagic seabirds known as PETRELS.

Albatrosses are best known for their exceptional wingspans—between 2 and 3.6 m (79 and 141 in)—and their majestic soaring flight. They are well adapted to life at sea and can travel for great distances without flapping their wings. They

only come ashore to breed. Some species do not begin breeding until they are seven years or older. Excluding the solitary sooty albatross, all the members of the Diomedeidae have a colonial nesting habit, which is determined by feeding preferences and competition for resources. They breed annually, except the wandering and royal albatross and grey-headed mollymawk, which nest every second year. All petrels, including albatrosses, lay a single egg per clutch.

Alexander Island A large island—386 km (240 miles) long and 80 km (50 miles) wide—that lies west of the base of the ANTARCTIC PENINSULA. It was discovered in January 1821 by Thaddeus BELLINGSHAUSEN, who named it Alexander I Land.

It was not until 1940 that a sledging party led by Finn RONNE discovered that it was an island joined to the mainland by a deep shelf of ice.

Alfred Wegener Institute Named after the German geophysicist, meteorologist and Arctic explorer, Alfred Wegener, who originated the hypothesis of CONTINENTAL DRIFT, the Institute coordinates polar research and expeditions, and conducts research in the Arctic and Antarctic regions and at temperate latitudes. Studies of global change are a central research focus of the Institute, which operates the research vessel *Polarstern*.

Algae Numerous species of unicellular (single-celled) algae are found in Antarctica. They make up the PHYTOPLANKTON, the group of simple plant organisms living on the surface of the ocean, and cryoplankton, the unicellular algae that live and reproduce in snow banks and GLACIERS.

Algae are unicellular or multicellular plants occurring in water or moist environments. DIATOMS are unicellular marine algae, and SEA-WEEDS are multicellular algae. Algae have chlorophyll for photosynthesis but lack true stems, roots and leaves.

Patches of cryoplankton are found in summer on snowfields near the Antarctic coast, particularly on the Peninsula. So-called RED SNOW is coloured by algae that contain green and red pigments. Where this occurs close to PENGUIN colonies, which supply enough nutrients for these algal populations to thrive, large mats of green algae such as *Nostoc commune* form. Algae are also found in the meltwater pools that form briefly in summer and in the water column and bottom communities of FRESHWATER LAKES.

Allan Hills To the west of the DRY VALLEYS, in South VICTORIA LAND. Concentrations of METE-ORITES have been found here, including a 32 g (1.1 oz) fragment believed to be part of the moon, which was discovered in January 1982.

Alph River Flows 20 or 30 km (12–19 miles) through a 10 m (33 ft) wide and 0.5 m (1½ ft) deep channel along the edge of the Koettlitz Glacier in VICTORIA LAND. See RIVERS.

American sheathbill (*Chionis alba*) See SNOWY SHEATHBILL.

Amery Berg One of the largest ICEBERGS recorded, it measured about 110 by 75 km (68 by 47 miles), and broke off the AMERY ICE SHELF in 1963. It drifted west and collided with the Fimbul Ice Shelf, where another large iceberg fractured off. It was tracked by SATELLITE for about 12 years, until it broke into small fragments.

Amery Ice Shelf The third largest ICE SHELF system in the Antarctic region, covering 1.5 million sq km (579,150 sq miles)—11 percent of the total Antarctic grounded and floating ice area—and occupying a large indentation in the Indian Ocean coastline of Antarctica, in the AUSTRALIAN ANTARC-TIC TERRITORY. It extends inland from Prydz and MacKenzie Bays more than 320 km (198 miles) to

Above: *Upwelling ocean currents rich in dissolved nutrients and long summer sunlight hours create ideal conditions for algal development, and huge blooms envelop the underside of Antarctic sea ice—sometimes referred to as an upside-down prairie.*

the foot of LAMBERT GLACIER. In 1963 a large chunk of ice sheared off the shelf. Known as the AMERY BERG, it was one of the largest icebergs recorded.

Amundsen, Roald (1872–1928) Norwegian explorer, born south of Oslo, Norway. He began studying medicine, but his main ambition was Arctic exploration and he trained by skiing and trekking long distances. After working in the Arctic on sealing vessels, he joined the 1897–99 BELGIAN-ANTARCTIC EXPEDITION to Antarctica, led by Adrien de GERLACHE, as second mate on the *BELGICA*. Amundsen was among those who made the first Antarctic sledging journey. He formed a close friendship with Frederick COOK, who described Amundsen as 'the biggest, the strongest, the bravest ...' of the crew.

From 1903–06 he sailed in the Arctic, became the first person to navigate the north-west passage and returned to Norway a hero. Amundsen then planned to be first to reach the North Pole, and he borrowed the FRAM from Fridtjof NANSEN for the purpose. However, in 1909, when he heard that Robert Peary had reached the North Pole, he altered his destination to the SOUTH POLE.

The *Fram* entered the BAY OF WHALES on 14 January 1911 on what had become known as the NORWEGIAN ANTARCTIC EXPEDITION. A hut was erected at a place Amundsen called FRAMHEIM. After a winter spent in preparations, he set off for the South Pole with Helmer Julius HANSSEN, Olaf BJAALAND, Oscar WISTING and Helge Sverre HASSEL. They took four sledges, drawn by 13 dogs each, killing the dogs for food as they went. When

Above: Roald Amundsen, first to reach the South Pole on 14 December 1911.

Below: The original Amundsen-Scott South Pole Station was buried by snow and replaced in 1975 by the geodesic dome. Like its predecessor, this too is being replaced as the pressure of the mounting snow makes the base unsustainable.

they passed the southernmost point previously reached by Ernest SHACKLETON, Amundsen recalled, 'No other moment of the whole trip affected me like this. The tears forced their way to my eyes.' At 3 pm on 14 December 1911 the team became the first to reach the South Pole. The return journey went smoothly and they arrived back at Framheim on 25 January 1912 with 11 of the 52 dogs they had set out with.

On his return to Norway, Amundsen built the *Maud*, in which he navigated the north-east passage around Siberia, only the second person to do so. In 1926, along with Lincoln ELLSWORTH, he made the first trans-Arctic flight across the North Pole in an airship called the *Norge*. Amundsen disappeared two years later while attempting to rescue the Italian Umberto Nobile, who had crashed his airship in the Arctic.

Amundsen Sea Off the coast of MARIE BYRD LAND in West Antarctica, it was named after Roald AMUNDSEN by a 1928–29 Norwegian expedition. The sea was mapped by the 1939–41 UNITED STATES ANTARCTIC SERVICE EXPEDITION.

Amundsen-Scott South Pole Station Established at the SOUTH POLE in November 1956 by the USA as part of INTERNATIONAL GEOPHYSICAL YEAR. The station's name honours Roald AMUNDSEN and Robert SCOTT, who reached the Pole within a few weeks of each other in the summer of 1911–12. The station's first chief scientist was Paul SIPLE. The original station was buried by snow. Rebuilt in 1975 as a geodesic dome, which sits astride six TERRITORIAL CLAIMS, it has been redeveloped since.

Research is conducted at the base into astronomy, astrophysics, biomedical studies, glaciology, geophysics, meteorology and upper atmosphere physics.

The base holds 28 in winter and 130 or more in summer. Like its predecessor, it is slowly being buried beneath the accumulating snow and construction began in 2002 on a new station, to be built on stilts.

Animals Many of Antarctica's animals are endemic to the region; all are uniquely adapted to survive the extreme conditions they encounter. By definition an animal is any organism capable of voluntary movement, which possesses specialized sense organs and ingests complex organic substances—in other words, anything that moves, feels and eats. Antarctica's animals range from simple CORALS and SEA ANEMONES to more complex INVERTEBRATES such as KRILL, TARDIGRADES and SQUID, and VERTEBRATES, including BIRDS, FISH and WHALES.

Antarctic and Southern Ocean Coalition (ASOC) Founded in 1977, when the CONSULTATIVE PARTIES of the ANTARCTIC TREATY began negotiating a regime to manage living marine resources in the SOUTHERN OCEAN. The Antarctic and Southern Ocean Coalition (ASOC) monitors activities and coordinates environmental advocacy about Antarctica and the Southern Ocean. A loose coalition of over 230 environmental groups in 49 countries, ASOC is an umbrella organization that facilitates the exchange of information among groups, but does not direct their activities.

The Antarctic Project was founded in 1982 as a secretariat for ASOC. Based in Washington DC, it is the only environmental group working full-time on Antarctica and has the goal of holding governments legally, politically and morally accountable for their actions in Antarctica and the Southern Ocean. ASOC has observers at meetings of the Commission for the CONVENTION ON THE CONSERVATION OF ANTARCTIC MARINE LIVING RESOURCES, the ANTARCTIC TREATY CONSULTATIVE MEETING, and the INTERNATIONAL WHALING COMMISSION.

The objective of ASOC is to protect the last unspoiled wilderness, preserving its value to science and keeping it as the last frontier for human endeavours. ASOC initially campaigned to make Antarctica a WORLD PARK. Its major activities are a special implementation campaign to ensure that the MADRID PROTOCOL is implemented by the signatory governments, and a campaign against WHALING and illegal, unregulated and unreported FISHING in the Southern Ocean.

Antarctic bottom water The coldest, most saline water in the global ocean system, it drives circulation in the SOUTHERN OCEAN. Laden with oxygen, it seeps slowly north from Antarctica, aerating and cooling the world's oceans. Most Antarctic bottom water is produced in the WEDDELL SEA, but some is produced in the ROSS SEA, off WILKES LAND.

Exposure to ICE SHELVES 'supercools' sea water to below freezing point. As SEA ICE is formed, the sea water is 'distilled', with the ice taking up only about 15 percent of the salt, leaving dense, salty brine that sinks towards the ocean floor. This water flows off the Antarctic CONTINENTAL SHELF and moves north, along with cold ANTARCTIC SURFACE WATER. These cold bands of water mix with warm, southerly flowing water at the ANTARCTIC CONVERGENCE. Here, some of the bottom water rises and begins to flow back towards Antarctica.

Antarctic Circle A geographical boundary that circles Antarctica at about 66.6°S. It is the southernmost point at which the sun can be seen on midwinter's day (21 June): south of the circle it is dark 24 hours a day in winter. The circle also marks the most northerly point at which the sun is visible for 24 hours a day on midsummer's day (21 December), when the sun is at its highest position above the horizon. Crossing the Antarctic Circle is considered to be a symbolic point of entry into Antarctic waters. On 17 January 1773, James COOK and the crews of the *Resolution* and *Adventure* were the first to cross this significant line.

Antarctic Coastal Current See EAST WIND DRIFT.

Antarctic cod (Nototheniidae) The most common FISH type in Antarctica. Antarctic cod are sluggish fish with large heads and tapered bodies, which live on the sea bottom, usually in the shallower coastal waters of the Antarctic continent and the SUBANTARCTIC ISLANDS. In order to cope with the freezing conditions, some species produce a GLYCOPROTEIN compound similar to ANTIFREEZE in

Above: The sluggish giant Antarctic cod (Dissostichus mawsoni), *with its large head and tapered body, is the most common fish in Antarctic waters.*

their blood. This blocks the formation of potentially fatal ice crystals in their body fluids.

Antarctic cod belong to the suborder NOTOTHENIOIDEI, all of which lack the swim-bladders present in most fish to control buoyancy. Because all Antarctic cod were once bottom-dwellers (DEMERSAL FISH) and did not need to float, swim-bladders were unnecessary.

The giant Antarctic cod, *Dissostichus mawsoni*, also known as the ANTARCTIC TOOTHFISH, is the largest fish in Antarctica, reaching up to 120 kg (265 lb) in weight and growing to over 2.2 m (7 ft) long.

Antarctic continent Representing 9 percent of the Earth's continental surface, Antarctica covers almost 14 million sq km (5.4 million sq miles); it is twice as big as Australia, and the entire USA could easily fit within its boundaries. The name Antarctica means 'opposite the Arctic'. It is the most isolated continent on the globe, lying 950 km (589 miles) from South America, 2300 km (1426 miles) from Tasmania, 2200 km (1364 miles) from New Zealand and 3600 km (2232 miles) from Africa.

The TRANSANTARCTIC MOUNTAINS split the continent into EAST (Greater) ANTARCTICA and WEST (Lesser) ANTARCTICA. At either end of the mountain range are two wide basins: the ROSS SEA lies along the Pacific Ocean sector, and the WEDDELL SEA faces the Atlantic. A vast, perpetual ice cover smothers both East and West Antarctica, and spills out over the SOUTHERN OCEAN so that the coastline is fringed with ICE SHELVES. The ANTARCTIC PENINSULA, a finger of land extending north from West Antarctica towards the southern tip of South America, breaks up the nearly circular outline of the continent.

Despite its isolation, Antarctica has many interrelationships with the global ATMOSPHERE and oceanic systems and it is believed that first

indications of CLIMATE change will come from this icy southern continent. The giant Antarctic ICE SHEET, for example, drives Southern Hemisphere weather and has the potential to dominate future sea levels.

Averaging about 2300 m (7544 ft) above sea level, Antarctica is the world's highest continent (Asia is next, averaging about 900 m/3000 ft above sea level). If its icy surface layer were stripped away, however, Antarctica would probably average little more than about 500 m (1640 ft) above sea level. The bare bones, or BEDROCK, of the continent are undulating and mountainous. The relief would range from 4897 m (16,062 ft) above sea level at VINSON MASSIF to more than 2539 m (8328 ft) below sea level in the BENTLEY SUBGLACIAL TRENCH.

The first landing on the continent may have occurred on 7 February 1821, by John DAVIS, an American sealer. The first confirmed landing was on 24 January 1895, when the Norwegian whaling ship *Antarctic* landed a party at Cape ADARE in the northern Ross Sea. The party included Leonard Kristensen, Carsten BORCHGREVINK and Henryk BULL, who considered stepping onto the Antarctic mainland to be 'both strange and pleasurable', although he thought the crew would have preferred to find a right whale 'even of small dimensions'.

Antarctic Convergence Also known as the POLAR FRONT, the Convergence is a roughly circular belt of water about 40 km (25 miles) wide lying between latitudes 48°S and 60°S in the SOUTHERN OCEAN. Not a fixed boundary, the Convergence shifts with the seasons. It is often considered the true boundary of Antarctica, as it marks a definite change in ocean temperatures and chemical composition. The Convergence forms where cold north-flowing ANTARCTIC BOTTOM WATER and ANTARCTIC SURFACE WATER meet warmer water flowing south from the Atlantic, Indian and Pacific Oceans; this produces a sharp temperature change of up to 2°C (3.6°F).

Travelling south across the Convergence, sailors notice a new bite in the WIND. In a calm weather crossing, eerie 'frost smoke'—water vapour produced from the cooling of warm air—rises from the ocean surface. Captain Reginald Ford, who sailed over the Convergence with Robert SCOTT's 1901–04 NATIONAL ANTARCTIC EXPEDITION, described the experience: '... we have been sailing in an unknown sea, shrouded in fog, and surrounded on all sides by snow and ice. The mist brings with it a sense of dread and peril—the birds vanish, and the silence becomes more noticeable—the cold seems to grow in intensity. The ice, which has been glistening in the light of the powerful sun, becomes cheerless and full of hidden dangers.'

The term 'Convergence' came into general use after Dr G Wüst, scientist on the METEOR EXPEDITION, published his theories on ocean circulation.

Antarctic Divergence A zone around the Antarctic continent at approximately 65°S where opposing winds and water masses collide. It separates EAST WIND DRIFT from WEST WIND DRIFT. The nutrient-rich upwelling of CIRCUMPOLAR DEEP WATER occurs in this area, when that southward-flowing water mass meets the north-flowing ANTARCTIC SURFACE WATER. Above the Divergence, the CIRCUMPOLAR TROUGH marks a similar atmospheric boundary between easterly and westerly airflow.

Antarctic **Expedition (1893–95)** A Norwegian commercial WHALING expedition. Financed by Svend FOYN and led by Henryk BULL, the intention was to search out right WHALES. A steam whaler, the *Kap Nor*, was refitted and renamed *Antarctic* and captained by Leonard Kristensen. The expedition planned to explore three areas: the South Atlantic and southern Indian Oceans regions; the New Zealand SUBANTARCTIC ISLANDS, and the ROSS SEA.

Only fin whales were sighted in the Atlantic and Indian Ocean sectors. However, at Îles KER-GUELEN, 3000 seals were killed and their oil and skins sold later at Melbourne, Australia for £3,000. The ship ran aground at CAMPBELL ISLAND and returned to Melbourne for repairs. Dr William BRUCE was to have joined the expedition here as scientist; in the event, Carsten BORCH-GREVINK, who was taken on as an ordinary crewman, conducted scientific investigations on the voyage. The *Antarctic* crossed the ANTARCTIC CIRCLE on Christmas Day 1894. Cape ADARE was reached on 16 January 1895 and eight days later the first confirmed landing on the Antarctic continent occurred; Borchgrevink and Alexander von Tunzleman, who was taken on in New Zealand, both claimed they were the first to step ashore (however, research by R K Headland indicates this was not the first landing on the continent; it is believed that an American whaler landed on the continent in the mid-19th century). Rock samples were taken at the landing point and collections made of lichen, seaweed and jellyfish. Although botanical and geological findings

were impressive, and the confirmation of James Clark ROSS's belief that in summer the water was ice-free inside the PACK ICE of great importance, as a commercial venture the expedition was a failure.

Antarctic fulmar (*Fulmaris glacialoides***)** Also known as southern fulmars, these birds are often found over and around the fringes of PACK ICE and they have a circumpolar flight distribution. Rather large birds—48 cm (19 in) long—they look like oversize, pale bluish-grey gulls, although in flight they glide more than gulls. They have pink bills with blackish tips and blue nasal tubes. The birds feed on CRUSTACEANS, FISH and KRILL, and breed in large colonies on the Antarctic coast, and on islands south of the ANTARCTIC CONVERGENCE. Nests comprising stone chips and pebbles are constructed on rock ledges or scree slopes in October; laying takes place in early December and chicks fledge in March.

Antarctic fur seal (*Arctocephalus gazella***)** The only eared seal that lives in the polar waters south of the ANTARCTIC CONVERGENCE. Much more agile on land than phocid (or 'earless') seals, they are shallow divers, catching KRILL, FISH and SQUID at depths of about 60 to 80 m (195–260 ft); they are believed to dive mainly at night. Males are larger than females at around 2 m (7 ft) in length and weigh around 125 to 200 kg (275–440 lb). Fur seals have dark grey bodies with lighter coloured throats and bellies. Their skin consists of a coarse

layer of guard hairs beneath which is a thick, silky under-pelt about 2.5 cm (1 in) long that provides insulation (their blubber is not as thick as that of phocid seals).

Breeding takes place on the rocky shores of SOUTH GEORGIA and neighbouring islands from late October to January, where the males establish breeding territories containing about 10 females. They return to the same sites each year. The single pups conceived the previous year are born within a few days of the females arriving, and about seven days later the females mate again. The pups, which weigh about 5.5 kg (12 lb) at birth, are nursed for 110 to 115 days and grow rapidly; they are sometimes preyed upon by LEOPARD SEALS and possibly by KILLER WHALES. Male fur seals can live about 15 years, and females up to 23 years.

The population is believed to be over 1 million, about half that estimated at the start of commercial SEALING in the late 18th century.

Antarctic hairgrass (*Deschampsia antarctica***)** The most common of the two FLOWERING PLANTS that grow below 60°S. Antarctic hairgrass thrives in the soils created by LICHENS and MOSSES. It is found in small, low patches in sheltered spots on the ANTARCTIC PENINSULA as far south as 68°S, and growing much more vigorously on the SOUTH ORKNEY and SOUTH SHETLAND ISLANDS. The grass has small, inconspicuous blooms.

Some botanists believe hairgrass to be a relic of early FLORA that was destroyed during the ICE

*Below: One of just two flowering plants found in Antarctica, Antarctic hairgrass (*Deschampsia antarctica*) grows in clumps as far south as 68°S.*

AGES and has been reintroduced to the Antarctic continent from the northern islands.

Antarctic Heritage Trust (AHT) New Zealand-based organization formed in 1987 to restore and protect HISTORIC SITES, particularly in the ROSS SEA region. The Trust is the guardian of the expedition HUTS of Robert SCOTT and Ernest SHACKLETON on ROSS ISLAND, and that of Carsten BORCHGREVINK at Cape ADARE, which is the first building to be erected on the Antarctic continent. It has also documented artefacts associated with the huts.

Antarctic Ice Cap See ICE SHEET.

Antarctic pearlwort (*Colobanthus quitensis*) One of only two FLOWERING PLANTS that grow below 60°S, Antarctic pearlwort has tiny, inconspicuous pink blooms and forms a small, compact cushion about 25 cm (10 in) across. It is less widespread and usually less abundant than ANTARCTIC HAIR-GRASS (*Deschampsia antarctica*). Both are found on the SOUTH SHETLAND and SOUTH ORKNEY ISLANDS and the ANTARCTIC PENINSULA, and are most frequent at low altitudes and on the warmer sites of north- or west-facing slopes. Antarctic pearlwort needs nutrients, water and soil to grow. Colonizing LICHENS and MOSSES provide the soil needed for it to become established and the guano of BIRDS often provides essential nutrients.

Antarctic Peninsula The northern part of WEST ANTARCTICA, the Peninsula is an extension of the TRANSANTARCTIC MOUNTAIN chain that arches into the SOUTHERN OCEAN. One of the richest breeding grounds on the Antarctic continent, the Peninsula reaches to within 970 km (600 miles) of South America and sits astride a dramatic CLIMATE divide. On the eastern edge, it is exposed to the vast, frozen WEDDELL SEA. On the western coast the climate is warmed by the Southern Ocean and buffeted by fierce WINDS.

The land of the Peninsula was formed in submarine troughs that were filled with sediment about 220 million years ago. This was uplifted and eroded, and the next layer was laid down beside it in a second trough, about 180 million years ago. The rocks that have been folded and uplifted since this time are rich in FOSSILS and are similar to those found in South America. In fact, the tip of the Peninsula can be linked to South America via an OCEAN RIDGE system.

Antarctic petrel (*Thalassoica antarctica*) One of Antarctica's most beautiful birds, these conspicuously marked, chocolate-and-white PETRELS have wide wings with trailing edges, and their rumps and tails are white. Antarctic petrels are 43 cm (17 in) long with wing-spans of 102 cm (40 in).

They are circumpolar summer breeders, breeding along the coasts of the Antarctic continent and as far as 250 km (155 miles) inland. Breeding colonies are densely packed and situated on cliffs, steep rocky slopes and crevices, as well as on NUNATAKS. The adults start arriving about a month before breeding begins in November and chicks are fledged in March. During the rest of the year, flocks of Antarctic petrels keep close to the

Above: *Antarctic skuas (*Catharacta maccormicki*) nest on open rocky ground.*

PACK ICE, although they are sometimes found north of the ANTARCTIC CONVERGENCE in winter. They feed on small SQUID, KRILL and FISH, especially Antarctic herring, caught through shallow dives, dips and surface-seizing.

Antarctic prion (*Pachyptila desolata*) The most polar of all the PRIONS, Antarctic prions are medium sized, with bluish-grey plumage, black 'M'-shaped markings from wing-tip to wing-tip and black-tipped tail feathers, dark elongated eye patches, broad blue bills and pale blue feet with cream webs. Breeding takes place on several Antarctic and southern SUBANTARCTIC ISLANDS. The largest population—of 22 million birds—is on SOUTH GEORGIA. Distribution is circumpolar, and they often form vast flocks, especially when migrating or feeding. Their multi-purpose bill allows them to catch quite large KRILL and small SQUID by surface-seizing; when these are scarce, they can effectively sieve the water for tiny amphipods using the small lamellae that fringe their bill's upper mandible. During the winter they move into subantarctic and temperate waters.

Antarctic shag (*Phalacrocorax bransfieldensis*) The only CORMORANT that breeds on the ANTARCTIC PENINSULA. Around 10,000 pairs of these black-and-white shags are found in over 50 known colonies on the Peninsula, where they are out-numbered only by PENGUINS. They also breed on SOUTH SHETLAND and ELEPHANT ISLANDS. Their breeding success varies, depending on food availability; when food is scarce they are forced to forage further afield, away from the ice. They are named 'bransfieldensis' after Edward BRANSFIELD, believed to be the second person to sight the Antarctic Peninsula, on 30 January 1820.

Antarctic skua (*Catharacta maccormicki*) Similar in appearance to the SOUTHERN SKUA, but smaller, Antarctic skuas have three plumage phases: light, intermediate and dark. They range widely around Antarctica and have been seen flying over the SOUTH POLE.

They lay two eggs but, because of the extreme conditions of their southern habitat and the aggressive nature of the stronger chick, often the weaker one perishes through cold and starvation.

Antarctic skuas begin arriving at their breeding grounds on snow-free coasts of the Antarctic continent around October, the same time as ADÉLIE PENGUINS start to settle into their colonies. The skuas often form their territories on the edges of the penguin colonies, giving them plenty of opportunities to plunder penguin eggs and attack young chicks. Like other skuas, Antarctic skuas also feed extensively on FISH and on carrion such as dead SEALS and BIRDS. They also breed on the SOUTH SHETLAND ISLANDS.

Young birds migrate to the Northern Hemisphere, as far as Greenland.

Antarctic surface water Originating close to the coast of Antarctica, it flows north until it encounters warmer south-flowing surface water at the ANTARCTIC CONVERGENCE—there, the Antarctic surface water, richer in oxygen and nutrients and thus denser than the northern water, sinks and mixes with underlying water.

At the ANTARCTIC DIVERGENCE, the point where the north-flowing surface water collides with the warmer, south-moving, fast-flowing CIRCUMPOLAR DEEP WATER, there is an area of water mass upwelling.

The surface water has a far lower SALINITY than ANTARCTIC BOTTOM WATER.

Antarctic tern (*Sterna vittata*) Antarctic terns are found and breed on the ANTARCTIC PENINSULA and on many islands throughout the SOUTHERN OCEAN, including most of the SUBANTARCTIC ISLANDS.

They are very similar in size and appearance to ARCTIC TERNS, with white plumage, black caps and red bills and legs, although their bodies are slightly longer (about 40 cm/16 in). Beautifully proportioned, with wing-spans double the length of their bodies, these graceful birds have a circumpolar distribution, and the population has been estimated at around 90,000 pairs. They feed mainly on KRILL.

Breeding colonies are located relatively close to the sea. When preyed upon by SKUAS, the terns join together to defend their colonies. The chicks leave their nests a few days after hatching and find themselves a hiding place. Unlike many seabirds, Antarctic terns do not always return to the same breeding area each year.

Antarctic toothfish (*Dissostichus mawsoni*) Weighing in at about 120 kg (265 lb) and reaching up to 2.2 m (7 ft) in length, the Antarctic toothfish is the largest FISH in Antarctica. Also known as the giant ANTARCTIC COD, it gets its Latin name *mawsoni* from Australian explorer Douglas MAWSON. One of two species of toothfish, the Antarctic toothfish is found in the high-latitude region close to the Antarctic continent. Its relative, the PATAGONIAN TOOTHFISH (*Dissostichus eleginoides*), lives in subantarctic waters on shelves around islands and submarine banks.

Toothfish belong to the NOTOTHENIOIDEI, the dominant suborder of fish in the SOUTHERN OCEAN. They are long-lived—the maximum recorded age is about 45 years—and slow-growing. It has been estimated they gain only 1 kg (2.2 lb) in weight and 2.5cm (1 in) in length per year. Toothfish reach sexual maturity when they are between 70 cm (2⅓ ft) and 95 cm (3 ft) long, at about eight to 10 years old.

Like many of its notothenioid relatives, the toothfish lacks a swim-bladder for buoyancy, but it is considered to be virtually neutrally buoyant: this is because it has cartilage in its skeleton instead of bone, which saves weight, and because its body contains large deposits of triglyceride, a lipid (or fat) that is lighter than seawater. Almost 10 percent of the toothfish's bodyweight is fat.

The blood of the Antarctic toothfish contains GLYCOPROTEIN, which acts as an ANTIFREEZE, inhibiting the formation of ice crystals in the subzero water temperatures.

The toothfish is unique among notothenioids in having rod-dominated retinae in its eyes, an adaptation for vision in conditions of very low light.

Toothfish eggs and larvae are pelagic, floating near the sea surface where the larvae feed on ZOOPLANKTON. The adults are permanent members of the mid-water community, living at depths of around 300 to 500 m (984–1646 ft) where they eat smaller fish, larger CRUSTACEANS and SQUID. Toothfish are known to be eaten by SPERM and KILLER WHALES and WEDDELL SEALS, but are usually too large to be eaten by other predators.

Antarctic Treaty *See following pages.*

Antarctic Treaty Consultative Meeting An Antarctic Treaty Consultative Meeting (ATCM) is a meeting of the states that are CONSULTATIVE PARTIES to the ANTARCTIC TREATY. ATCMs are held under the auspices of Article IX of the Antarctic Treaty, which allows meetings 'for the purpose of exchanging information, consulting together on matters of common interest pertaining to Antarctica, and formulating and considering, and recommending to their Governments, measures in furtherance of the principles and objectives of the Treaty.'

Observers from NON-CONSULTATIVE PARTIES to the treaty, non-governmental organizations such as the ANTARCTIC AND SOUTHERN OCEAN COALITION, and intergovernmental organizations like the INTERNATIONAL WHALING COMMISSION may also be invited to the meetings but do not have a vote.

The first ATCM was held in Canberra in 1961, then every two years after that until 1993, when the meetings began to be held annually. Special Consultative Meetings (SCMs) have also been held at irregular intervals since 1977 to discuss issues such as treaty membership and the creation of new agreements, such as the 1991 MADRID PROTOCOL.

Because decisions are made by reaching a consensus, each member state effectively has a veto. Before they can be implemented, decisions must be ratified by the state governments. This decision-making process means that it can take years to reach agreement on an issue; such agreements are often reached by negotiation, bargaining and compromise.

ATCM decisions are usually in the form of recommendations. These can lead to new regulations governing activity in Antarctica or result in the negotiation of a new convention or treaty. The ATCM has a central role in the coordination of the different elements of the ANTARCTIC TREATY SYSTEM (ATS), which facilitates the exchange of information and discussion of the issues affecting Antarctica and the SOUTHERN OCEAN.

Antarctic Treaty System The Antarctic Treaty System (ATS) is a convenient way of referring to the different organizations and agreements that have been developed by the CONSULTATIVE PARTIES of the ANTARCTIC TREATY to manage Antarctica.

Antarctic Trough See CIRCUMPOLAR TROUGH.

Antarctica New Zealand Formed on 1 July 1996 to develop and manage NEW ZEALAND's activities in Antarctica; formally known as the New Zealand Antarctic Institute. It is located at the International Antarctic Centre in Christchurch, coordinates New Zealand's Antarctic research programmes and operates SCOTT BASE on ROSS ISLAND. Antarctica New Zealand represents New Zealand internationally on Antarctic matters.

Antártida Argentina Between 1943 and 1947 ARGENTINA defined a TERRITORIAL CLAIM to a wedge-shaped sector of the Antarctic continent between 25°W and 74°W, below 60°S, encompassing 1,230,000 sq km (475,000 sq miles) of land and ice.

Like CHILE, Argentina makes reference to the 1494 TREATY OF TORDESILLAS in claiming that it inherited historic rights to SOVEREIGNTY over Antarctic territory from Spain. This territorial claim overlaps partially with that of Chile's and completely with the British claim, and is also linked to the dispute with BRITAIN over the FALKLAND ISLANDS that led to WAR in 1982. Previously, Argentina had made formal claims to the SOUTH ORKNEY ISLANDS in 1925, and to the Falkland Islands as a whole in 1937—both of which had been claimed by Britain.

Argentina was one of the original signatories to the ANTARCTIC TREATY, and it has never relaxed its sovereignty claims.

The six bases maintained by Argentina are aimed at settlement as well as science, improving their sovereignty claim. The 'colony' at HOPE BAY at the tip of the ANTARCTIC PENINSULA is the closest to a normal human settlement in Antarctica. It was here, at ESPERANZA STATION, in January 1978 that Emilio de Palma was born, the first Antarctic BIRTH.

Antifreeze Antarctic organisms as diverse as MOSS, SPRINGTAILS and FISH have evolved a variety of antifreeze compounds to combat subzero temperatures and potentially fatal freezing of living tissues.

Many Antarctic fish, particularly those belonging to the suborder NOTOTHENIOIDEI, synthesize GLYCOPROTEINS in their liver, which then circulate in the blood preventing the formation of ice crystals. Glycoproteins are molecules made of repeating units of sugar and amino acids and work by binding to the sides of ice crystals as they form, preventing further growth.

On land, INVERTEBRATES such as MITES and SPRINGTAILS avoid freezing by supercooling; they keep their body fluids liquid at temperatures as low as −35°C (−31°F) with the aid of a variety of compounds, including polyalcohols (eg, glycerol), alcohol derivatives of sugars (mannitol) and sugars themselves (fructose, trehalose). Antifreeze substances are also produced in mosses and some ALGAE.

Antipodes Islands Named the 'Penantipodes' in 1800 by their discoverer Captain Henry Waterhouse, the Antipodes lie in the Pacific Ocean at 50°S, between BOUNTY and CAMPBELL ISLANDS. The group consists of a main island, the Windward Islands, the Leeward Islands and Bollons Island.

On the main island, dark volcanic cliffs rise to an exposed plateau of grassland, densely covered with TUSSOCKS. The Windward Islands are surrounded by volcanic cliffs that have been undercut by the sea, and the coast is laced with narrow caves. Bollons, the second largest island, is the eroded rim of an extinct volcano.

The islands are breeding sites for great numbers of sea BIRDS, including the WANDERING ALBATROSS. During the 19th century, sealers decimated the FUR SEAL populations.

*Above: Of all Antarctica's seasonal visitors, Arctic terns (*Sterna paradisaea*) are the greatest travellers, migrating between the polar regions to benefit from the summer bounty at both extremes of the Earth.*

Anvers Island Discovered in 1898 by Adrien de GERLACHE, Anvers Island is 190 km (118 miles) long and mountainous, and is the largest island in the PALMER ARCHIPELAGO, which lies off the northwest coast of the ANTARCTIC PENINSULA. The USA's PALMER STATION is located on the island's southwest coast. The site of a major OIL spill in 1989, the Argentinian supply ship *Bahia Paraiso* sank near the station. Its nearly submerged hull is still visible.

Large numbers of the WINGLESS MIDGE are found among ANTARCTIC HAIRGRASS and ANTARCTIC PEARLWORT at Biscoe Point, which has been designated a SITE OF SPECIFIC SCIENTIFIC INTEREST.

Arctic and Antarctic Research Institute RUSSIA's oldest and largest research institution concerned with the polar regions.

Established in 1920, it was initially concerned with research and trade in the Arctic region. Formerly the Arctic Research Institute, it was renamed in 1958, when it became responsible for the organization and coordination of Russia's Antarctic science programme.

Arctic tern (*Sterna paradisaea*) Of all the BIRDS, these TERNS are among the greatest travellers. They breed near the Arctic Circle then, as the northern winter approaches, fly to the PACK ICE of Antarctica, a one-way journey of around 20,000 km (12,400 miles) that takes up to four months. This migration allows the terns to take advantage of longer days and, therefore, longer feeding opportunities in both hemispheres. After three months feeding—like most terns, their main dietary component is FISH and CRUSTACEANS—they

begin the long flight back to their northern breeding grounds, leaving the young first- and second-year birds behind.

Breeding starts around the end of May. One to three eggs, usually coloured blue or brown, and occasionally pink, are laid in shallow depressions on the ground. Once hatched, the chicks can fly after about three weeks but remain dependent on their parents for much longer.

Small and compact—they are 36 cm (14 in) long, with wing-spans double their length—their wings are narrow and set forward on the body to offset their long tails.

Arçtowski, Henryk (1871–1958) Polish scientist and explorer, born in Warsaw. Arçtowski studied in Paris, and joined Adrien de GERLACHE's 1897–99 BELGIAN ANTARCTIC EXPEDITION. Arçtowski produced the first detailed account of the physical geography and geology of Antarctica and described 'the way geological surveys had to be carried out' in the ANTARCTIC PENINSULA: 'A few strokes of the oars brought us to the beach amid cries of "Hurry up, Arçtowski!" I gave a hammer to Tellefsen, with orders to chip here and there down by the shore, while I hurriedly climbed the moraine, picking up specimens as I ran, took the direction with my compass, glanced to the left and the right, and hurried down again at full speed to get a look at the rock *in situ*; meanwhile Cook had taken a photograph of the place from the ship.' Arçtowski was one of the first to recognize that WINDCHILL could be as dangerous to the human body as just cold alone.

In his book, *Through the First Antarctic Night*, the expedition's doctor Frederick COOK

described Arçtowski 'interviewing' a sea leopard: 'The animal sprang from a new break in the ice onto the floe, upon which [Arçtowski] had a number of delicate meteorological instruments, and without an introduction, or any signs of friendship, the animal crept rapidly over the snow and examined Arçtowski and his paraphernalia with characteristic seal inquisitiveness. ... Arçtowski made warlike gestures, and uttered a volley of sulphureous Polish words, but the seal didn't mind that. ... Now and then its lips moved, and there was audible a weird noise, with signs we took to be the animal's manner of inviting its new companion to a journey under the icy surface, where they might talk the matter over out of the cold blast of the wind, in the blue depths below.' Over an entire year, Arçtowski made daily meteorological observations; his findings were recorded in an appendix to Cook's book, along with another appendix on 'The bathymetrical conditions of the Antarctic regions.' In 1919 Arçtowski was appointed professor of geology at Luwow University, Poland, and in 1939 he moved to the USA, where he remained for the rest of his life.

Arçtowski Station Arçtowski Station, POLAND's first permanent Antarctic base, was established in ADMIRALTY BAY, KING GEORGE ISLAND, in 1977 and has operated continuously since. It was established in part so that Poland could become the 13th CONSULTATIVE PARTY to the ANTARCTIC TREATY. The station has an OVERWINTERING population of up to 20 people. Research is conducted into oceanography, geology, geomorphology, glaciology, meteorology, climatology, seismology, magnetism and ecology.

Antarctic Treaty

The Antarctic Treaty was signed by 12 states on 1 December 1959 and has operated continuously since it entered into force on 23 June 1961. The treaty was the result of long negotiations at the WASHINGTON CONFERENCE. The treaty is relatively brief, consisting of 14 articles and a short preamble.

Article I of the treaty states that 'Antarctica shall be used for peaceful purposes only'. All activities of a military nature, including the establishment of military bases or fortifications, testing weapons, and carrying out military manoeuvres, are prohibited. Article V also specifies that nuclear explosions and the disposal of radioactive waste are prohibited in Antarctica, but allows the peaceful use of nuclear energy. The use of military personnel and equipment is permitted for scientific research and other peaceful purposes, such as providing logistical support, equipment and supplies to the bases.

Article II calls for the freedom of scientific investigation in Antarctica that had occurred during the INTERNATIONAL GEOPHYSICAL YEAR to continue. This is promoted in Article III, in which it is agreed to exchange information about planning for scientific programmes, to exchange scientists across expeditions and stations, and for scientific observations to be exchanged and be made freely available.

The problematic issue of TERRITORIAL CLAIMS is addressed in Article IV, which effectively retains the legal *status quo* of 1959 by stating: 'No acts or activities taking place while the present Treaty is in force shall constitute a basis for asserting, supporting or denying a claim to territorial sovereignty in Antarctica or create any rights to sovereignty in Antarctica. No new claim, or enlargement of an existing claim, to territorial sovereignty in Antarctica shall be asserted while the present treaty is in force.'

This does not resolve the problem of DISPUTED CLAIMS—as Australian diplomat Keith Brennan noted, 'the Treaty did not freeze territorial claims, it merely put them on the back burner to keep warm.' Article IV is crucial to the treaty's success because it accommodates the different interests of the states claiming territory, the states which might want to claim territory in the future, and the states that do not recognize any territorial claims in Antarctica.

Article VI applies the provisions of the treaty to the area below 60°S latitude, including all ICE SHELVES, but it does not affect the rights of states under international law on the high seas within that area. Article VIII provides for jurisdiction over scientists, observers and support staff to be retained by the state of which they are nationals. This is important in helping to preserve the 'freeze' on SOVEREIGNTY issues in Article IV. One of the major loose ends of the treaty is that nationals of third-party states and other private citizens are not covered by this article and jurisdiction over them is unclear. To help solve problems such as this, the treaty members are required to 'consult together with a view to reaching a mutually acceptable solution.' Article XI provides a mechanism for resolving disputes that allows the parties to refer a case to the International Court of Justice. No such disputes had occurred by 2002.

The treaty does not contain any enforcement provisions, but Article VII establishes a simple and practical observation scheme. Treaty members are allowed to appoint observers, who at any time can freely inspect all 'stations, installations and equipment within those areas, and all ships and aircraft at points of discharging or embarking cargoes or personnel in Antarctica.' The exchange of scientists between national research programmes also acts as a *de facto* system of inspection. Inspections also allow observers to see if compliance with other regulations of the ANTARCTIC TREATY SYSTEM, such as the environmental rules of the MADRID PROTOCOL, are being correctly observed. The USA has carried out the most inspections; only a few other members have conducted inspections, mainly because they are expensive and time consuming. No violations of the Antarctic Treaty had been detected by 2002.

Article IX requires the CONSULTATIVE PARTIES to meet at suitable times and places to exchange information, consult together, and recommend measures that further the principles and objectives of the treaty. This is an important article because it allows the treaty to evolve over time. ANTARCTIC TREATY CONSULTATIVE MEETINGS (ATCMS) are now held every year. During the treaty negotiations, agreement could not be reached on administrative arrangements and the treaty has never had a secretariat. Although most members now agree that a secretariat would be useful, they have been unable to agree on specifics, including its location.

Article XII allows the treaty to be modified by the unanimous agreement of the consultative parties at an ATCM. Provision is also made for holding a review conference any time after 30 years from the date the treaty first entered force (post 1991), at which a majority decision could change the treaty. No such formal review has been held, in part because the negotiations for the Madrid Protocol during 1989–91 served to allow a comprehensive review of the Antarctic Treaty.

The 12 original signatories to the Antarctic Treaty are: ARGENTINA, AUSTRALIA, BELGIUM, CHILE, FRANCE, JAPAN, NEW ZEALAND, NORWAY, SOUTH AFRICA, USSR, BRITAIN and the USA. (The former USSR is now represented by RUSSIA.) These 12 have been joined by 15 other states which have demonstrated their interest by 'conducting substantial research activity' in Antarctica: BRAZIL, Bulgaria, CHINA, Ecuador, Finland, GERMANY, INDIA, ITALY, the Netherlands, POLAND, Peru, the Republic of Korea, SWEDEN, Spain and Uruguay. All these states are consultative parties.

As at May 2000, a further 17 states had acceded to the treaty without becoming consultative parties: AUSTRIA, Canada, Colombia, Cuba, Czech Republic, Democratic Peoples Republic of Korea, Denmark, Greece, Guatemala, Hungary, Papua New Guinea, Romania, Slovak Republic, Switzerland, Turkey, Ukraine and Venezuela. These states are known as NON-CONSULTATIVE PARTIES.

Below: Delegates attend the Washington Conference, which resulted in the signing of the Antarctic Treaty, in December 1959.

Above: The flags of the first 12 signatory countries of the Antarctic Treaty fly at the South Pole.

Right: In January 1999 New Zealand hosted a meeting for delegates from 24 Treaty member nations, dubbed 'Ministerial on Ice', at Scott Base. It was aimed at drawing attention to Antarctic environmental issues and encouraging policies, which more actively protect the region. The entourage visited various locations, including the Dry Valleys, several bases, and (right) Castle Rock on Hut Point Peninsula, Ross Island.

Above: New Zealand artist Nigel Brown painted his impressions of Antarctica in 1998.

Argentina Geographically, Argentina is one of the closest countries to Antarctica. It first demonstrated its interest in Antarctica in 1904, when it took over management of ORCADAS STATION from the SCOTTISH NATIONAL ANTARCTIC EXPEDITION on Laurie Island in the SOUTH ORKNEYS, and it made formal claims to the South Orkneys in 1925, and to the Falkland Islands Dependencies (FID) as a whole in 1937. Although Argentina was involved in WHALING in the SOUTHERN OCEAN, its activity on the Antarctic continent remained low until World War II, when expeditions were sent to the ANTARCTIC PENINSULA and the SOUTH SHETLAND ISLANDS in 1942 and 1943.

Between 1943 and 1947 Argentina defined a TERRITORIAL CLAIM to ANTÁRTIDA ARGENTINA, a wedge-shaped sector of the continent between 25°W and 74°W, below 60°S, encompassing 1,230,000 sq km (475,000 sq miles) of land and ice. This territorial claim overlaps partially with that of CHILE's and completely with the British claim, and is also linked to the dispute with BRITAIN over the FALKLAND ISLANDS. The Argentinian government has taken a high level of interest in Antarctica and in maintaining its territorial claim. President Lastiri and the Argentinian Cabinet spent August 1961 at Marambio Base, which was declared the provisional seat of government for the month.

Large expeditions in 1946–47 and 1947–48 established bases in the Antarctic Peninsula, which the Argentinians call San Martin Land. Between 1951 and 1956 the INSTITUTO ANTÁRTICO ARGENTINO was established to develop scientific and technological activities in Antarctica.

A participant in the INTERNATIONAL GEOPHYSICAL YEAR, Argentina was one of the original 12 CONSULTATIVE PARTIES to the ANTARCTIC TREATY. During treaty negotiations, it opposed the dumping of NUCLEAR WASTE in Antarctica, attempts at greater internationalization of the continent, and the concept of binding arbitration of Antarctic disputes through the International Court of Justice. The treaty has been useful in reducing tensions with Britain and Chile, although Argentina and Britain fought a short WAR over the Falklands in 1982 and a dispute with Chile over some islands in the DRAKE PASSAGE was not resolved until 1984.

Argentina maintains six year-round bases, one of the largest national efforts in Antarctica. It has some regional economic interests, notably in FISHING in the South Atlantic sector of the Southern Ocean. Some Argentinian fishing operators have been involved in the illegal, unregulated and unreported TOOTHFISH FISHING. Argentina benefits from the expanding interest in Antarctic TOURISM, Ushuaia being the main departure point. Science remains an important activity, and a wide range of programmes in earth, biological and atmospheric sciences is conducted.

Argentine Islands Islands off the west coast of the ANTARCTIC PENINSULA discovered by Jean-Baptiste CHARCOT's 1908–10 FRENCH ANTARCTIC EXPEDITION and named in recognition of the assistance given to the expedition by ARGENTINA. FARADAY STATION on Galindez Island, which was transferred from BRITAIN to the UKRAINE and renamed VERNADSKY STATION in 1988, is the oldest operational station in the Peninsula area. British scientists discovered the OZONE HOLE at this base in 1985.

Arnoux's beaked whale (*Berardius arnouxii*) One of the TOOTHED WHALES, it is also called the Southern fourtooth whale and some researchers consider it to be the same species as BAIRD'S BEAKED WHALE (*B. bairdii*). They range south as far as the PACK ICE, and grow to 10 m (32 ft) in length and 7 to 8 tonnes (7–8 tons) in weight. As with many members of the beaked whale family (Ziphiidae), little is known about their reproductive cycle, lifespan, diet, behaviour and complete range.

Artists Before the invention of PHOTOGRAPHY, voyages of exploration relied on artists to record newly discovered places and species. The first images of the Antarctic region were by William HODGES, artist on James COOK's 1772–75 voyage. When Ernest Goupil, official artist on Jules-Sébastien DUMONT D'URVILLE's 1837–40 circumnavigation, died from dysentery, he was replaced as artist by the expedition surgeon Louis Le Breton; drawings by both artists were published as lithographs in the 10-volume *Atlas Pittoresque*, accompanying d'Urville's narrative of the voyage. Other official artists included Frank Stokes on the

ANTARCTIC EXPEDITION, who sailed north with 150 Antarctic sketches before the rest of the members were shipwrecked, and George Marston, on the two expeditions led by Ernest SHACKLETON.

Because Antarctica is now recognized for its cultural as well as scientific value, many artists now visit the continent as part of national 'Artists in Antarctica' programmes. One of the earliest was in 1964, when Australian painter Sidney Nolan flew to MCMURDO BASE as the guest of the US ANTARCTIC PROGRAM. Nolan exhibited Antarctic works in New York and London the following year.

Astrolabe Island Home to colonies of CHINSTRAP PENGUINS, this small island is located in Bransfield Strait at the western tip of the ANTARCTIC PENINSULA. The island was discovered by Jules-Sébastien DUMONT D'URVILLE in 1838, and named after his ship.

Astronomical research Some of the earliest voyages to the Antarctic regions were for the purpose of astronomical research. Edmond HALLEY's second voyage to the Southern Hemisphere in 1699–1700, which reached 52°24'S, just south of the POLAR FRONT, was to test whether celestial observations were an accurate method of determining longitude at sea.

William WALES and William Baly, astronomical observers on James COOK's second voyage, published the *Original Astronomical Observations, made in the course of a Voyage towards the South Pole, and Round the World* in 1777.

I M Simanov, professor of astronomy, was a member of Thaddeus BELLINGSHAUSEN's 1819–21 voyage.

Today, such mysteries as the origins of galaxies and the fate of the universe are being investigated at the SOUTH POLE. A world centre for astronomy, it is ideal because of its high altitude, high LATITUDE, clear skies and the long period of WINTER darkness. Telescopes at the Pole are able to detect very faint objects at long distances in the night sky.

Antarctic astronomical research programmes did not begin in earnest until the 1960s. The USA's NATIONAL SCIENCE FOUNDATION established an astrophysical research centre based in an area called the DARK SECTOR at the South Pole, where the unique atmospheric qualities are preserved. In higher latitudes, water vapour absorbs infrared radiation, but at the Pole most radiation from deep space reaches the surface. Invisible to the naked human eye, it is detected by telescope and the data obtained are used to identify young galaxies.

Scientists are searching for ripples in background microwave radiation emitted from the Big Bang, and have recently found that the universe is expanding at an accelerating rate, not shrinking as was once thought. Astronomers at the Pole are also studying tiny subatomic 'ghost' particles called 'neutrinos', which constantly bombard the surface of the Earth, and are so small that they fly through solid objects but are filtered by ice, to learn more about power sources of galaxies, eruptions from stars and life cycles of supernovae.

Above: *Lieutenant Edward Evans observes an occultation of Jupiter in 1911 during the 1910–13 British Antarctic Expedition.*

Atlantic Ocean South of the eastern Americas and Africa, the Atlantic Ocean merges with the SOUTHERN OCEAN at the northern tip of the ANTARCTIC PENINSULA and stretches about 30° longitude, embracing TRISTAN DA CUNHA and GOUGH ISLANDS, as well as the SOUTH ORKNEYS, SOUTH SHETLANDS, SOUTH SANDWICH ISLANDS, SOUTH GEORGIA and the world's most isolated island BOUVETØYA.

Atmosphere The atmosphere is the Earth's shield, surrounding and protecting it from the sun. It is divided into layers with different temperatures and different physical properties. Antarctica's WEATHER is produced in the 'troposphere'; the lowest 10 km (6 miles) of the atmosphere, where heat and energy move freely, and the air is dynamic and turbulent. The troposphere is capped by the 'stratosphere', a stable lid on the atmosphere below.

The stratosphere acts as an OZONE 'trap', and is a key element in the 'GREENHOUSE EFFECT' and the OZONE HOLE. In summer, ultraviolet SOLAR RADIATION is absorbed by ozone, which warms the stratosphere and sets up a weak easterly airflow. As winter sets in and Antarctica is plunged into darkness, the TEMPERATURE of the stratosphere drops by as much as 50°C (90°F) and its WIND strengthens and reverses direction. It spirals around the SOUTH POLE, forming the 'CIRCUMPOLAR VORTEX', a whirlpool of wind that persists over winter and seals off the Antarctic stratosphere. The sun's return weakens, then destroys the vortex, causing rapid heating of the stratosphere. A chain reaction that eats away ozone, and leads to the formation of the 'hole' over Antarctica, begins.

Above the stratosphere lie the 'mesosphere', reaching to 90 km (56 miles), and the 'thermosphere', reaching to about 700 km (430 miles). Beyond this again is the upper atmosphere, or 'geospace', a vast area extending to over 50,000 km (32,240 miles) above the surface, where the sun's atmosphere and magnetic field interact with those of the Earth. At the outer surface of geospace is the 'magnetosphere', which contains the Earth's magnetic field, and the 'ionosphere', an electrically charged layer that absorbs ULTRAVIOLET RADIATION from the sun in a chemical reaction called ionization. This process splits gas molecules into electrons and ions and creates ionized gases that reflect radio waves, making long-distance communication in Antarctica possible.

'Plasma'—ionized gas emitted from the sun—travels towards Earth in a 'SOLAR WIND' that hits the magnetosphere at about 3 million km (1.8 million miles) per hour. Over most of the Earth the solar wind compresses the outer surface of the magnetosphere, and stretches it out in a long comet-like tail. But over the South Pole, magnetic field lines thread back together and arc down towards the surface, funnelling the solar wind down through the magnetosphere. Charged plasma penetrates deep into the polar atmosphere at this 'magnetic cusp', generating spectacular AURORA or 'southern lights' and other OPTICAL PHENOMENA. Radio waves, generated from natural sources, such as lightning, and from artificial sources on the Earth's surface, are also carried in magnetic field lines, and reenter the atmosphere above Antarctica.

Major 'eruptions' on the surface of the sun pelt the magnetosphere with plasma, and cause magnetic storms in geospace. The plasma is accelerated towards the surface by huge voltages generated in the magnetosphere; the energy from the solar wind increases from 10,000 megawatts to 15 million megawatts during a storm. Disrupting the magnetic field, even down to the surface of Antarctica, the storms play havoc with radio communications and navigation systems; they can deflect SATELLITES off course or damage them with intense radiation and may even cause power cuts.

*Above: One of the smaller cormorant species, the Auckland Islands cormorant (*Phalacrocorax colensoi*) nests in tussock grasses.*

Atmospheric research Antarctica's geographical position makes it an ideal vantage point for the observation of atmospheric processes. It is the only place on Earth where there is a land mass at polar latitudes on which scientists can place instruments to observe the ATMOSPHERE. Also, data are not complicated by interference from industrialization or other human activities.

The upper atmosphere, or geospace, is an important focus in Antarctic research, and Antarctica is ideal for the observation of radio waves; the atmosphere is virtually transparent to many wavelengths. The radio waves that reach the continent's surface carry information about the sun's energy and the way it is transmitted through the layers of the atmosphere. This has practical implications in the study of magnetic storms that affect radio communications, navigation and SATELLITES. Information is collected from ground-based data, satellite data and complex computer modelling. There is a network of automatic OBSERVATORIES that collect year-round data from geospace in remote regions of Antarctica, and several permanent observatories have been established to study the ionosphere.

After evidence of the existence of the OZONE HOLE was presented in 1985, there was a great increase in polar chemistry research in Antarctica. Changes in OZONE concentration in the Antarctic stratosphere are monitored by measuring ULTRA-VIOLET RADIATION (which is absorbed by ozone in the stratosphere); if the amount reaching the surface increases, then ozone levels must be decreasing. Spectrometers also measure levels of other atmospheric gases such as nitrogen dioxide, and the minute amounts of UV light emitted from the

moon and planets that reach Antarctica over winter. In the summer months, conditions are often ideal for ballooning experiments: the circulation of WIND around the stratosphere allows BALLOONS to circumnavigate the continent, gathering information on OPTICAL PHENOMENA. Satellites also provide continuous data and imagery of ozone concentrations, and laser systems can detect the formation of polar stratospheric clouds, in which ozone destruction takes place.

Auckland Islands The largest group of islands in the New Zealand subantarctic region, discovered in 1840 by Captain Abraham Bristow on his whaling ship *Ocean*. Four main islands—Adams, Auckland, Disappointment and Enderby—are made up of two large VOLCANOES: Ross Volcano, centred on Disappointment Island, and the Carnley Volcano, which makes the rim of Carnley Harbour on Auckland Island. Sheer cliffs, rippled by old lava flows, border the western and southern coasts, and tower up to about 300 m (1000 ft). In the path of the ROARING FORTIES, this dramatic landscape is swept by powerful winds and waves. Serrated by deep fiords that have been carved by glacial action, the eastern coasts are relatively sheltered. There are two large, well-protected harbours, Carnley and Ross, on the eastern shores of Auckland Island.

The islands have a rich array of flora, closely related to plants on the New Zealand mainland. The largest subantarctic forest grows on the east and south coasts of the Auckland Islands. The islands are home to the largest breeding colony of WANDERING ALBATROSS, about 50,000 WHITE-CAPPED MOLLYMAWKS, and yellow-eyed penguins.

In summer, HOOKER'S SEA LIONS converge on Enderby Island to breed. The ENDERBY ISLAND SETTLEMENT, established in 1849 and abandoned several years later, was the only permanent settlement attempted on these islands, apart from short-lived Maori habitation and two failed sheep farmers. The New Zealand government declared the islands a nature reserve in 1934.

The Auckland Islands are known as a 'ship's graveyard'. At least eight vessels have sunk along the west coast, including the famous *General Grant*, and more than 100 lives have been lost in SHIPWRECKS. In 1880, the New Zealand government began setting up huts for CASTAWAYS containing basic supplies and, until 1927, made annual trips to the islands in search of shipwrecked crews. The most recent shipwreck was that of the yacht *Totorore* that sank in South Bay in 1999; no trace has been found of the crew.

Auckland Islands cormorant (*Phalacrocorax colensoi*) One of the smaller and most vulnerable cormorants, they are 63 cm (25 in) long, and are found only in the AUCKLAND ISLANDS, where they breed in colonies among the TUSSOCK grasses. The population of about 5000 birds is threatened by pigs and CATS.

Aurora Ernest SHACKLETON purchased the *Aurora* for the ROSS SEA section of his IMPERIAL TRANS-ANTARCTIC EXPEDITION of 1914–17. The ship was 40 years old and had sailed to Antarctica with the AUSTRALASIAN ANTARCTIC EXPEDITION of 1911–14.

It left Sydney on 15 December 1914 under the command of Captain J R Stenhouse, and reached MCMURDO SOUND on 24 January 1915. The ship was intended to serve as a base for the party, and was moored to the shore at Cape EVANS. During a blizzard on 6 May the *Aurora* broke free and, over the next 10 months, drifted 1900 km (1025 naut. miles) north, and eventually reached Port Chalmers, in New Zealand. When repairs were completed, the *Aurora* returned south, this time with Shackleton, recovered from his heroic trek from ELEPHANT ISLAND to SOUTH GEORGIA, on board. They reached Cape Evans on 10 January 1917. The stranded men, equipment and specimens were loaded and on 17 January the ship sailed north again, steaming into Wellington Harbour, New Zealand, on 9 February, where it was sold for £10,000 to pay off the crew and an outstanding loan. *Aurora* was last heard of in mid-1917 when it left Newcastle bound for Chile with a load of coal and was lost without trace.

Aurora australis Also referred to as the 'southern lights', the aurora is a manifestation of complex processes that occur in the upper ATMOSPHERE. The aurora was known to Maori of New Zealand as 'Tahu-nui-a-rangi', or 'the great burning of the sky'.

Like its Northern Hemisphere counterpart the aurora borealis, the aurora australis has its origins in the SOLAR WIND, a cloud of charged particles blowing radially off the sun. At times of enhanced solar activity, manifested by increased numbers of sunspots, the velocity of the solar wind increases. On its night side, the Earth's magnetic field lines enclosed in the magnetosphere (see ATMOSPHERE) are stretched out by this supersonic flow into a 'tail', into which the solar wind particles can enter. Because the particles are both positively and negatively charged, an electrical field—or potential—is established across the tail of the magnetosphere. This causes the charged particles to be accelerated down the field lines and into oval-shaped auroral regions above the north and south polar regions—oval-shaped because they represent the 'footprint' of the distorted tail field lines mapping down to the Earth.

As the energetic electrons enter the atmosphere, they ionize nitrogen molecules (that is, a secondary electron is ejected), which then emit light strongly in the ultraviolet region. The secondary electrons continue downwards, exciting nitrogen and oxygen gases, which then also emit light at characteristic wavelengths (colours)—a process akin to that seen in mercury, sodium and neon gas fluorescent tubes.

In the aurora, the dominant colour visible is greenish white, due to the excitation of atomic oxygen; this is also the first colour usually seen, partly because the colour sensitivity of the eye peaks in the green part of the spectrum. The light emissions typically occur at heights above the Earth of from 100 to 1000 km (62–620 miles), depending on the altitude distribution of the atmospheric gases, and on the energy and sign of the incoming particles. During intense auroral events, these very high emissions have been seen from the Equator.

Early in the evening the aurora appears as smooth arcs of light, which later may develop bars of light, called rays, that are aligned along the magnetic field lines. These 'curtains' become more active, first swaying, then bending and flowing—a crimson red sometimes seen at the bottom of a curtain is caused by excitation of nitrogen molecules. Around midnight, the rays may appear to spread out from a central focal point, called a 'corona', the result of a parallax effect when looking directly up the magnetic field lines. Later, the aurora will decay, sometimes leaving a faint background glow, through which the occasional light may flicker.

The first recorded sighting of this OPTICAL PHENOMENON was on 17 February 1774, from 57°8'S, 80°59'E, on James COOK's final voyage to the Pacific. In March 1820 Thaddeus BELLINGSHAUSEN sighted an aurora '... in all its magnificence and brightness ... The whole vault of the heavens except 12° or 15° from the horizon was covered with bands of rainbow colour which, with the rapidity of lightning, traversed the sky in sinuous lines from south to north, shading off from colour to colour.' Eleven years later, on 3 March 1831, John BISCOE and his crew were in the SOUTHERN OCEAN on a night when the aurora appeared: 'Nearly the whole night, the Aurora Australis showed the most brilliant appearance, at times rolling itself over our heads in beautiful columns, then suddenly forming itself as the unrolled fringe of a curtain, and again suddenly shooting to the form of a serpent, and at times appearing not many yards above us ...'

Because the aurora is basically a discharge phenomenon, electric currents as large as 1 million amperes can flow along the auroral curtains at heights of about 100 km (62 miles), sometimes resulting in a disruption of power distribution systems on the Earth below.

Aurora Australis The first BOOK produced in Antarctica. The 120-page book with cover and lithographs by George Marston was printed at the sign of 'The Penguins', at Cape ROYDS, by Ernest JOYCE and Frank WILD in 1908 during the BRITISH ANTARCTIC EXPEDITION. The *Aurora Australis* was edited by Ernest SHACKLETON and contained contributions from many expedition members.

Below: Also known as the 'southern lights', the aurora australis phenomenon is created by charged particles high in the atmosphere. This view was photographed using a long exposure, giving rise to the apparent movement of the stars.

Australasian Antarctic Expedition (1911–14) Australian and New Zealand exploratory and scientific expedition, led by Douglas MAWSON. This was the first expedition to establish RADIO contact in Antarctica and the first to use an AIRCRAFT, albeit one that did not fly: a Vickers REP monoplane was taken, but its wing was damaged in an accident in Adelaide, Australia, before departure and it was only utilized as a tractor-sledge. The expedition mapped the part of Antarctica nearest Australia, and carried out geological, biological, magnetic and meteorological studies.

The expedition ship, the *AURORA*, captained by John King DAVIS, sailed from Hobart, Australia, on 2 December 1911. It dropped five men at MACQUARIE ISLAND to establish a radio relay station between Antarctica and Australia. On arrival at the Antarctic coastline, a base at Cape DENISON was established in COMMONWEALTH BAY. The *Aurora* dropped a second group of eight men—led by Frank WILD—further along the coast on the SHACKLETON ICE SHELF at QUEEN MARY LAND.

Cape Denison turned out to be one of the windiest places on Earth. Throughout March and April 1912 the wind blew for days on end at speeds between 96 and 128 km (60 and 80 miles) per hour. 'The winds have a force so terrific as to eclipse anything previously known in the world. We have found the kingdom of blizzards,' Mawson wrote. Erection of the radio mast began in April 1912 but, because of poor weather, was not finished until October. Two-way communication between Cape Denison and Macquarie Island was finally achieved in February 1913.

In November 1912 conditions improved slightly and five separate groups set off to explore: three headed east, one went south towards the SOUTH MAGNETIC POLE, and one went west. Mawson led Lieutenant Belgrave Ninnis and Dr Xavier Mertz eastwards to explore the GEORGE V LAND coast. The different groups produced important information about new lands, with detailed scientific measurements, and the western group discovered the first METEORITE in Antarctica. All groups, except Mawson's, returned by 15 January 1913, when the *Aurora* was scheduled to leave.

Mawson's group was in trouble. On 14 December Ninnis fell down a CREVASSE to his death, together with a SLEDGE carrying most of the food and supplies. About 584 km (315 naut. miles) from Cape Denison, the two remaining men turned back, killing their DOGS for food. Mertz died on 7 January—years later it was discovered that he had been poisoned by the Vitamin A from the dogs' livers.

Mawson trekked the final 185 km (100 naut. miles) alone with very little food, surviving a fall down a crevasse and a BLIZZARD. He arrived at Cape Denison on 8 February to see the *Aurora* sailing away. Captain Davis had waited as long as possible but the winter ice was closing in. However, a six-man search party had stayed behind; they spent another winter in the Antarctic until the ship picked them up the following spring. They returned to Australia on 5 February 1914.

Frank Wild's western party, which had been picked up by the *Aurora* on 23 February 1913, explored the coastline on the Shackleton Ice Shelf. A party sent eastwards reached the Denman Glacier, and others crossed the Helen Glacier, discovering large PENGUIN rookeries.

Australia Australia's earliest involvement in Antarctica was through participation in British expeditions. Douglas MAWSON, who had taken part in the 1907–09 BRITISH ANTARCTIC EXPEDITION, led the 1911–14 AUSTRALASIAN ANTARCTIC EXPEDITION and the 1929–31 BRITISH-AUSTRALIAN-NEW ZEALAND ANTARCTIC RESEARCH EXPEDITION (BANZARE), which charted the East Antarctic coastline from ENDERBY LAND to GEORGE V LAND.

BANZARE prepared the way for a 1933 British TERRITORIAL CLAIM that was then placed under the control of Australia in June 1933. The AUSTRALIAN ANTARCTIC TERRITORY is the largest claim in Antarctica, covering 6,500,000 sq km (2,509,650 sq miles) in two wedges from 45°E and 136°E, and 142°E and 160°E, below 60°S. The area between the two sectors, TERRE ADÉLIE, is claimed by FRANCE. The AUSTRALIAN NATIONAL ANTARCTIC RESEARCH EXPEDITIONS (ANARE) was created in 1947.

Australia participated in the INTERNATIONAL GEOPHYSICAL YEAR, and was one of the original 12 CONSULTATIVE PARTIES to the ANTARCTIC TREATY. The first ANTARCTIC TREATY CONSULTATIVE MEETING (ATCM) was held in Canberra in 1961, and the permanent secretariat of the CONVENTION ON THE CONSERVATION OF ANTARCTIC MARINE LIVING RESOURCES (CCAMLR) is located in Hobart, Tasmania.

Australia operates bases in Antarctica and the SUBANTARCTIC ISLANDS, and has a continuing research programme. By choosing not to ratify the CONVENTION ON THE REGULATION OF ANTARCTIC MINERAL RESOURCE ACTIVITIES (CRAMRA) in 1989—partly because of sovereignty issues, and partly because of strong environmental lobbying—Australia played a crucial role in the debate over Antarctic MINING. It has a significant FISHING industry around its subantarctic islands, and has increased maritime surveillance there to combat 'pirate' TOOTHFISH FISHING. Australia has also been involved in Antarctic TOURISM through overflights of the continent and as a departure point for CRUISES.

Australian Antarctic Division The Australian Antarctic Division was established in 1948 to direct AUSTRALIA's Antarctic programme and is based at Kingston, near Hobart, Tasmania. It has over 300 staff and four permanent stations in Antarctica, conducts scientific research and represents Australia internationally on Antarctic issues.

Australian Antarctic Territory A 1933 British TERRITORIAL CLAIM was transferred to Australia in June 1933. The Australian Antarctic Territory is the largest territorial claim in Antarctica, covering 6,500,000 sq km (2,509,650 sq miles) in two wedges from 45°E and 136°E, and 142°E and 160°E, below 60°S. The 1954 Australian Antarctic Territory Act provides for the governance of the Australian Antarctic Territory (AAT).

An original CONSULTATIVE PARTY to the ANTARCTIC TREATY, Australia has always maintained its SOVEREIGNTY claims. 'The Australian Antarctic sector is of vital importance to Australia. For strategic reasons, it is important that this area, lying as it does so close to Australia's backdoor, remains under Australian control,' foreign minister Richard Casey said in 1953. One real fear was the establishment of MILITARY BASES in Antarctica. During the INTERNATIONAL GEOPHYSICAL YEAR, the USSR established several bases in the territory Australia claims.

Australia declared an Exclusive Economic Zone (EEZ) for its territorial claim in Antarctica in 1979, but has not taken any steps to enforce it.

Above: ANARE station on Macquarie Island.

Australian National Antarctic Research Expeditions (ANARE) Douglas MAWSON, an influential proponent of Antarctic exploration, encouraged the establishment of the ANARE programme in 1947, which is administered by the AUSTRALIAN ANTARCTIC DIVISION. Headed by Philip LAW from 1949 to 1966, ANARE set up several bases on HEARD and MACQUARIE ISLANDS in 1947–48 and the oldest permanently occupied scientific base south of the Antarctic Circle, MAWSON BASE in MAC.ROBERTSON LAND in 1954.

Austria Austria's involvement in Antarctica has been limited: the traditional focus of its polar efforts has been in the Arctic.

A proposed Austrian Antarctic expedition, under the leadership of Dr Felix König (who had participated in the 1911–12 GERMAN SOUTH POLAR EXPEDITION), was halted by the outbreak of World War I in 1914, as the expedition party was readying to leave Trieste. Austria acceded to the ANTARCTIC TREATY in 1987 and signed the MADRID PROTOCOL in 1991.

b

Bacteria The most abundant organisms on Earth. Each litre (1.76 pints) of sea water is estimated to contain about a billion bacteria. Bacteria are minute and have been found everywhere that life can exist. A small number can cause disease in other organisms, but most are decomposers, feeding on waste and dead organisms. They are an important component of Antarctic lake sediments, where they are involved in decomposition and recycling of nutrients. Bacteria are also found in ICE SHEETS at the North and South Poles and have been recovered and germinated from centuries-old deposits of Antarctic ice and snow.

Baird's beaked whale (*Berardius bairdii*) One of the TOOTHED WHALES found in Antarctic waters, it is sometimes considered to be the same species as ARNOUX'S BEAKED WHALE (*B. arnouxii*), although it is larger, growing to 12 m (39 ft) and 13 to 15 tonnes (13–15 tons) in weight. As with many members of the beaked whale family (Ziphiidae), little is known about their reproductive cycle, lifespan, diet, behaviour and complete range.

Baleen whale Five of the six species of baleen whales that occur in Antarctic waters—BLUE, FIN, SEI, HUMPBACK and MINKE—are also found in the Northern Hemisphere, although the populations are separate and do not mix. The exception is the SOUTHERN RIGHT WHALE, which, as the name suggests, is found only in southern waters.

Baleen whales have no teeth, but instead feed using a collection of slender triangular plates of horny material (baleen) that grow from their upper jaws. Baleen is very similar to human fingernails and the plates are fringed with bristles, through which the whales filter ZOOPLANKTON, particularly KRILL, from the water. Baleen whales feed by taking large quantities of sea water into their mouths, then, using their huge tongues (in the right whale the tongue is the size of a small car) they push the water through the 'comb' of baleen. Any food in the water is trapped inside the comb, and the water is squirted out the sides of the mouth. Although they all eat some of the same food, each species targets krill at different stages of development and therefore direct competition is minimal.

Baleen whales are found in Antarctic waters in summer, taking advantage of the abundant food available then. Breeding takes place in warmer tropical or subtropical waters: migration distances can exceed 10,000 km (over 6200 miles). Migrations occur sequentially, with the blue whales arriving first, followed by the fins, hump-

Right: A large hydrogen-filled balloon was the first aircraft to ascend in Antarctica when Robert Scott's National Antarctic Expedition used it to survey the ice shelf.

backs and finally the sei whales; there is segregation within species too, according to size and feeding requirements, with pregnant females and older males generally arriving early. Time spent in Antarctic waters can span four to six months, depending on species and class within species. It is thought the smaller minke whales may, in fact, winter in the cold Antarctic waters.

Schools of minkes and fins can number up to a 100 in the feeding grounds, although groups between 10 and 25 are more common. The other baleen species tend to congregate in threes and fours. The average reproductive cycle is two to three years; gestation is between 10 and 12 months, and birth occurs in the warmer winter grounds. Most baleen whales are physically mature at around age 25, and their life expectancy can be up to 90 years.

Balleny, John (dates unknown) British sealer and explorer. Earliest knowledge of Balleny is from 1798 when he was in London and part-owner of a brig. He was master of coastal traders until 1831, and was appointed skipper of the schooner *Eliza Scott*, one of the SEALING vessels owned by the ENDERBY BROTHERS. With the 54-tonne (54-ton) cutter *Sabrina*, under the command of Captain Freeman, he sailed south from NEW ZEALAND in 1838. After breaking through PACK ICE in February 1839, Balleny discovered a chain of five high volcanic islands, which he named after himself; he named individual islands after partners in Enderby Brothers' firm, the largest of which, Young Island, rises to a height of 1220 m (4000 ft). Continuing southwards, the two ships took a sample of rock from an ICEBERG and came within sight of part of the Antarctic continent they called the Sabrina Coast. During a storm

shortly afterwards the *Sabrina* was lost. Although he returned with only 178 seal skins, Charles Enderby praised Balleny for his work. Balleny's journal and charts were of great benefit to James Clark ROSS, who was about to set sail when the *Eliza Scott* returned. Little is known of Balleny's later life except that he sailed trading vessels between London and India for some years.

Balleny Islands Located on the ANTARCTIC CIRCLE, north of OATES LAND, the Balleny Islands are the rugged tips of giant VOLCANOES that reach up from 3 km (2 miles) below the ocean surface. English whaling captains John BALLENY and Thomas Freeman made the first landing south of the Antarctic Circle on the islands in 1839. Physicist Louis BERNACCHI, sailing past the Ballenys in 1899 on the SOUTHERN CROSS EXPEDITION, wrote 'I can imagine no greater punishment than to be "left alone to live forgotten and die forlorn" on that desolate shore.'

Balloons Piloted by Robert SCOTT, the first balloon to be launched in Antarctica flew from the ROSS ICE SHELF on 4 February 1902. Named *Eva*, this was the first AIRCRAFT to fly above the continent. From an unsteady perch, in a 'very inadequate' basket, Scott surveyed the landscape from 240 m (787 ft), and noted the progress of the sledging party that had left that day. On the second flight, Ernest SHACKLETON took the first aerial photographs of the continent. Today, private hot-air balloons are occasionally seen in Antarctica's skies.

Balloons have been used routinely for low-level meteorological observations since Scott's 1901–04 expedition, when meteorologist George Clark Simpson used them to determine the air TEMPERATURE above ground level. During the

Above: Antarctica's bedrock, most of which is buried deep beneath the ice cap, has been raised to the surface in the Transantarctic Mountains.

INTERNATIONAL GEOPHYSICAL YEAR, balloons were used to collect data from the upper atmosphere, providing a way of making continuous observations at altitudes of up to 40 km (25 miles) above Antarctica.

Since 1988, scientists have launched a series of balloons into the CIRCUMPOLAR VORTEX to collect valuable high-altitude data for ASTRONOMICAL and ATMOSPHERIC RESEARCH programmes.

Banks, Joseph (1743–1820) British botanist, born in London. The son of a wealthy Lincolnshire family, educated at Oxford University, Banks became interested in botany while a schoolboy. In 1766 he travelled to Labrador and Newfoundland to collect specimens, and in 1768 was appointed by the ROYAL SOCIETY to accompany James COOK on his first voyage to the Pacific. Banks assembled a team of two assistant botanists, two artists, four servants, two greyhounds, a boat, and all the necessary scientific instruments; by the time the *Endeavour* returned in 1771, five of his human entourage had died. Banks's botanical discoveries on that voyage were

as well publicized as Cook's navigational and exploratory achievements.

Chosen to sail to Antarctica on the *Resolution* in 1772, the wealthy Banks spared no expense refitting the ship to suit his own requirements (which included a team of 15); when Cook had the top-heavy additions removed, Banks resigned.

Instead, he went on a botanical expedition to Iceland. Banks was elected president of the Royal Society in 1778, an office he held for the next 41 years, and was appointed director of the Royal Botanical Gardens.

Bay of Whales The base from which Roald AMUNDSEN made his successful attempt to reach the SOUTH POLE in 1911. It was a broad cove on the edge of the ROSS ICE SHELF, due north of ROOSEVELT ISLAND at 160°W, formed by ice breaking away from the shelf; in turn, it disappeared in the 1950s when more of the shelf broke away, taking the Bay of Whales with it. The Bay of Whales was named by Ernest SHACKLETON in January 1908.

The Bay of Whales was also favoured by Richard BYRD, whose 1928–30 expedition built a

base, LITTLE AMERICA, inland from the bay. Byrd returned twice with other expeditions. In 1935 the Bay was the endpoint of Lincoln ELLSWORTH's successful trans-Antarctic flight from the WEDDELL SEA.

Beardmore Glacier One of the largest known valley GLACIERS, it is 200 km (124 miles) long. Descending 2200 m (7216 ft) from the POLAR PLATEAU, it flows north into the ROSS ICE SHELF. It provided a route for Ernest SHACKLETON's 1908 attempt to reach the SOUTH POLE: 'we have now traversed nearly one hundred miles of crevassed ice, and risen 6000 ft ... and we have an extended view of the glacier and mountains,' Shackleton wrote. Robert SCOTT followed the same route in 1912. Recently, scientists have found petrified wood and FOSSILS of ferns and CORAL within the glacier—evidence of the CLIMATE in Antarctica once being much warmer.

Bedrock The basement of Antarctica is very different from the icy exterior of the continent. After World War II, scientists began to probe beneath

the ice with RADAR, and found a remarkable hidden landscape.

The ancient foundations of EAST ANTARCTICA date back at least 3000 million years, to the Precambrian era. The basement ROCK is composed of granites, gneisses and schists that have been heated and compressed (metamorphosed) into a solid continental shield, 40 to 50 km (25 to 30 miles) thick. This bedrock sits well above sea level, and is similar to that of Australia, India, Africa and South America, to which Antarctica was once joined as GONDWANA. In some areas the bedrock has been covered with younger sedimentary material, including limestone and sandstone.

WEST ANTARCTICA joins East Antarctica at the junction of the TRANSANTARCTIC MOUNTAINS, which traverse the continent. The western region is much smaller in area and is made up of a number of crustal fragments, which would form an archipelago of islands if the Ice Cap did not fuse them together. The bedrock of West Antarctica is mainly less than 600 million years old—much younger than the East Antarctic shield. Buckled and folded, it forms a mountainous landscape of Himalayan proportions. However, in some places the terrain is below sea level, often by as much as

2500 m (8200 ft). The mountains that do protrude above the surface of the ice include the highest on the continent, the VINSON MASSIF—4897 m (16,066 ft) above sea level on the SENTINEL RANGE.

Belgian Antarctic Expedition (1897–99) Belgian expedition, the first to winter in the Antarctic, led by Adrien de GERLACHE. The expedition was funded from Belgium, the ship, the BELGICA, a whaler previously named *Patric*, was built in Norway, and expedition members came from a number of countries including Poland, Russia, Romania, Norway and America. They included Roald AMUNDSEN as second mate, Henryk ARĊTOWSKI as geologist, and Dr Frederick COOK as surgeon. The aim of the expedition was to locate the SOUTH MAGNETIC POLE.

The expedition arrived in Antarctic waters on 20 January 1898, late in the season. Two days later a gale hit the *Belgica*, sweeping a sailor, Carl Wiencke, overboard to his death. The ship sailed along the coast of GRAHAM LAND, on the west of the ANTARCTIC PENINSULA, which had not been visited for 60 years. The passage between the coastline and the belt of islands that include ANVERS and the SOUTH SHETLANDS was later named GERLACHE STRAIT. Between 23 January and 12 February 1898 de Gerlache made 20 landings on these islands, which the expedition charted and named, and samples of rocks, lichens, mosses and insects were collected.

On the last day of February the *Belgica* entered PACK ICE and became trapped in the BELLINGSHAUSEN SEA. The expedition was ill-equipped and unprepared to deal with an Antarctic winter. The men suffered dreadfully, physically and psychologically, and their suffering was exacerbated by difficulties of communicating in different languages. 'The curtain of blackness which has fallen over the outer world of icy desolation has also descended upon the inner world of our souls,' Cook wrote. 'Oh for that heavenly ball of fire! Not for the heat ... but for the light—the hope of life.' On 5 June Lieutenant Emile Danco died.

Once the sun returned, research resumed and sledge parties were sent to explore the ice drift. On the last day of the year they sighted open water 640 m (2000 ft) ahead of the ship. The men spent a month attempting to cut a channel and free the ship, and were close to success when the wind changed and the channel closed up—the prospect of a second Antarctic winter caused panic. Eventually, on 14 March 1899, the ship was freed.

The expedition took the first photographic images of the Antarctic, kept consecutive meteorological records over an Antarctic winter, which included a minimum temperature of –43°C (–44°F), and produced a detailed description (by Arċtowski) of the physical geography and geology of the Antarctic Peninsula.

Belgian Scientific Research Programme on the Antarctic Established in 1985 with the aim of actively contributing to global research on climate

Right: Russian explorer Thaddeus Bellingshausen.

and to a science-based conservation and management of Antarctica.

Belgica A three-masted Norwegian-built whaler, the 250 tonne (250 ton) *Patric* was purchased by Adrien de GERLACHE for the 1897–99 BELGIAN ANTARCTIC EXPEDITION. Refitted and renamed the *Belgica*, the ship sailed from Antwerp on 16 August 1897 and reached Antarctic waters on 20 January 1898.

By 2 March the *Belgica* became trapped in PACK ICE on the western coast of the ANTARCTIC PENINSULA at about 71°S, and was not freed until the ice broke up the following March. Over the 12 months it was imprisoned, the ship drifted more than 17 degrees of longitude.

Belgium The involvement of Belgium in Antarctica has been episodic. The Belgian government partially funded the 1897–99 BELGIAN-ANTARCTIC EXPEDITION, the first to OVERWINTER below the ANTARCTIC CIRCLE. During INTERNATIONAL GEOPHYSICAL YEAR it built King Baudouin Base, which closed in 1961. Joint Dutch-Belgian research was conducted from 1964 to 1967, and later Belgian scientists worked at SANAE BASE.

One of the original 12 CONSULTATIVE PARTIES to the ANTARCTIC TREATY, Belgium has no TERRITORIAL CLAIMS in Antarctica. Most Belgian Antarctic scientific investigations are cooperative programmes conducted with other states.

Bellingshausen, Thaddeus (1778–1852) Russian sailor and explorer, born on the island of Ösel in Estonia. Bellingshausen joined the Russian Navy when he was 10 and sailed as a naval officer on the first Russian round-the-world voyage captained by Adam Ivan Krusenstern in 1803–06.

In 1819 he was selected to lead the Russian expedition to Antarctica with the 500 tonne (500 ton) sloops *Vostok* and *Mirnyi*, the latter under the command of Mikhail Petrovich LAZAREV, in what was destined to be one of the great voyages of dis-

covery. In January 1820 the expedition crossed the ANTARCTIC CIRCLE and a fortnight later, at 67°S, are believed to have sighted the Antarctic continent, the first to do so. The expedition retreated north for the winter and returned to the SOUTHERN OCEAN in December. During a circumnavigation of Antarctica, at 69°S Bellingshausen discovered PETER I ØY and, later, ALEXANDER ISLAND; the area between the two is now known as the BELLINGS-HAUSEN SEA. On 15 December he became the first to capture an EMPEROR PENGUIN before sailing on to explore around the SOUTH SHETLAND and SOUTH SANDWICH ISLANDS, previously discovered by James COOK. Here, he observed the wholesale slaughter of SEALS, which he believed were doomed to extinction. In this area, one morning when the fog cleared, Nathaniel PALMER's whaler *Annawan* was sighted between the two Russian ships.

Although little acknowledged in his lifetime, Bellingshausen's was a great achievement. He was a superb navigator and an acute observer, adding considerably to knowledge about Antarctica. He returned to Russia with a great quantity of specimens and scientific data on ICE, WEATHER, atmospheric pressure, WIND, OCEAN CURRENTS and volcanic activity. His account of the voyage was published 10 years later. In 1831 he was promoted to the rank of admiral, and in 1839 he was appointed governor of the port of Kronstadt in Russia.

Bellingshausen Base The Russian Antarctic Research Station at Bellingshausen on KING GEORGE ISLAND, named after Thaddeus BELLINGS-HAUSEN who explored the region in 1820. It was established in February 1968 and has operated continuously since.

Bellingshausen Sea Lies off the coast of Antarctica, southwest of Cape Horn between the ANTARCTIC PENINSULA and the AMUNDSEN SEA. It remains impounded in PACK ICE for the entire year. Unlike that in the WEDDELL SEA, the ice drift does not follow a definite pattern. The ship *Antarctic*, for example, followed a meandering, aimless course when locked in the ice throughout the 1898–99 winter. The sea is named after Thaddeus BELLINGSHAUSEN. Major islands in the sea include ALEXANDER ISLAND, PETER I ØY and CHARCOT ISLAND.

Benthos The Greek word 'benthos' means 'depth of the sea'. A remarkable benthic—or seabed—community teeming with life lies far beneath the great expanses of ice surrounding the Antarctic continent. Here, in these dark cold waters, is one of the most stable marine systems in the world, shaped by three key factors: water temperature, anchor ice formation and annual PLANKTON bloom.

Minerals from sediments deposited by melting ICEBERGS enrich the benthos—it is estimated that 500 million tonnes (500 million tons) of sediment are deposited on the sea floor every year—and, along with the high concentration of oxygen found in very cold water and organic nutrients produced by PHYTOPLANKTON blooms during the long summer daylight hours, create an ideal medium for benthic life.

Benthic organisms include SPONGES, BRY-OZOANS, SEA ANEMONES and CORALS, which are filter-feeders anchored to the seabed. MOLLUSCS, CRUSTACEANS and ECHINODERMS are some of the many varied organisms that move over them. Growing larger and more slowly than temperate counterparts, the benthic organisms have low metabolic rates. Sponges can grow to enormous sizes, and some may be several centuries old. Some LIMPET species also live for over a 100 years. The GIANT ISOPOD (*GLYPTONOTUS ANTARCTICUS*) occupies the niche taken by crabs in other ocean environments (there are no crabs in Antarctica).

Bentley Subglacial Trench Located on the ICE SHEET in WEST ANTARCTICA, the trench is 2539 m (8328 ft) below sea level, the lowest point on the lowest continent on Earth.

Berkner Island The 322 km (200 mile) long and 137 km (85 mile) wide island separates the RONNE and FILCHNER ICE SHELVES in the WEDDELL SEA. Covered by ice, the island was discovered during the INTERNATIONAL GEOPHYSICAL YEAR by an American party led by Finn RONNE. It was named after Lloyd V Berkner, physicist on BYRD'S FIRST EXPEDITION, and is also known as Hubley Island or the Berkner Ice Rise.

Below: Named in recognition of the celebrated explorer, Bellingshausen Base is located on King George Island.

Above: Sponges such as this large specimen (Suberites montiniger) *make up an important component of the Antarctic benthic (sea-floor) community.*

Berlin, Mount Located in MARIE BYRD LAND, Berlin is the only VOLCANO in WEST ANTARCTICA that is still active.

Bernacchi, Louis (1876–1942) Belgian physicist and meteorologist, born in Brussels. His family moved several times, eventually settling in Tasmania, Australia. He studied astronomy, magnetism and meteorology at the Melbourne Observatory, and in 1897 was appointed scientist with the 1897–99 BELGIAN ANTARCTIC EXPEDITION, but this was abandoned when it was thought the *BELGICA* was lost. In May the same year he went to England to join Carsten BORCHGREVINK and his *SOUTHERN CROSS* EXPEDITION. Bernacchi was in charge of meteorological observations and photography and worked with William COLBECK, the magnetic observer.

After the expedition arrived at Cape ADARE, Bernacchi found 'numerous pieces of quartz ... like the auriferous quartz I have often seen in Australia,' he wrote in his journal; this find supported the theory of GONDWANA and CONTINENTAL DRIFT. Bernacchi's meteorological observations, made at two-hourly intervals over an entire year, provided the first detailed account of the CLIMATE of the Antarctic continent.

He joined the 1901–04 NATIONAL ANTARCTIC EXPEDITION, during which he continued meteorological research, recorded detailed observations of AURORA AUSTRALIS and, with Ernest SHACKLETON, was co-editor of the *SOUTH POLAR TIMES*.

Bicycles Ridden at many of the bases, bicycles are used at the SOUTH POLE to commute from the station to the DARK SECTOR, where astronomical observations are carried out. A standard mountain bike has been modified to include quick-release seat buttons that can be operated with gloves on; toothed-edged pedals for insulated boots; outrigger skis; metal disc brakes, and resin wheels, moulded in two pieces so that the bike 'floats' over the snow.

Biological Investigations of Marine Antarctic Systems and Stocks (BIOMASS) A major international oceanographic research programme, BIOMASS was organized through the SCIENTIFIC COMMITTEE ON ANTARCTIC RESEARCH (SCAR) and the Food and Agricultural Organization of the UNITED NATIONS in the late 1970s, with the objective of gaining 'a deeper understanding of the structure and dynamic functioning of the Antarctic marine ecosystem as a basis for future management of potential living resources.'

The First International Biological Experiments (FIBEX) was held in 1980–81, and the Second International Biological Experiments (SIBEX I and II) in 1983–85. These expeditions involved ships and scientists from 11 different countries. As many as 16 vessels attempted to collect simultaneous information about WEATHER, the OCEAN, and ANIMALS from different parts of the Southern Ocean. There was a focus on species, such as KRILL, that had potential to be harvested. The data

collected are stored electronically through the BRITISH ANTARCTIC SURVEY in Cambridge, and are available for use by scientists from all participating countries. BIOMASS ended in 1992.

Biological research Early biological research focused on collecting and classifying the unique species of the Antarctic continent and the SOUTHERN OCEAN. After World War II, biologists began to consider ways in which life in the Antarctic adapted to extreme environmental conditions. This approach was pioneered in the late 19th century by German biologists Victor Hensen and Karl Brandt, who had postulated that plankton populations in the polar seas were much greater than in warmer tropical waters, mainly because of high nitrate and phosphate concentrations near the surface.

Little biological research was carried out on continental Antarctica until the 1960s. When Vivian FUCHS was appointed director of the Falkland Islands Dependencies Survey (later the BRITISH ANTARCTIC SURVEY) in 1958, he commenced a programme of biological research in inshore communities on Signy Island in the SOUTH SHETLANDS.

After the INTERNATIONAL GEOPHYSICAL YEAR, the number of coastal scientific stations increased and scientists began to access the LAKES of the DRY VALLEYS. A major international biological research programme, carried out in the 1960s and based on SOUTH GEORGIA, found that the long, wet subantarctic growing season stunted vegetation growth (in contrast to the short Arctic summers which limited reproduction and plant development).

Other areas of recent biological interest have been the physiology of Antarctic FISH, including their ANTIFREEZE, SEAL numbers, cold desert ecosystems in the DRY VALLEY region, and the ecology of freshwater and saline LAKES.

Increased human activity in Antarctica has led to research into the impact of TOURISM on PENGUIN behaviour and the effects of effluent from bases on the undersea ecology, while the effects of CLIMATE change and the OZONE HOLE on Antarctic species remain a significant focus.

Birds The few families of marine birds—PENGUINS, CORMORANTS, PETRELS, TERNS and SKUAS—that inhabit the Antarctic region are dependent on the sea for food. Each spring, in a dramatic end to the long, dark polar winter, millions of seabirds return to the coasts of Antarctica to breed. By autumn, most of these birds have left the continent, flying north as the PACK ICE builds up. Only EMPEROR PENGUINS remain through the winter.

Antarctic birds have features that help conserve body heat: special waterproof plumage; unique, short, overlapping feathers that trap insulating air, a layer of subcutaneous fat (which in penguins is 2 cm (nearly 1 in) thick at the start of the breeding season), large body size, small appendages (to minimize heat loss) and a lack of bare skin. These features enable them to extend their range southwards to exploit the rich food supplies of the SOUTHERN OCEAN and to establish breeding colonies on the Antarctic continent and the SUBANTARCTIC ISLANDS.

Births On 7 January 1978, Silvia Morello de Palma, wife of Chilean Army captain Jorge de Palma, gave birth to Emilio Marcos at ESPERANZA STATION on the ANTARCTIC PENINSULA. This was the first birth in Antarctica. Since then a number of babies have been born at this Chilean base, as part of the country's Antarctic TERRITORIAL CLAIM.

Biscoe, John (1794–1843) British sailor and explorer, born in Middlesex, England. Biscoe joined the British Navy in 1812 and, after serving in the American War of Independence, joined the merchant marine in voyages to the Caribbean and Asia.

In 1830 he was appointed by the ENDERBY BROTHERS to lead a SEALING expedition in the brig *Tula* and the cutter *Lively*. The expedition sailed first to three of the SOUTH SANDWICH ISLANDS but on finding no SEALS continued south and east. From the top of his foremast, at 69°S and 45°E, Biscoe saw black mountain tops above an ice-covered land, which he named ENDERBY LAND. Hopes of making a landing were abandoned by mid-March 1831, by which time the two ships had become separated.

Although both were battered in wild storms and heavy seas, and the crews were severely affected with scurvy, the two ships managed to return to Hobart, Australia, where they wintered

*Petrels, such as the South Georgia diving petrel (*Pelecanoides georgicus*) (above) and mottled petrel (*Pterodroma inexpectata*) (below), comprise the greatest biomass of all bird species in the Antarctic region.*

over and met Jules-Sébastien DUMONT D'URVILLE and Charles WILKES.

The next summer, the expedition returned to the SOUTHERN OCEAN. On 14 February 1832 Biscoe discovered ADELAIDE ISLAND, and a group of small islands, the Biscoe Islands. From here they sailed to the SOUTH SHETLANDS, then on to the FALKLAND ISLANDS, where the *Lively* was wrecked and most of its crew deserted.

In 1837 Biscoe moved to Sydney, Australia, then Hobart. When he became seriously ill, a public appeal was launched for him and his family to return to England but he died at sea.

Bismarck Strait Stretch of water separating GRAHAM LAND from the PALMER ARCHIPELAGO. Discovered and charted in 1873 by Eduard DALLMAN on the whaling ship *Grönland* during the first steam-powered voyage in Antarctic waters.

Bjaaland, Olav (1872–1961) Norwegian skier and explorer, born in Morgedal, Telemark. The 1902 Nordic ski champion, he was also a talented musician and carpenter, skills that were utilized on the 1910–12 NORWEGIAN ANTARCTIC EXPEDITION.

Over the 1911 Antarctic winter at FRAMHEIM, Bjaaland remodelled the sledges, reducing their weight by one-third in order to make them faster, and prepared 10 sets of skis for the use of the polar party. With Sverre HASSEL, he also built a sauna.

He was a member of the party that reached the SOUTH POLE for the first time, on 14 December 1911, under the leadership of Roald AMUNDSEN: 'it's as flat as a lake at Morgedal and the skiing is good,' he wrote. He surprised the rest of the party by producing a case full of cigars to accompany their celebratory dinner—of a little seal meat—at the South Pole. On his return to Norway, he lived on his farm at Morgedal.

Black-backed gull (*Larus dominicanus*) See KELP GULL.

Black-bellied storm petrel (*Fregatta tropica*) These medium-sized STORM PETRELS have wing-spans measuring 46 cm (18 in) and are 20 cm (8 in) long. Sooty brown on their upper bodies, with white rumps and stomachs that have black lines running down the middle (from which they get their name), their underwings are white with broad black margins.

Black-bellied storm petrels have erratic flight patterns, flying low over the sea and sometimes using their feet against the water surface. They feed on free-living barnacle larvae, small euphausiids, amphipods and occasional FISH, which they dip to catch.

The females lay solitary eggs in burrows during November to February. Breeding colonies comprise loose groups formed on several SUB-ANTARCTIC ISLANDS and some Antarctic islands. They have a circumpolar distribution and a population estimated at 150,000 breeding pairs.

Black-browed mollymawk (*Thalassarche melanophrys*) This MOLLYMAWK is medium-sized with a bright yellow, pink-tipped bill and distinctive black brows. Its underwings are lined with black, especially along the leading edge, and its wing-span is 2.4 m (8 ft).

The black-browed mollymawk breeds annually on many of the SUBANTARCTIC ISLANDS. Its nest is cone-shaped and built of mud and vegetation. It ranges widely throughout the SOUTHERN OCEAN, feeding on FISH, SQUID and CRUSTACEANS and is most abundant on CONTINENTAL SHELF waters, but can be seen close to the edge of the Antarctic ICE during the austral summer. Like the WANDERING ALBATROSS and ROYAL ALBATROSS, this bird is increasingly in danger from long-line FISHING.

Black icebergs Unusual ICEBERGS in which rocks and debris are incorporated, first observed by James WEDDELL on his 1822–24 voyage. He saw 'a pinnacle of an iceberg so thickly incorporated with black earth as to present the appearance of a rock.'

Blizzards With gale-force WINDS and blowing SNOW, Antarctic blizzards may last for days at a time. Visibility is often reduced to zero, and it may be hard to distinguish objects at an arm's distance. Winds greater than 5 m (16 ft) per second can whip up snow and grit, and in a full-scale blizzard, speeds of 150 km (93 miles) per hour with gusts exceeding 190 km (118 miles) per hour race across the landscape, tearing away anything they can carry.

In March 1902, during the NATIONAL ANTARCTIC EXPEDITION, a party caught in a blizzard lost all sense of direction in the whirling snow; although they stopped just short of a precipice, one man slipped past and disappeared to his death. According to expedition member Captain Reginald Ford, 'To be out in a blizzard is a really trying experience. You try and turn your back to the wind, but it is hopeless, it seems to surround you. The snow doesn't fall in flakes as in an ordinary snowstorm at home, but is driven by the wind into hard fine particles like sand, and this icy sand attacks you everywhere. The particles force their way into your nostrils and your eyes fasten to your eyelashes and then freeze, so that before you can open your eyes you must first rub off the ice.'

Blue petrel (*Halobaena caerulea*) Small and dark-crowned, the birds superficially resemble PRIONS, but have distinctive square, white-tipped tails, white foreheads and short black bills, dark tapering-patterned necks, and white underparts. Their wings have dark, narrow 'M' markings extending from wing-tip to wing-tip, and wing-spans are about 62 cm (24 in).

Extremely abundant birds, blue petrels breed in dense colonies, nesting in thick TUSSOCKS on many islands including SOUTH GEORGIA, MACQUARIE and Îles KERGUÉLEN. Blue petrels have a circumpolar distribution and are seen almost everywhere from the PACK ICE to the more southern parts of Africa, Australia and South America. Their flight is swift and erratic and they feed at night, taking KRILL and amphipods from the sea's surface.

Blue whale (*Balaenoptera musculus*) The largest living mammal, the blue whale is believed to be the largest animal that has ever existed. The biggest recorded specimen weighed nearly 200 tonnes (200 tons) and was over 30 m (100 ft) long.

These BALEEN WHALES are bluish-grey in colour, and they have close to a 100 black throat pleats. Early whalers used to call them 'Sulphur Bottoms' because their bellies are often covered with yellow layers of DIATOMS. Not particularly social animals, they usually travel alone or in small groups of around three. Blue whales, and the smaller subspecies known as pygmy blue whales (*Balaenoptera musculus brevicauda*), are found in both the Northern and Southern Hemispheres, although the populations do not mix.

They have a circumpolar distribution, and migrate great distances from their warmer winter breeding grounds at about 40°S latitude to summer in the plankton-rich Antarctic waters. Here they feed on vast quantities of KRILL: in a single day they can eat over 8 tonnes (8 tons). Like all baleen whales, blue whales are filter feeders, and use their baleen plates as sieves to trap krill. The first of the baleens to arrive (in late spring), they will have all departed by April, having spent about four months feeding intensively.

They find little suitable food in winter breeding waters, living off their store of blubber. Mating takes place in winter in the warmer northern waters and gestation lasts for 12 months. The calf weighs around 2.5 tonnes (2.5 tons) at birth and is suckled for about eight months. In the last stages of nursing, the calves can put on up to 100 kg (220 lb) a day. They become sexually mature aged about five. The females have a two- to three-year reproductive cycle, and generally give birth to a single calf. The population of blue whales is unknown, but some estimates place numbers at fewer than 2000. The population was devastated by commercial WHALING during the first half of the 20th century. In the 1930–31 season alone, 30,000 whales were killed in one small area around SOUTH GEORGIA. Blue whales are classified as endangered.

Above: Antarctic blizzards can last for days at a time. Visibility is often reduced to zero: it can be hard to distinguish objects even at an arm's distance.

*Above: Blue eyed cormorants (*Phalacrocorax atriceps*) breed in large flocks, nesting along coastlines.*

Blue-eyed cormorant (*Phalacrocorax atriceps*) Also known as imperial shags, there are a number of subspecies: the KERGUELEN CORMORANT and the imperial cormorant. Colonies range from the west coast of South America to 68°S on the ANTARCTIC PENINSULA.

They eat FISH, diving to depths up to 100 m (33 ft) to obtain them. With wing-spans of around 1.1–1.25 m (43–49 in), they have blue-black backs and caps of the same colour, white breasts and throats, dark brown bills and pink feet. These sociable birds forage and breed in large flocks, nesting along coastlines, sometimes among PENGUIN colonies. Nests are made of SEA-WEED, and bound together by guano. Eggs are laid in October and take about five weeks to hatch. Newborn chicks are naked, with not even a sprinkling of down, and they are brooded by both parents for three weeks, then looked after for another three weeks. Some chicks and eggs are taken by SKUAS and SHEATHBILLS. LEOPARD SEALS are a potential threat to fledglings reaching water for the first time.

Books During the 1908 winter at Cape ROYDS, members of the BRITISH ANTARCTIC EXPEDITION wrote and produced the *AURORA AUSTRALIS*, the first book published in Antarctica.

The earliest books about Antarctica were personal journals, accounts of voyages and detailed reports of scientific investigations. They include such works as James COOK's journals, James WEDDELL's *Voyage Towards the South Pole* (1825), the five-volume *The Narrative of the United States Exploring Expedition* (1845) and 20 volumes of scientific reports from the same expedition, and Joseph HOOKER's lavishly illustrated *The Botany of the Antarctic Voyage* (1847).

Many heroic first-hand narratives of survival in an inhospitable environment were bestsellers when first published, and some have become classics of their kind: in 2001 Aspley CHERRY-GARRARD's *The Worst Journey in the World* was voted No 1 on *National Geographic*'s list of the 10 greatest adventure stories of all time.

Today, books on Antarctica are wide-ranging in subject, encompassing fiction (mainly thrillers and adventure stories), history, politics, biography, science and natural history.

Borchgrevink, Carsten (1864–1934) Norwegian scientist and explorer, born in Oslo. At the age of 24 he emigrated to Australia, and worked in the outback before taking up a teaching position in New South Wales. On hearing that Henryk BULL was recruiting for his 1893–95 ANTARCTIC EXPEDITION he applied for a scientific position. Enlisted as a crewman, Borchgrevink collected specimens of LICHEN growing on POSSESSION ISLAND, thereby dispelling earlier ideas that PLANT life could not exist in Antarctica, and is usually credited as being the first person known to have stepped onto the Antarctic continent itself, on 24 January 1895 at Cape ADARE. Back in Australia, Borchgrevink attempted to obtain funds for a further expedition; although unsuccessful, he did secure the services of Louis BERNACCHI, a scientist at the Melbourne Observatory. In England Borchgrevink received financial support from millionaire publisher George Newnes for the 1898–1900 SOUTHERN CROSS EXPEDITION.

The sturdy Norwegian-built whaler *Pollux* was purchased and renamed the *Southern Cross*, and the expedition meticulously planned. It landed at Cape Adare on 17 February 1899. After the British flag was raised, the ship unloaded and a prefabricated hut was erected. The *Southern Cross* left and 10 men remained at Cape Adare to set up their scientific equipment. The long winter passed unpleasantly, with tension among Borchgrevink and the scientists and the death of zoologist Nicolai Hanson from an unknown illness—the first man to be buried on the Antarctic Continent.

Below: Carsten Borchgrevink, leader of the 1898–1900 Southern Cross Expedition.

Although Borchgrevink was not a popular leader, the expedition was well organised. It was the first to deliberately winter over, and a sledging party reached 78°50'S, the first such journey on the ROSS ICE SHELF. Borchgrevink made a voyage to the West Indies in 1902, then retired to Norway.

Botanical research Early botanical research was carried out on the SUBANTARCTIC ISLANDS by pioneers such as Joseph HOOKER, who travelled to Antarctica with James Clark ROSS aboard *Erebus* in 1839–43. At Cape ADARE Henryk BULL's 1894–95 expedition found the first LICHEN growing on the continent itself, and in 1898 Emil Racovitza, zoologist on the 1897–99 BELGIAN ANTARCTIC EXPEDITION, discovered the first of only two FLOWERING PLANTS on the continent. Recent studies have considered the colonization processes of plants on the Antarctic mainland. The breakup of ICE SHELVES and the spread of plants south along the ANTARCTIC PENINSULA, for example, suggest that regional warming is occurring.

Bottlenose whale See SOUTHERN BOTTLENOSE WHALE.

Bottom water See ANTARCTIC BOTTOM WATER.

Bounty Islands The largest island of the 22 in the Bounty group is less than 1 km (just over ½ mile) wide. Unlike other New Zealand SUBANTARCTIC ISLANDS (CAMPBELL, AUCKLAND and ANTIPODES ISLANDS), the Bountys are not volcanic but are solid BEDROCK outcrops scattered over the ocean surface. MOSSES and LICHEN, the only vegetation, cling to the rock surfaces, and the islets provide nesting grounds and perches for many species of BIRDS. The islands were discovered in 1788 by Captain William Bligh and named after his ship. Soon after this, SEALING fleets arrived and, by 1831, when John BISCOE visited the islands, the seals had been almost wiped out.

Bouvet de Lozier, Jean-Baptiste-Charles (1705–86) French captain and explorer. The French search for TERRA AUSTRALIS INCOGNITA began in 1738 when Bouvet sailed into the South Atlantic Ocean looking for a suitable base for his employer, the French East India Company. On 1 January 1739 he came upon a headland, which he named Cape Circumcision and which he believed to be part of a continental land mass. However, although the latitude was correct, the longitude was so far out it was not until 1898 that the headland—in fact, a small island—was rediscovered. Since 1930 it has been annexed to NORWAY and named BOUVETØYA in his honour. Bouvet sailed 2000 km (1240 miles) along the PACK ICE before heading home.

Bouvet Island See BOUVETØYA.

Bouvetøya Also known as Bouvet Island, this 54 sq km (21 sq mile) ice-covered island in the southern Atlantic Ocean is the most isolated on Earth—not counting a small island to the south west, Larsøya, the nearest land is more than 1600 km (992 miles) away. A VOLCANO at the tip of the island is extinct, but there is still a lot of geothermal activity: the below-ground temperature has been measured at 25°C (77°F). GLACIERS block access to the south and east coasts and steep cliffs barricade entry from the north, making the island virtually inaccessible by boat. However, this did not prevent sporadic visits by sealers in the 19th century. In 1928 Bouvetøya was claimed by NORWAY. It was declared a nature reserve in 1971, and a small research station was set up in 1997.

An orbiting satellite recorded a brief, intense burst of light on 22 September 1979 to the west of Bouvetøya, and scientists at Australian research stations detected radioactive debris soon after this (see CORE SAMPLES). The evidence points to a thermonuclear bomb test, but no country has ever admitted responsibility.

Bowers, Henry Robertson (1883–1912) British naval lieutenant and explorer, born in Greenock, Scotland. He became a cadet on the *Worcester* aged 14, then in 1899 joined the merchant navy. In 1905 he was assigned to the Royal Indian Marine Service, and was stationed in Burma and Ceylon for the next five years. Bowers joined the 1910–13 BRITISH ANTARCTIC EXPEDITION. Known as 'Birdie' because of his beaky nose, he was eternally cheerful and was an exceptional navigator and organizer. Expedition leader Robert SCOTT noted, 'In the transport department ... I find that Bowers is the only man on whom I can thoroughly rely to carry out the work without mistake. ...'

In the winter of 1911 he walked with Edward WILSON and Aspley CHERRY-GARRARD to Cape CROZIER to find EMPEROR PENGUIN eggs—the first overland journey attempted during the Antarctic winter. Bowers reached the SOUTH POLE with Scott on 17 January 1912. He had been included in the Pole party at the last minute; as a result, he had no skis, unlike the others. He died on the return journey from starvation and exhaustion, around 29 March 1912. Some of the photographs he took during the polar journey and at the South Pole were exhibited at the Fine Art Society Exhibition of Herbert PONTING's photographs in London in 1913: '... they are without doubt the most tragically interesting in existence. The films from which they were made ... lay beside the dead body of the leader for eight months before they were found,' a note in the exhibition catalogue explained.

Bransfield, Edward (c. 1795–1852) British naval officer. Little is known of Bransfield's early life. He was master of the British Navy ship HMS *Andromache* on the Pacific coast of South America in the first decades of the 19th century.

When William SMITH reported his discovery of the SOUTH SHETLAND ISLANDS, Smith's brig, the *Williams*, was chartered by the British Admiralty and placed under Bransfield's command with Smith as pilot. Bransfield's instructions were to secure the islands in the name of King George III, to survey their anchorages and harbours, and to seek out possible locations for British naval bases. They sailed from Valparaíso in CHILE on 20 December 1819 and first sighted land on 18 January 1820. Turning eastwards through the Bransfield Strait on 30 January they saw land to the southwest which they named 'Trinity Land', the northern tip of the ANTARCTIC PENINSULA. For many years this was believed to be the first sighting of the Antarctic continent, although it seems more likely that the honour goes to Thaddeus BELLINGSHAUSEN who recorded seeing an ICE SHELF on 27 January 1820, three days earlier.

Sailing northwards, Bransfield charted the northern coast of D'URVILLE ISLAND before moving on to ELEPHANT ISLAND, and on 4 February reached Clarence Island where they landed and took formal possession. Nine days were spent surveying the area before turning southeast, but they were unable to get further than 64°50'S because of the dense PACK ICE. On his return to England, Bransfield resigned from the navy. His charts are now housed in the Hydrographic Department of the British Admiralty.

Brash ice Accumulations of floating ICE made up of fragments not more than 2 m (6½ ft) across. They are the wreckage of other forms of ice.

Brazil Brazil's interest in Antarctica increased in the 1960s and 1970s as a result of private efforts. Although the Brazilian government had considered making a TERRITORIAL CLAIM, in 1975 it chose instead to accede to the ANTARCTIC TREATY, which prohibits new claims but allowed Brazil to participate in decision-making about Antarctica. Brazil undertook its first Antarctic expedition in 1982.

In the following year it became a CONSULTATIVE PARTY to the Antarctic Treaty, but it did not establish a research station until 1984. Brazil is a member of the Commission for the CONVENTION ON THE CONSERVATION OF ANTARCTIC MARINE LIVING RESOURCES (CCAMLR) and a signatory to the MADRID PROTOCOL.

Below: Brazil's Commandante Ferraz Station is located on King George Island, part of the South Shetland Archipelago.

Above: The British Antarctic Expedition team celebrate the 43rd birthday of Captain Robert Scott (centre) with a dinner on 6 June 1911.

Britain The first British discoveries in Antarctica were in 1772–75, when James COOK successfully circumnavigated the then unknown Antarctic continent and was probably the first to cross the ANTARCTIC CIRCLE. The expedition landed at SOUTH GEORGIA and discovered the SOUTH SANDWICH ISLANDS, which were claimed for Britain. In 1776, on his final voyage, Cook travelled to the PRINCE EDWARD ISLANDS and to the bleak Îles KERGUÉLEN, which he called the Islands of Desolation. A British merchant captain, William SMITH, discovered the SOUTH SHETLANDS in February 1819 and it is thought that Edward BRANSFIELD may have sighted the ANTARCTIC PENINSULA early the next year (albeit three days after Thaddeus BELLINGSHAUSEN), giving Britain a claim to the discovery of Antarctica. In 1834 Britain occupied the FALKLAND ISLANDS, and established a crown colony there in 1841. Later expeditions—from that of 1838–41 led by James Clark ROSS to Ernest SHACKLETON's 1914–17 IMPERIAL TRANSANTARCTIC EXPEDITION—were concerned with SCIENCE and furthering Britain's imperial goals through discovery, EXPLORATION, occupation, COMMERCE and administration.

In 1908 and 1917 Britain claimed South Georgia, the SOUTH ORKNEYS, the South Shetlands, and South Sandwich Islands, and roughly 17 percent of the Antarctic continent between 80°W and 20°W. This claim was organized as the Falkland Islands Dependencies (FID); since 1962, the claim, excluding the Falkland Islands sector, has been known as the BRITISH ANTARCTIC TERRITORY. In 1923 Britain also claimed the ROSS DEPENDENCY, later ceded to NEW ZEALAND, and in 1933 a third claim—the largest TERRITORIAL CLAIM, cov-

ering 6,500,000 sq km (2,509,650 sq miles)—was made in EAST ANTARCTICA then immediately ceded to AUSTRALIA.

Britain participated in the INTERNATIONAL GEOPHYSICAL YEAR and became one of the 12 original CONSULTATIVE PARTIES to the ANTARCTIC TREATY. The treaty reduced strategic tensions involved in Britain's rivalry with ARGENTINA and CHILE over the DISPUTED CLAIMS in Antarctica and the Falklands.

Britain is a member of the INTERNATIONAL WHALING COMMISSION (IWC), the CONVENTION ON THE CONSERVATION OF ANTARCTIC MARINE LIVING RESOURCES (CCAMLR) and the CONVENTION FOR THE CONSERVATION OF ANTARCTIC SEALS (CCAS) and a signatory to the MADRID PROTOCOL. In CCAMLR Britain has adopted a pro-conservationist stance.

Britannia Range Located west of the ROSS ICE SHELF, this mountain range was discovered by the 1901–04 NATIONAL ANTARCTIC EXPEDITION. It is bordered to the north by the Hatherton and Darwin Glaciers and to the south by the BYRD GLACIER. It includes Mount McClintock, which rises to 3492 m (11,454 ft) above sea level.

British Antarctic Expedition (1898–1900) See SOUTHERN CROSS EXPEDITION.

British Antarctic Expedition (1907–09) Private British exploratory expedition. Unable to gain official sponsorship, Ernest SHACKLETON turned to industrialist William Beardmore, who loaned £20,000 for an expedition aimed at reaching the SOUTH POLE. A further £5,000 was granted by the

Australian Government. Sailing in the NIMROD, and using what he had learnt on the 1901–04 expedition led by Robert SCOTT, Shackleton made a number of improvements in CLOTHING and TRANSPORT. He brought woollen clothing, fur-lined sleeping bags, Siberian huskies and Manchurian ponies, and motorized transport—the first on the Antarctic continent—in the form of an Arrol-Johnson car.

Because Scott refused to allow the expedition the use of his old HUT, a new base was established at Cape ROYDS. The men were not experienced dog-handlers and most of the huskies became weak and died, and, after the last of the horses was shot, lost or died (some perished after eating volcanic gravel), man-hauling became their means of transport. Among many achievements, the expedition made the first ascent of Mount EREBUS; reached the farthest south point ever—88°23'S—only 180 km (112 miles) short of the South Pole; and was the first to reach the SOUTH MAGNETIC POLE. During the long winter there were various contributors to the first BOOK to be published in Antarctica: AURORA AUSTRALIS, illustrated by George Marston, printed by Ernest JOYCE and Frank WILD, and bound by Bernard Day.

In order to repay the loans raised prior to his departure, Shackleton undertook a rigorous series of lecture tours throughout BRITAIN and the USA. His two-volume record of the expedition, *Heart of the Antarctic*, was published in 1909.

British Antarctic Expedition (1910–13) British expedition to the SOUTH POLE, led by Captain Robert SCOTT. The expedition had two aims: to be the first to reach the South Pole and to carry out a detailed scientific programme. Although Scott had returned to Britain a hero from the 1901–04 NATIONAL ANTARCTIC EXPEDITION, Ernest SHACKLE-TON had stolen some of his limelight with his own 1907–09 expedition, and Scott had difficulty raising sufficient funds for the venture. When the expedition ship, the TERRA NOVA, sailed in June 1910, it was uncertain whether the explorers and sailors could be paid. On the way to Antarctica, Scott received a cable from Roald AMUNDSEN, which read, 'Beg leave to inform you, FRAM proceeding Antarctica. Amundsen'. Three days after leaving New Zealand, already six weeks behind schedule, a fierce storm almost sunk the Terra Nova. It was then stuck in PACK ICE, which further delayed the expedition.

The ship arrived at Cape EVANS on 4 January 1911, where a hut was constructed and a party began laying food and fuel depots for the attempt on the South Pole the following summer. Snowfalls were heavier than expected and the ponies that had been brought to haul the sledges did not perform well. Consequently, the final depot, known as One Ton Depot, was laid 48 km (30 miles) short of the planned 80°S. Meanwhile, the Terra Nova took another party westwards along the ROSS ICE SHELF, where they found the Fram, Amundsen's ship, in the BAY OF WHALES. A third party explored the Koettliz Glacier region and the DRY VALLEYS.

In the winter of 1911 scientist Edward WILSON, along with Henry BOWERS and Aspley CHERRY-GARRARD, set out on a 105 km (65 mile) trek to Cape CROZIER to find EMPEROR PENGUIN eggs; Wilson hoped the chick embryos would help explain the origin of all birds. This was the first man-hauling Antarctic journey ever attempted during the depths of winter and was described by Cherry-Garrard as 'the worst journey in the world'. The three men survived conditions in total darkness with blizzards and temperatures as low as –60°C (–75°F). They returned with three pale-green eggs and some frozen chicks, but studies on them were inconclusive.

The following summer, a party of six, led by Victor Campbell and known as the Northern Party, carried out exploratory surveys of the VICTORIA LAND coastline. After wintering in a hut at Carsten BORCHGREVINK's old site at Cape ADARE and spending the spring and summer surveying the area, including the ADÉLIE PENGUIN colony, the group was transported by the Terra Nova to Evans Cove in TERRA NOVA BAY. However, the ship was not able to pick them up in mid-February as expected, and they were forced to spend the winter in an ice cave, living off SEALS and PENGUINS. The following spring they sledged back down the coast to Cape Evans.

The Southern Party set out in October 1911 for the South Pole. Scott experimented with a variety of transport methods: motor sledges, ponies, dogs and man-haulage. Eventually, they decided on man-hauling. Two support parties turned back, leaving the polar party: Scott, Wilson, Bowers, Lawrence OATES and Edgar EVANS. Bowers was included at the last minute, an extra man in a party with supplies for four. They reached the South Pole on 17 January 1912 to find that Amundsen had beaten them by one month.

On the return journey, all the men suffered from inadequate food, exhaustion and cold. Evans collapsed and died on 17 February. Oates walked out into the snow to his death on 15 March. The other three died only 20 km (12 miles) from One Ton Depot. The final entry in Scott's diary was on 29 March. He left a 'Message to the Public' in which he wrote: 'Had we lived, I should have had a tale to tell of the hardihood, endurance, and courage of my companions which would have stirred the heart of every Englishman. These rough notes and our dead bodies must tell the tale. ...' Their bodies were found by a search party the following November.

Despite the tragedy, the expedition carried out valuable scientific research in a number of fields.

British Antarctic Survey (BAS) The organization that coordinates and undertakes BRITAIN's

Antarctic research programme. At the end of World War II OPERATION TABARIN was renamed the Falkland Islands Dependencies Survey, which later became the British Antarctic Survey. Based at Cambridge, England, it holds the database for the international BIOLOGICAL INVESTIGATIONS OF MARINE ANTARCTIC SYSTEMS AND STOCK (BIOMASS) project. Although the BAS is a civilian organization, after the 1982 WAR in the FALKLAND ISLANDS, the British government increased its funding, thus reinforcing its presence in the Antarctic zone.

British Antarctic Territory In 1908 and 1917 Britain claimed SOUTH GEORGIA, the SOUTH ORKNEYS, the SOUTH SHETLANDS and SOUTH SANDWICH ISLANDS, and roughly 17 percent of the Antarctic continent between 80°W and 20°W. This claim was organized as the Falkland Islands Dependencies (FID). In 1923 Britain also claimed the ROSS DEPENDENCY, later ceded to NEW ZEALAND. A third claim—the largest TERRITORIAL CLAIM in Antarctica, covering around 40 percent of the mainland and stretching 6,500,000 sq km (2,509,650 sq miles) in two wedges from 45°E and 136°E, and 142°E and 160°E, below 60°S—was made in EAST ANTARCTICA in 1933, then immediately ceded to AUSTRALIA. At one stage Britain planned to annex the entire continent, but was prevented by claims made by FRANCE and NORWAY.

The claims of ARGENTINA and CHILE overlap with those of Britain—during World War II the British OPERATION TABARIN was designed to keep a watch on Argentinian and Chilean moves in Antarctica as well as German SUBMARINES. After the war there were several incidents involving the three rivals, including in February 1952 when Argentinian soldiers fired over the heads of a British scientific party at HOPE BAY on the northern tip of the ANTARCTIC PENINSULA.

In 1962 the British Antarctic Territory (BAT), consisting of the mainland section of GRAHAM LAND plus the South Orkneys and South Shetland Islands, was split off from the Falkland Islands Dependencies, which now include South Georgia and the South Sandwich Islands.

British-Australian-New Zealand Antarctic Research Expedition (BANZARE) (1929–31) Two international Antarctic expeditions in consecutive summers to carry out scientific research, both led by Douglas MAWSON. The British government lent the ship DISCOVERY, initially commanded by John King DAVIS, for the expedition.

The aims were three-fold: political—to make TERRITORIAL CLAIMS for the British; economic—to investigate possibilities for WHALING; and scientific—to map the coastline and make geological, meteorological and other studies. A small AIRCRAFT, a Gypsy Moth, was taken to make aerial surveys.

The Discovery sailed from Cape Town on 19 October 1929. A new coastline, which Mawson named MAC.ROBERTSON LAND, was discovered at the start of January by aeroplane. On 13 January 1930 a party landed on Proclamation Island and the British flag was raised, claiming sovereignty over ENDERBY LAND, KEMP LAND and

Below: Scott's pole party leaving the depot at 3°S on the 1910–13 British Antarctic Expedition.

Mac.Robertson Land. The next day Mawson met Captain Hjalmar RIISER-LARSEN, who was commanding a rival expedition from NORWAY. They agreed that 45°E would act as a dividing line between the two expeditions. In mid-March Captain Davis, worried that the vessel might run out of coal, insisted that the *Discovery* return to Australia: 'I am not going to take any risks for that bloody rubbishing business of raising the flag ashore,' Davis said.

On his return, Mawson immediately recommended a second trip to examine the coastline between GEORGE V LAND and his new discoveries. The second BANZARE voyage left Hobart on 2 November 1930. The *Discovery* was now captained by K N MacKenzie. New radio equipment was installed and an aeroplane was again used for aerial surveys. *Discovery* reached Cape DENISON on 4 January 1931. The ship then sailed east along the coast, naming and claiming land along the way.

After its return to Australia on 19 March 1931, Mawson summarized the achievements of BANZARE: 'The work conducted ... through the two summer seasons ... resulted in the amassing of an immense amount of data regarding the region lying south of Australia and the Indian Ocean ... Long stretches of new coast line were discovered. ... perhaps the most important outcome of the expedition is that ... the presence of a real continent within the ice has been finally established. ...'

The reports of the BANZARE expeditions cover work in the fields of geography, geology, oceanography, botany and zoology. Two volumes of data, comprising 2700 pages, were published; volume one in 1937 and volume two in 1975.

British Graham Land Expedition (1934–37) British exploratory expedition, led by John RYMILL in the ship *Penola*. The aim was to explore the eastern side of the ANTARCTIC PENINSULA and to sledge west to the WEDDELL SEA through one of the straits reported by Hubert WILKINS and Lincoln ELLSWORTH. Among the expedition's equipment was a single-engine de Havilland Fox Moth AIRCRAFT.

Their chosen location for a base in the ARGENTINE ISLANDS was abandoned for a point farther south on one of the Debenham Islands. SLEDGING trips were undertaken on ADELAIDE ISLAND and ALEXANDER ISLAND, where much important surveying work was carried out. A new body of water—which they named King George VI Sound—was discovered. The sledge journeys along the edge of the Antarctic Peninsula proved the non-existence of any strait linking to the Weddell Sea; what Wilkins and Ellsworth had seen from the air had been glaciers. As well as ornithological work, other important discoveries were FOSSILS of ferns and other early Cretaceous vegetation; these proved that a dramatic climatic change had taken place in the distant past.

British Imperial Antarctic Expedition (1920–22) Private four-man expedition, led by John Cope, surgeon on Ernest SHACKLETON's 1914–17 expedition: 'a nice fellow but hopeless at organizing anything,' according to Hubert WILKINS, who reluctantly joined the expedition as aviator. Other members were 19-year-old geologist Thomas Bagshawe, and surveyor M C Lester, who was not much older. Cope's initial objective was to fly over the SOUTH POLE; however, with insufficient funds to buy one plane, he decided to extend the survey work of Otto NORDENSKJÖLD from a base at HOPE BAY in DECEPTION ISLAND.

On Christmas Eve 1920, the four men were taken to Hope Bay by whaler. Finding it iced over, the captain continued on to Andvard Bay on the Danco Coast, which they reached on 12 January 1921. Hindered by the location—they were cut off from the WEDDELL SEA region by mountains—and lack of equipment, in late February Cope decided to return north on a whaler to find a ship to take them to Hope Bay. The frustrated Wilkins also left. Bagshawe and Lester volunteered to winter over, sheltering in a makeshift HUT at Waterboat Point and a beached life boat. Towards the end of winter, the pair undertook some scientific observations, including measuring tide levels and recording numbers of PENGUIN eggs. Bagshawe and Lester were picked up by the whaler *Graham* on 13 January 1922.

British Imperial Transantarctic Expedition (1914–17) See IMPERIAL TRANSANTARCTIC EXPEDITION.

British National Antarctic Expedition (1901–04) See NATIONAL ANTARCTIC EXPEDITION.

British Whaling Expedition (1892–93) Private expedition in search of new WHALING grounds. By the 1890s, Arctic whaling was less profitable than previously. Scottish whalers organized an expedition to examine the possibility of hunting RIGHT WHALES in the WEDDELL SEA.

A fleet of four ships—all ice-resistant Arctic whalers armed with muzzle-loading harpoon guns, large-bore rifles that fired dum-dums, and rockets to attack the whales after harpooning—sailed from Dundee, Scotland, on 6 September 1892. Among the three doctors taken on as naturalists and scientific observers was William BRUCE. Although no RIGHT WHALES were sighted, the ships returned with a full cargo of SEAL oil and skins. The only exploratory work was done by Captain Robertson in the *Active* and included the discovery of DUNDEE ISLAND and surveying of the southern coast of JOINVILLE ISLAND.

Bruce was particularly frustrated by the lack of facilities and opportunities for the scientists to observe natural phenomena and conduct investigations. He wrote that, 'The scientific work of the expedition was not done in very favourable circumstances; commerce was the dominating note. A great deal more might have been done for the geology and biology of the Antarctic Regions if some opportunities had been afforded me.' The expedition's artist W G Burn Murdoch was equally frustrated: 'We turned from the mystery of the Antarctic with all its white-bound secrets still unread,' he said.

Brittle stars (Ophiuroidea) The most mobile ECHINODERM, brittle stars get their name from their ability to avoid capture by throwing away their arms, then growing new ones. At least 35 species of brittle stars are believed to live in Antarctica.

The brittle star *Ophiacantha antarctica* is circumpolar in distribution, is the most abundant and widely distributed echinoderm in the ROSS SEA and plays an important role in the biological balance of the BENTHOS, feeding on DIATOMS, COPEPODS and other tiny ZOOPLANKTON. It probably feeds by manipulating its flexible arms and long, erect, thin arm spines to capture food on or near the sea bottom. By contrast, the larger brittle star *Ophiosparte gigas*, with arms up to 17 cm (over 6½ in) long, is an active predator of large prey, especially other brittle stars, and will even cannibalize its own species. Other prey of *O. gigas* include MOLLUSCS, CRUSTACEANS and SPONGES. The

Below: *Brittle stars (Ophiuroidea) derive their name from their uncanny ability to discard their arms as a defence mechanism.*

Most modern Antarctic buildings (right) are constructed of insulated refrigeration panelling, and are heated year-round by diesel-powered generators. In contrast, Ernest Shackleton's wooden hut at Cape Royds (above) was heated by a large, American-made coal range, using fuel brought by the expedition. Blubber may have been used to supplement the fuel.

smaller brittle star *Ophionotus victoriae* responds to contact with the menacing *Ophiosparte gigas* by fleeing if it can—otherwise its arms are clipped off one by one and eaten.

Bruce, William (1867–1921) Scottish doctor, scientist and explorer, born in London. Bruce studied medicine at Edinburgh and pursued his interest in natural history. During 1892–93 he was doctor and scientific observer aboard *Balaena*, one of the four ships involved in the BRITISH WHALING EXPEDITION to the SOUTHERN OCEAN; on that voyage he performed his first operation (on a sailor who had been hit on the head with an iron block), and collected samples in the FALKLAND ISLANDS. Disappointed that the expedition was focused on COMMERCE, Bruce determined to return to Antarctica for genuine scientific study, and took part in two expeditions to the Arctic, where he met Fridtjof NANSEN.

After three years of planning and fund-raising, Bruce's SCOTTISH NATIONAL EXPEDITION left from the Clyde on 2 November 1902. Becoming ice-bound, they wintered over in Scotia Bay at Laurie Island in the SOUTH ORKNEY ISLANDS. When the weather improved and the ship was freed, Bruce explored the WEDDELL SEA; sailing beside a high ice-barrier for 250 km (150 miles), which he believed skirted solid land. He named the area COATS LAND after his patron James Coats. The expedition returned home on 21 July 1904.

Bruce continued his polar researches, making a total of 10 Arctic journeys. He received an honorary LLD from the University of Aberdeen in recognition of his services to science. He remained interested in Antarctica and was generous with advice to later explorers, notably Wilhelm FILCHNER, Ernest SHACKLETON and Douglas MAWSON. His ashes were scattered over Antarctic waters.

Bryozoans Encrusting animals that are found on hard surfaces throughout the world's oceans, along with SPONGES. Bryozoans are a particularly diverse component of the Antarctic BENTHOS. Sometimes known as 'lace corals', 'sea mats' or 'moss animals', they are colonial organisms, superficially similar to CORAL but with a much more complex anatomy. Bryozoan colonies are composed of individual organisms called 'zooids', which usually have ciliated tentacles to propel food particles towards their mouths. Predators include NUDIBRANCHS, SEASTARS and SEA URCHINS. Scientists on a BRITISH ANTARCTIC SURVEY supply ship recently discovered the first free-floating bryozoan colonies recorded. Found drifting in the WEDDELL SEA, these 'brown golf balls' also represent the first recorded hollow bryozoan colony. It is not yet known whether the colonies represent a totally new species of bryozoan or are, in fact, a mobile juvenile form of a bottom-dwelling species which, as it increases in size, sinks to the sea floor and becomes attached to a substrate.

Buildings The early explorers' HUTS were built mainly of wood, with several layers of boards and some insulating material—often seaweed—to protect against the cold. Although stoves burned constantly, in the extreme conditions it was difficult to keep indoor temperatures above freezing. BLIZZARDS could force large amounts of snow through gaps in walls, as well as burying the hut and its occupants under large snowdrifts. After World War II, Antarctic buildings became more sophisticated. Permanent facilities for science and research were required. Most bases have their own power stations and sophisticated insulation, buildings are consolidated into blocks rather than isolated huts, and indoor temperatures are maintained at a comfortable level.

A major concern with modern Antarctic buildings is preventing FIRES. The isolation of many bases, and conditions that make fire-fighting difficult and survival without shelter limited, have made fire a major threat. Now most buildings are made of fire-resistant materials and many bases have emergency shelters separated from the main buildings for use in case of fire.

Bull, Henryk Johan (1844–1930) Norwegian whaler and explorer. Bull travelled to AUSTRALIA in the early 1890s to establish a WHALING operation in the SOUTHERN OCEAN. Unsuccessful, he returned to NORWAY where he obtained backing from the Arctic whaling pioneer, Sven FOYN, who provided him with a barque-rigged whaler that was renamed *Antarctic* and sailed under the command of Captain Leonard Kristensen.

The expedition reached Îles KERGUÉLEN in December 1893 and remained there for two

Below: The contribution of Admiral Richard Byrd to Antarctic exploration is commemorated at McMurdo Station.

months. Although no whales were found, which Bull attributed to incompetence on the part of Kristensen, 3000 SEALS were killed. After a further fruitless hunt for whales around CAMPBELL ISLAND, Bull took on scientists and some new crew, including Carsten BORCHGREVINK, in Melbourne, Australia.

The *Antarctic* sailed for Campbell Island before continuing southwards to the Southern Ocean. Despite striking an ICEBERG (and having to return to Port Chalmers, New Zealand, for repairs), on 24 January 1895 the expedition reached Cape ADARE, where they made the first confirmed landing on the Antarctic continent. They spent only 90 minutes ashore, gathering LICHEN specimens and erecting a marker in Norwegian colours with details of the ship and date. The expedition returned to Melbourne in March 1895.

As a commercial venture, the expedition was a failure. However, the discovery of PLANT life on the continent and the confirmation of James Clark ROSS's belief that in summer the water was ice-free inside the PACK ICE were of great importance. Bull was involved in later sealing expeditions to southern waters, and in 1906 was shipwrecked for two months on Îles CROZET.

Bunger Hills Also known as Bunger Oasis and Bunger Lakes, Bunger Hills is a 780 sq km (301 sq miles) snow-free area just inland from the SHACKLETON ICE SHELF in WILKES LAND. It is covered with low, rounded hills and surrounded on all sides by walls of ice rising nearly 120 m (394 ft) high and includes a number of meltwater ponds and LAKES. Bunger Hills was discovered by US Navy pilot David Bunger on 11 February 1947 during a photographic flight for OPERATION HIGHJUMP.

Burks, Cape A promontory on the east side of the entrance to Hull Bay on the coast of MARIE BYRD LAND. Cape Burks, which was discovered and mapped by the crew of USS *Glacier* in January 1962, is named for Lieutenant-Commander Ernest Burks, the first person to set foot on the Cape. Cape Burks is the site of the Russian RUSSKAYA BASE.

By-catch Incidental catches of seabirds and FISH species in the course of longline FISHING for tuna and other species. It is estimated that in the SOUTHERN OCEAN longline techniques killed between 20,000 and 70,000 ALBATROSSES, 5000 to 11,000 GIANT PETRELS and 80,000 to 180,000 WHITE-CHINNED PETRELS in the four years between 1996 and 1999. Despite efforts by members of the CONVENTION ON THE CONSERVATION OF ANTARCTIC MARINE LIVING RESOURCES, this problem is not under control.

Byrd, Richard Evelyn (1888–1957) American admiral and explorer, born into a famous Virginian family. Credited with adding a larger region to the map of Antarctica than any other explorer, he was also largely responsible for modern-day USA involvement in the continent.

Byrd entered the US Navy at 20 and learned to

fly in World War I. Together with Floyd Bennett, he made the first flight over the North Pole in 1926, then decided to fly over the SOUTH POLE—the main objective of BYRD'S FIRST EXPEDITION of 1928–30 to Antarctica. He established the expedition base LITTLE AMERICA to the east of the BAY OF WHALES and, with pilot Bernt Balchen, radio operator Harold June and photographer Ashley McKinley, achieved what was believed to be the first flight over the South Pole, on 28–29 November 1929, in a three-engined Ford monoplane, the *Floyd Bennett*. (Historians have recently shown that Byrd could not have flown over the Pole.). Byrd returned to a hero's welcome in the USA on 19 June 1930: 'One of the true pleasures of being an explorer, was the pleasure of return,' he wrote.

BYRD'S SECOND EXPEDITION of 1933–35 was much larger, added considerably to scientific and geographical knowledge about Antarctica, and pioneered the successful use of TRACTORS. Byrd spent the 1934 winter alone in a weather station, 198 km (123 miles) inland from Little America, monitoring the weather and, after becoming seriously ill with carbon monoxide poisoning, was rescued after four and a half months.

In 1939–41 Byrd returned to Antarctica, this time as leader of the US government-funded UNITED STATES ANTARCTIC EXPEDITION, which was intended to consolidate previous American exploration in the Antarctic and make further land surveys. Scientific work and surveys were carried out from two base stations, Little America II near the former base, and one 3200 km (2000 miles) to the east on Palmer Peninsula.

By this time Byrd—then promoted to admiral—was the most experienced Antarctic explorer alive. He commanded the 1946–47 OPERATION HIGHJUMP, the largest expedition ever mounted in the Antarctic, during which he made a flight over the South Pole.

His final journey to Antarctica was on the 1955–57 OPERATION DEEPFREEZE, the USA's two-part contribution to the INTERNATIONAL GEOPHYSICAL YEAR. Byrd headed the first stage, which established an airfield at MCMURDO SOUND to support the establishment of a base at the South Pole.

Byrd died in the USA in March 1957. He had had idealistic views about the future of Antarctica, writing: 'Let there be no boundaries here. Let the Antarctic stand as a symbol of peace and a beacon for the world.'

Byrd Glacier A 135 km (84 mile) long, 24 km (15 mile) wide GLACIER that drains from the POLAR PLATEAU to the ROSS ICE SHELF at Barne Inlet. Almost as wide as the English Channel, the Byrd Glacier is the largest glacier feeding into the Ross Ice Shelf. It discharges 18,000 km (635,688 ft) of ICE each year.

Byrd Polar Research Centre Based at Ohio State University and named for Richard BYRD, the Byrd Polar Research Centre conducts research into the global CLIMATE, with emphasis on the movement of Antarctica's ICE SHEETS, the circulation of Antarctic storm systems, and interactions between ICE and ATMOSPHERE.

Byrd Station Byrd Station was established in January 1957 by the USA as part of INTERNATIONAL GEOPHYSICAL YEAR, and of OPERATION DEEPFREEZE. The station, which was located in MARIE BYRD LAND and built under the ground, is named after Richard BYRD.

On 29 January 1968, the first ICE CORE to reach Antarctic BEDROCK was drilled with a cable-suspended, electromechanical rotary drill at Byrd Station. Other research conducted here included upper atmosphere physics, meteorology, geophysics and glaciology. The station was replaced by the Byrd Surface Camp in 1972, in part because snow was threatening to crush it. The Byrd Surface Camp served primarily as a refuelling site for LC-130 AIRCRAFT travelling to and from remote field sites in WEST ANTARCTICA. Byrd Surface Camp operates as a summer-only base and personnel provide weather observations to MCMURDO STATION and maintain the skiway as an alternative landing site for EMERGENCIES.

Byrd's First Expedition (1928–30) Private American expedition led by Richard BYRD; officially known as the United States Antarctic Expedition. It marked the beginning of the widespread use of modern technology in Antarctic exploration and the start of recent USA involvement. One of the best prepared expeditions up to that time, the main aim was to fly over the SOUTH POLE, but SCIENTIFIC RESEARCH also formed an important part. Funds were donated by J D Rockefeller Jr and Edsel Ford, and the *New York Times* paid US$60,000 for exclusive news coverage.

The expedition set out with two SHIPS, three AIRCRAFT, 50 men and 95 DOGS. On leaving the USA, Byrd had told the press, 'The dog sledge must give way to the aircraft; the old school has passed.' He also said they were going to 'a new place ... where men will not strut because there are no women about.'

The expedition reached the ROSS ICE SHELF on Christmas Day 1928. Byrd established a base, LITTLE AMERICA, to the east of the BAY OF WHALES and 14 km (8½ miles) inland. Within two weeks a small village had been constructed, dominated by 20 m (65 ft) tall radio towers.

It has long been believed that the expedition made the first flight over the South Pole, on 28–29 November 1929; however, recent research shows that Byrd could not have flown over the Pole. Nevertheless, it was an extraordinary accomplishment. When the Ford monoplane was 'over the Pole', Byrd dropped an American flag and radioed Little America, a message that was picked up by chance in New York and immediately broadcast by loudspeaker in Times Square, New York.

The expedition also carried out pioneering work in geology. Samples of sandstone and coal, discovered by geologist Laurence Gould on Mount Fridtjof Nansen, proved that the mountains were not volcanic as had previously been thought, but had been formed by movements of the Earth's crust. 'No symphony I have ever heard, no work of art before which I have stood in awe ever gave me quite the thrill that I had when I reached out ... and picked up a piece of rock to find it sandstone,'

Above: *Established by the USA in January 1957, Byrd Station, situated in Marie Byrd Land, is named in honour of explorer Richard Byrd.*

Gould wrote. 'It was just the rock I had come all the way ... to find.' On their return journey his party found a cairn and a note left by Roald AMUNDSEN on his way back from the Pole.

Byrd's Second Expedition (1933–35) Private American scientific expedition led by Richard BYRD; officially known as the United States Antarctic Expedition. Byrd's intention was to examine the uncharted and unclaimed areas between the 80° and 150° West meridians. The expedition sailed in two ships, the *Jacob Ruppert* and the *Bear of Oakland*, with a variety of VEHICLES, including a heavy Cletrac TRACTOR, three Citröen tractors and two Ford snow-mobiles, three AIRCRAFT, 153 DOGS, and 56 men in the OVERWINTERING party. Three cows taken to provide fresh milk were housed in purpose-built heated barns. Regular RADIO broadcasts were transmitted to the USA for the first time.

The ships arrived in the BAY OF WHALES on 17 January 1934, then had a difficult passage through the PACK ICE, a route they named 'Misery Trail'. A base was established at LITTLE AMERICA, on the site used by BYRD'S FIRST EXPEDITION in 1928–30, and 10 buildings were constructed. Some of the stores had to be moved further inland when cracks started appearing in the ice shelf around Little America.

An inland base, Bolling Advance Weather Station, was established 198 km (123 miles) from Little America, at 80°0S'S. The original intention was for three men to spend winter there to monitor WEATHER conditions, but Byrd decided to winter there alone. He left Dr Thomas Poulter, the

expedition's senior scientist, in charge of Little America with instructions that it be run as a republic: 'We have no class distinctions as in civilization.'

Byrd was left in the inland hut, measuring four paces by three, on 28 March 1934 and stayed for four and a half months. Apart from the extreme cold, carbon monoxide fumes from a faulty stove and the radio generator were slowly poisoning Byrd, and he collapsed suddenly at the end of May. Not wanting other lives endangered in a rescue attempt, he did not mention his condition during his regular radio contacts with Little America. Eventually, the base radio operator suspected he was ill, and a rescue party set out in early spring—in the guise of a meteor observation trip.

After three attempts, Poulter and two others arrived on 11 August. It was another two months before Byrd was well enough to fly back to base.

The expedition proved conclusively that there was no strait connecting the ROSS SEA and WEDDELL SEA, thereby establishing that Antarctica was a single continent. The depth of the continental ice cap was measured, more life forms were found, much meteorological data were collected, and new mountains (including the HORLICK RANGE) and a large plateau were discovered. Tractors were used successfully in the Antarctic, covering 845 km (530 miles) on one depot-laying trip. A SLEDGING party found FOSSILS of leaves and tree trunks, and seams of coal in the Supporting Party Mountains in the QUEEN MAUD MOUNTAINS. In all, over 1 million sq km (386,000 sq miles) were surveyed by air.

C

Campbell Island Formed from an extinct VOL-CANO, this SUBANTARCTIC ISLAND covers 114 sq km (44 sq miles) in the southern Pacific Ocean. Its craggy hills are clothed in TUSSOCK, and steep volcanic cliffs drop away to the sea. A number of birds, including the SOUTHERN ROYAL ALBATROSS, GREY-HEADED MOLLYHAWK and NORTHERN GIANT PETREL, nest on the island, which also supports a variety of PLANTS.

Captain Frederick Hasselborough, in the ship *Perseverance*, discovered the island on 4 January 1810 and named it after his sealing company. Farming was more successful on Campbell Island than early attempts on the AUCKLAND ISLANDS. In 1895, J Gordon released 200 sheep and built a homestead in Tucker Cove, in Perseverance Harbour. This farm operated until 1931, when the remaining 4000 sheep and a few cattle were abandoned. The New Zealand government declared the island a reserve in 1954, and in 1990 the last wild sheep were removed. Along with farming, two shore-based WHALING ventures operated on

the island between 1909 and 1916. Over this time more than 60 SOUTHERN RIGHT WHALES were killed, but numbers have recovered, and every winter they return to Northwest Bay to mate.

The first scientific observations were made by James Clark ROSS's 1839–43 expedition, which carried out plant and animal surveys. A French expedition visited Campbell Island twice between 1873 and 1875 to observe the transit of Venus, although on both occasions heavy cloud prevented any astronomy.

Campbell Island cormorant (*Leucocarbo campbelli*) A vulnerable species, Campbell Island cormorants are confined to the small volcanic island lying between the Subtropical and ANTARCTIC CONVERGENCES. The population is estimated at around 8000 birds. They grow to about 63 cm (25 in) and have wing-spans of 105 cm (41 in).

Cape petrel (*Daption capense*) Also known as the cape pigeon or pintado petrel. Cape petrels are distinctly coloured—the Spanish word 'pintado' means 'painted'—with speckled backs and wings, mostly white underparts, and black beaks and feet. They are 39 cm (15 in) long and their wings span 86 cm (33 in).

Cape petrels commonly breed on the coasts of the Antarctic continent and on many of the SUB-ANTARCTIC ISLANDS as far north as the CROZETS,

nesting on rock ledges and using small stones and gravel as building materials. Laying begins about November and the single chick fledges in March.

Their range is circumpolar, they avoid the PACK ICE and are commonly seen off the coasts of AUSTRALIA, South America and SOUTH AFRICA. Large flocks of cape petrels follow ships, feeding off anything edible that is thrown overboard. They also feed on SQUID, FISH and CRUSTACEANS. They are extremely noisy when squabbling over food.

Cape pigeon (*Daption capense*) See CAPE PETREL.

Carbon dioxide (**CO_2**) Carbon dioxide makes up only a small proportion of the ATMOSPHERE—about 0.03 percent—but is a major contributor to the GREENHOUSE EFFECT. In fact, it is the most important human-produced greenhouse gas, and to understand global climate change, scientists study the ways that carbon dioxide enters and leaves the atmosphere.

The SOUTH POLE is an ideal pristine location to measure changes in atmospheric carbon dioxide concentrations, because data are not complicated by nearby human activity. The USA began measuring carbon dioxide concentrations at the South Pole in 1956, and now has one of the longest records of carbon dioxide concentrations available. Evidence from ICE CORE samples suggests hat

Below: Formed from an extinct volcano, Campbell Island is a nesting site for several bird species and was a centre for the whaling industry.

*Above: The cape petrel (*Daption capense*) is also known as the pintado ('painted') petrel, due to its distinctive markings.*

carbon dioxide levels have increased at least 25 percent over the past two centuries. A series of core samples from VOSTOK STATION on the POLAR PLATEAU showed a definite relationship between carbon dioxide and air temperature. Scientists extracted carbon dioxide that had been trapped in small bubbles in the ice for as long as 160,000 years, and found that carbon dioxide levels shot up at the end of the last ICE AGE as temperatures warmed, and plunged again at the beginning of the present glacial period. They also documented another recent rise in carbon dioxide levels, this time related to industrialization and the use of fossil fuels.

Humans have increased levels artificially in two ways: firstly, by releasing carbon dioxide (eg, by burning fossil fuels) and, secondly, by cutting down forests, which absorb the gas. The SOUTH-ERN OCEAN is a major 'sink' of carbon dioxide—it is estimated to absorb about 30 percent of the carbon dioxide discharged into the atmosphere. Scientists are still attempting to understand how carbon dioxide moves through the FOOD CHAIN to its ultimate burial at the sea floor.

Carmen Land When crossing the ROSS ICE SHELF in 1912, Roald AMUNDSEN named non-existent Carmen Land and estimated its position as being south of DRONNING MAUD LAND, and 100 km (62 miles) east of him and his party. Richard BYRD disproved the existence of Carmen Land on one of the first flights over Antarctica in 1929. Amundsen's mistake is believed to have been caused by a MIRAGE.

Cartography Although there were no definite sightings of Antarctica before 1820, the continent was 'mapped' as early as AD 150 by the Egyptian astronomer and geographer Ptolemy. Inspired by Ferdinand MAGELLAN's voyage round the world, in 1531 French mathematician Oronce Finé published a map—drawn from his imagination—of 'Terra Australis recenter invento sed nondum

plene cognia' ('the southern land newly discovered but not yet fully known'). James COOK was the first to chart expanses of the SOUTHERN OCEAN on his 1772–75 voyage. In the mid-19th century, explorers such as Thaddeus BELLINGSHAUSEN, John BISCOE, John BALLENY and James Clark ROSS filled in pieces of coastline. By the end of the century the general outline of the Antarctic continent was known. Later explorers discovered and charted sections of coast, and began mapping the contours of the interior.

Extensive aerial surveying—which revealed the extraordinary hidden topography of MOUNTAIN ranges that are covered by ICE, SUBGLACIAL LAKES and basins—began in the 1920s with Lincoln ELLSWORTH, Hubert WILKINS and Richard BYRD. Today, RADARS on aerial flights and SATELLITES build a picture of larger areas. A satellite known as RADARSAT, which scatters radar beams continually over Antarctica, has once again revolutionized mapping in Antarctica. It can operate through winter darkness, penetrate thick cloud cover and rotate in orbit, and it produces maps that are a mosaic of bright and dark areas. Finely powdered snow and smooth ice tend not to scatter radar and look dark on images. Coarse, uneven material such as weathered ice and ROCK, CREVASSES and SASTRUGI scatter the radar beam and thus look bright.

Casey Station AUSTRALIA took over the USA's Wilkes Station in 1959 and replaced it with Casey Station, located about 2 km (1 mile) south of the original base, on VINCENNES BAY in WILKES LAND. The station was a novel design for the time, with personnel and work quarters connected in one long tunnel-shaped building, that was raised up so that the wind would prevent snowdrifts from forming. Research is conducted here into atmospheric sciences, biology, geosciences, human impacts and medicine. Rebuilt in 1988, the present Casey Station is home to about 20 people over winter, and up to 70 during summer.

Castaways Members of several early expeditions became castaways on the Antarctic continent, most famously the ANTARCTIC EXPEDITION led by Otto NORDENSKJÖLD, which was marooned on SNOW HILL and at PAULET ISLANDS on the northeastern tip of the ANTARCTIC PENINSULA; Ernest SHACKLETON's 1914–17 expedition, which lost its ship, the ENDURANCE, in the WEDDELL SEA, wintered at ELEPHANT ISLAND and made the heroic journey to SOUTH GEORGIA to seek rescue; and Douglas MAWSON who, after losing his two companions in GEORGE V LAND, completed an arduous solo trek back to Cape DENISON—only to see his ship, the *Aurora*, steaming off in the distance and so had to remain on the continent (with the six men who had remained to search for his party) for another winter.

Many sealers and whalers were marooned in the SOUTHERN OCEAN. The crews from these stricken vessels often struggled for months on end, waiting for rescue. The New Zealand government began setting up castaway huts on its SUBANTARCTIC ISLANDS in 1880, containing food and blankets, and guns to hunt pigs, goats and rabbits. Until 1927, they made annual trips to the islands in search of castaways, until improved radio communications made this unnecessary.

Below: Between 1880 and 1927 the New Zealand government maintained 'castaway' huts on its subantarctic islands for marooned sailors. This hut is on Enderby Island.

Cephalopod A sophisticated class of marine MOL-LUSCS, including OCTOPUS and SQUID, that together constitute an important component of the Antarctic FOOD WEB. Scientists estimate there are about 100 million tonnes of squid in Antarctic seas, of which SPERM WHALES may consume as much as 50 million tonnes (50 million tons) each year.

Cephalopods have well-defined heads with highly developed eyes and a number of arms or tentacles. There are two orders: the Decapoda, including squid, which have 10 arms, and the Octopoda, to which the eight-armed octopus belongs. It is believed that about 70 species occur in Antarctic waters.

The shell normally present in molluscs is absent in squid and octopus, and usually reduced and internal in other cephalopods. Their mantles (outer skins), however, are thick and muscular. Cephalopods move by relaxing and contracting the mantle, allowing water to enter the mantle cavity then jetting it out through an exhalent siphon. This jet-propelling mechanism means that, unlike most molluscs, cephalopods are able to move at high speed. They are active, free-ranging predators and, in the case of squid, can reach gigantic proportions—*Architeuthis dux*, the giant squid, for example, reaches 18 m (60 ft).

Challenger Expedition (1872–76) British oceanographic research expedition, led by marine biologist Charles THOMSON, then professor of natural history at Edinburgh University. At Thomson's instigation, the ROYAL SOCIETY funded a global expedition to undertake OCEANOGRAPHIC RESEARCH. HMS *Challenger*, an ex-Royal Navy man-o'-war, was converted into a fully equipped scientific survey ship and George NARES was appointed captain.

The expedition left Portsmouth on 21 December 1872. They sailed down the Atlantic and called at TRISTAN DA CUNHA in October 1873. They established an observation station to view and record the transit of Venus at Îles KERGUÉLEN, and spent a month surveying the coast around the islands. Next they sailed southeast to HEARD ISLAND, and in late February 1874 reached south 72°22'E 66°33'S—becoming the first steamship to cross the ANTARCTIC CIRCLE. Naturalist John MUR-RAY recorded that, 'The *Challenger* has dredged up fragments of mica schists, quartzites, sandstones, compact limestones, and earthy shales, which leave little doubt that within the Antarctic Circle there is a mass of continental land quite similar in structure to other continents.' Thus, they were able to prove scientifically that Antarctica was a continent (rather than a group of islands) without having sighted it, by deducing that the continental rock and soil material had been carried away and dropped by icebergs.

The scientific work carried out during the *Challenger*'s four-year circumnavigation—a voyage that marked the beginnings of modern oceanography and covered 127,634 km (79,309 miles)—involved dredging the sea floor (in the course of which much new marine life was discovered), taking soundings, measuring water temperatures, and monitoring directions and rates of currents, as well as atmospheric, meteorological and magnetic observations. Results were published in 50 volumes.

Charcot, Jean-Baptiste (1867–1936) French scientist and explorer, born at Neuilly-sur-Seine. The son of famous neurologist Jean-Martin Charcot, Jean-Baptiste graduated in medicine in 1895. He had the 250 tonne (250 ton) vessel *Français* purpose-built for polar conditions, and on the 1903–05 FRENCH ANTARCTIC EXPEDITION, which was originally intended to rescue Otto NORDEN-SKJÖLD's ANTARCTIC EXPEDITION, he carried out extensive SCIENTIFIC RESEARCH, notably, charting the western side of the ANTARCTIC PENINSULA. On the second FRENCH ANTARCTIC EXPEDITION in 1908–10 on the *POURQUOI PAS?*, he continued surveying uncharted coast: his charts of the west of the Peninsula were used exclusively before the area was systematically explored by the British in 1935.

During World War I Charcot served with the French navy, and in 1926, at his instigation, the French government made a TERRITORIAL CLAIM to TERRE ADÉLIE. In his later years, Charcot carried out OCEANOGRAPHIC RESEARCH in the North Atlantic around the Hebrides and in the Greenland Sea, and land surveying on the eastern side of Greenland. In September 1936 the *Pourquoi pas?* was wrecked off the coast of Iceland and Charcot and all but one of the crew were drowned.

Charcot Island Once believed to be part of the Antarctic continent and known as Charcot Land, the island, which is located off ALEXANDER ISLAND in the BELLINGSHAUSEN SEA, was surveyed by Hubert WILKINS in January 1930.

Cherry-Garrard, Apsley (1886–1959) British explorer and writer, born in Bedford, England. He came from a wealthy family and studied history and classics at Oxford University. He donated £1,000 to the 1910–13 BRITISH ANTARCTIC EXPEDITION and, although he was initially turned down by Robert SCOTT, eventually joined the expedition as assistant zoologist.

During the expedition's first winter, Cherry-Garrard trekked 105 km (65 miles) with Edward WILSON and Henry BOWERS to the Cape CROZIER rookery to find EMPEROR PENGUIN eggs, which Wilson hoped would explain the origin of all birds. This was the first journey ever attempted during the depths of the Antarctic winter. It was described by Cherry-Garrard in his acclaimed best-selling book, *The Worst Journey in the World*. He wrote: 'This journey had beggared our language: no words could express its horror.'

Cherry-Garrard was involved in Scott's attempt to reach the SOUTH POLE, accompanying the polar party to the Upper Glacier Depot, and was disappointed not to be included in the final party. He was a member of the search party that found the bodies of Scott, Wilson and Bowers in November 1912.

Cherry-Garrard was invalided home in World War I. From then onwards, his physical and mental health was frail. 'Polar exploration is at once the cleanest and most isolated way of having a bad time which has been devised,' he wrote.

Chile Chile has been interested in Antarctica from the time of its independence from Spain in 1818. In 1940 its TERRITORIAL CLAIM was defined: TERRITORIO CHILENO ANTÁRTICO is a wedge-shaped sector running between 53°W and 90°W, down to the SOUTH POLE, with no northern boundary specified. The first permanent Chilean base was constructed in 1947 in the SOUTH SHET-LAND ISLANDS, and multiple bases were established by the late 1950s. The Chilean government has taken an active role in Antarctic issues: in 1947 it declared Chilean jurisdiction out to 320 km (200 miles) from the coastline, and the next year, the country's president, Gabriel Gonzaliz Videla, visited Chilean Antarctic bases. Thirty years later

Above: 'Birdie' Bowers (left), Edward Wilson (centre) and Apsley Cherry-Garrard (right) prior to setting out to Cape Crozier in 1911, the first overland winter journey undertaken in Antarctica.

and nearly two decades after Chile had signed the territorially-neutral ANTARCTIC TREATY, in January 1977, President Augusto Pinochet also visited Antarctica, and declared that Chile's claim was a continuation of its mainland territory.

Chile has always opposed efforts at internationalizing the Antarctic continent. In July 1948 it responded to USA proposals for such internationalization with the ESCUDERO DECLARATION. This welcomed scientific cooperation, suggested political neutrality of expeditions, and proposed a five-year suspension of discussions on SOVEREIGNTY. These principles were later adopted in part by the Antarctic Treaty.

A participant in the INTERNATIONAL GEOPHYSICAL YEAR, Chile is party to the 1972 CONVENTION FOR THE CONSERVATION OF ANTARCTIC SEALS, the 1982 CONVENTION ON THE CONSERVATION OF ANTARCTIC MARINE LIVING RESOURCES and the 1991 MADRID PROTOCOL.

Chile's scientific investigations are concentrated in the ANTARCTIC PENINSULA and the surrounding waters. The country has been involved in harvesting KRILL since the 1970s and in longline FISHING from the 1980s. Interest in developing TOURISM has been growing, and Punta Arenas is one of the GATEWAYS to Antarctica.

China China is a relatively new participant in Antarctica, the last major world power to become involved in the ANTARCTIC TREATY SYSTEM (ATS). It acceded to the ANTARCTIC TREATY in 1983 and became a CONSULTATIVE PARTY in 1985. The first major Chinese expedition to Antarctica was in November 1984.

China operates two year-round bases: the GREAT WALL BASE and the Zhongshan Station at Prydz Bay in PRINCESS ELIZABETH LAND. China is not a member of the CONVENTION ON THE CONSERVATION OF ANTARCTIC MARINE LIVING RESOURCES (CCAMLR), and it has been criticized for allowing shipments of illegally caught PATAGONIAN TOOTHFISH to be shipped through China to markets in the USA and JAPAN.

Chinstrap penguin (*Pygoscelis antarctica*) Chinstrap penguins are closely related to ADÉLIE and GENTOO PENGUINS. Although about the same height as the gentoo—they average about 76 cm (30 in)—they weigh considerably less at about 4.5 kg (10 lb). Like Adélies, the chinstraps' head and back are black, but they have white cheeks and a distinctive thin black line—like a chinstrap—running from the black cap to the chin. They feed mainly on KRILL, diving in the open sea and foraging among the PACK ICE.

Chinstraps are often found nesting with Adélies and gentoos on rocky coastal slopes, on the ANTARCTIC PENINSULA and SUBANTARCTIC ISLANDS in the South Atlantic. After the MACARONI, the chinstrap is the second most abundant SUBANTARCTIC penguin species: there are an estimated 4 million breeding pairs, almost half of these on the SOUTH SANDWICH ISLANDS. A small population breeds on BALLENY ISLANDS in the ROSS SEA. In summer, females lay two eggs in nests formed out of pebbles and 37 days later the chicks fledge.

Above: Chile's Carvajal Station, on Adelaide Island.

Chlorofluorocarbons Once considered completely safe and non-toxic, chlorofluorocarbons (CFCs) are a major cause of OZONE depletion over Antarctica, and also one of the most effective GREENHOUSE GASES. They have been used since the 1960s in the industrial production of everything from air-conditioners and refrigerators to aerosols, solvents and computer chips. About 1 million tonnes are produced each year, mainly in the Northern Hemisphere. In the mid-1970s, scientists found that CFCs, after remaining intact and inert for up to 50 years in the lower ATMOSPHERE, eventually infiltrate the stratosphere. Here, they are bombarded with ULTRAVIOLET RADIATION and break down, releasing chlorine, which in turn attacks ozone, causing the OZONE HOLE.

CFCs are not absorbed by the ocean or the soil, and there is no way of removing them from atmospheric circulation. Their chemical stability, which was once thought to make them neutral and safe, actually makes them one of the most dangerous artificial substances to be released into the atmosphere. The signing of the MONTREAL PROTOCOL in 1987 was a major step towards the reduction of CFCs—a commitment on the part of a number of major CFC-producing countries to limit their use. The protocol has been regularly amended in the light of improved understanding of ozone depletion. However, despite the reduction in CFCs released into the atmosphere, the long lifespan of the gases may mean that they continue to deplete ozone levels over Antarctica for many more years. It is estimated that if all CFC and related chemical production were to cease immediately, it would still take until the year 2050 for the ozone hole to disappear.

Below: Chinstrap penguins (Pygoscelis antarctica) are named for the thin black line that runs from their black caps to their chins.

Christensen, Lars (1884–1965) Norwegian whaling magnate and explorer, born at Sandefjord. His family was involved in WHALING and by the time Christensen was 23, he owned his own whaling company, which worked in Alaskan and South American waters. By 1920 his small firm had grown into one of the world's largest whaling companies, and he turned his attention to the SOUTHERN OCEAN. Between 1927 and 1937 Christensen organized nine separate Antarctic expeditions. He led two of these himself, and each combined whaling with EXPLORATION, specifically to claim territory for NORWAY. Over these 10 years, BOUVETØYA, PETER I ØY and DRONNING MAUD LAND were all added to Norwegian TERRITORIAL CLAIMS.

Christensen-sponsored expeditions carried out extensive surveys of Dronning Maud Land, and charted the area between the WEDDELL SEA and ENDERBY LAND. Christensen died in New York at the age of 81.

Circumnavigation To 1997, there had only been 10 known circumnavigations of Antarctica. The first was by James COOK on his 1772–75 voyage, followed by Thaddeus BELLINGSHAUSEN in 1819–20 and John BISCOE in 1830–33. After a century-long gap, there were four circumnavigations in the 1930s: Hjalmar RIISER-LARSEN and Nils Larsen between 1930 and 1939; the DISCOVERY EXPEDITION in 1931–33; Lars CHRISTENSEN and Klarius Mikkelson in the *Thorshavn* in 1933–34, and a second *Discovery* circumnavigation in 1937–39.

Circumpolar current See WEST WIND DRIFT.

Circumpolar deep water Warm, nutrient-rich water that lies beneath ANTARCTIC SURFACE WATER. When this south- and upward-flowing water mass collides with the north- and downward flowing surface water at the ANTARCTIC DIVERGENCE, the circumpolar deep water wells up towards the surface. It was first noted by William WALES and William Baly, astronomers sailing with James COOK on the 1772–75 voyage, when they dropped thermometers overboard and measured a significant increase in TEMPERATURE below the surface.

The water flows around the continent in an anti-clockwise direction and reaches speeds greater than 45 cm (18 in) per second, which is fast enough to scour the sea floor where it flows across the CONTINENTAL SHELF. The upwelling of circumpolar deep water extends from approximately 250 m (820 ft) to 4000 m (13,000 ft) below the ocean surface. The nutrients carried by the water support large quantities of PHYTOPLANKTON that forms the basis of the Antarctic marine ECOSYSTEM. After approximately two years at the Divergence, the water will begin to flow back to coastal Antarctica between layers of Antarctic surface water and ANTARCTIC BOTTOM WATER.

Circumpolar Trough A band of low air pressure, located between latitudes 60° and 70°S; also called the Antarctic Trough. It separates two broad bands of atmospheric circulation around Antarctica: a zone of westerly WINDS to the north and POLAR EASTERLIES to the south, close to the continent. The meeting of air masses at the Trough causes frequent storms; they usually last for a few days, then the weather clears, only to be replaced by another storm system.

At similar latitudes in the SOUTHERN OCEAN, the ANTARCTIC DIVERGENCE marks a surface boundary between different surface water masses. The only seasonal change in the position of the Trough is a small movement away from the continent in winter, and back in summer, on a half-yearly cycle.

Circumpolar vortex A whirlpool of WIND that spirals westwards around the SOUTH POLE over winter and seals off the Antarctic stratosphere. In mid-winter the vortex is at its strongest, reaching speeds of more than 100 m (328 ft) per second and is known as the 'polar night jet'. See ATMOSPHERE.

Clarie Coast Also called 'Wilkes Coast', the Clarie Coast is one of several named coasts on the edge of TERRE ADÉLIE. It is enclosed by the Pourqoi Pas? Glacier to the west and the Blodgett Iceberg Tongue to the east.

French explorer Jules-Sébastien DUMONT D'URVILLE discovered and named the sheer ice-covered coast after the wife of the captain of the *Zélee* on 30 January 1840 during his 1837–40 expedition.

Claims See TERRITORIAL CLAIMS.

Climate Three basic climatic regions can be distinguished in Antarctica: the interior, the coast and the ANTARCTIC PENINSULA. The interior is characterized by extreme cold and by light snowfall; it is a POLAR DESERT. In the interior, at the SOUTH POLE, WINTER lasts from March until October and TEMPERATURES average –57°C (–71°F). The coastal areas have milder temperatures because they are near sea level, and are influenced by WEATHER systems that form over the SOUTHERN OCEAN, which bring warmer, moister air to the coast, higher PRECIPITATION and more CLOUD FORMATION. The Antarctic Peninsula also has a relatively warm and moist coastal climate, but is more variable; when cold air flows north from the continent, temperatures can plummet.

Records of climate change from FOSSILS and from ice and sediment CORE SAMPLES have revealed that conifer and beech tree FORESTS grew in Antarctica only 2 million years ago. The continent drifted to its present near-polar position over the last 100 million years, and scientists believe that the climate has changed dramatically over this time (see ICE AGE).

Climatological research See METEOROLOGICAL RESEARCH.

Clothing Field parties are equipped with special clothing designed to withstand the COLD of Antarctica. The most effective clothing systems involve multiple layers, which are more effective at trapping heat than one thick layer. Such a system also allows people to regulate their body temperature by removing or adding layers. The modern fabrics used are able to 'breathe', and release moisture from perspiration, rather than allowing it to freeze against the skin.

For summer fieldwork, appropriate clothing is a base thermal layer, fleece overalls and jacket, and a thick, windproof outer shell (trousers and jacket). Sturdy work boots are needed around the bases, and for dealing with heavy EQUIPMENT; double-insulated mountaineering boots are used for SKIING and work on rock, and knee-length muklaks are efficient in cold, dry snow. Essential accessories include GOGGLES and glasses, to protect eyes against the harsh sunlight, which can quickly cause SNOWBLINDNESS. Gloves, balaclavas, hats and thick socks protect the extremities from FROSTBITE. Most of this gear is issued to people on their way to the ice and returned when they leave.

Roald AMUNDSEN's childhood in Norway gave him a considerable advantage in preparing clothing for his polar expedition. He had ex-Navy blankets made into felted suits, and 250 reindeer skins turned into clothes modelled on those worn by the Netchelli Eskimos. These weighed about half as much as wool, but protected the wearer from temperatures as low as –60°C (–76°F). But, like all clothing on early expeditions, they were not quick-drying and could not breathe, and when they became wet, the moisture froze on the fabric.

Below: Dressing for Antarctic conditions involves layers of clothing. Essential accessories include goggles or sunglasses to prevent snowblindness.

Above: Clouds over inland Antarctica are rare, due to the dry atmosphere; those that do form are thin and composed mainly of ice crystals.

Cloud formations Clouds only form regularly in areas where air masses absorb sufficient water vapour from the ocean surface. Heavy, low stratocumulus cloud forms over the SOUTHERN OCEAN, particularly during summer, when cyclonic WEATHER systems and associated fronts bring warm, humid air southwards. Strong WINDS that blow onshore, from the ocean to the coast, can pile this cloud over coastal regions.

In summer, billowing cumulus clouds may form over areas of ROCK that absorb heat, such as MOUNTAIN ranges, NUNATAKS and OASES. Because cumulous clouds need to be heated from below to form, they are rarely seen over ICE and SNOW surfaces. Hogsback clouds also form over mountains when there is strong high-level air flow; they are often the first sign that bad weather is on its way.

Cloud formation is limited over inland Antarctica, which is known for its arid CLIMATE. The air in this zone is very clear, and the only clouds are thin and composed of ice crystals; they have a very low moisture content and do not reach very high into the ATMOSPHERE.

Coal Antarctica—the coldest, windiest, driest and highest continent on Earth—was once covered in forests of *GLOSSOPTERIS* and other associated deciduous conifers: their remains have been fossilized in coal seams of Permian age (about 245 to 286 million years ago) in the TRANSANTARCTIC MOUNTAINS.

Coastal islands Windswept, bleak outposts of the Antarctic mainland, they include ALEXANDER, THURSTON, ROSS, ROOSEVELT and BERKNER ISLANDS. There are few coastal islands, given the size of the Antarctic continent. Most have been submerged by the sea, or overridden by vast ICE SHEETS. The climatic conditions on these islands vary little from those of the continent itself. Below-freezing

TEMPERATURES occur all year round, and average air temperatures are often lower than –9°C (15°F). The islands are heavily glaciated, and in winter are encased in thick PACK ICE.

Coats Land Discovered by William BRUCE's SCOTTISH NATIONAL ANTARCTIC EXPEDITION in March 1904 and named after the expedition's sponsors. Coats Land is bounded by the WEDDELL SEA to the north, DRONNING MAUD LAND to the northeast and the SHACKLETON RANGE to the southwest.

Colbeck, William (1871–1930) British naval officer and scientist, born at Kingston-upon-Hull. Colbeck went to sea at the age of 15. After a number of years in the British Navy and merchant navy, he joined the 1898–1900 SOUTHERN CROSS EXPEDITION as cartographer, magnetic observer and navigator. Before departure Colbeck and

Louis BERNACCHI took a course in the use of new instruments at Kew Observatory.

Antagonism between Colbeck and expedition leader Carsten BORCHGREVINK developed during the journey south, and on two occasions at Cape ADARE Borchgrevink dismissed then reinstated him. On 24 July 1898, Colbeck nearly burnt the Cape Adare hut down when papers near a candle he had lit caught alight; the burnt curtains are still in the hut. In February 1900, Colbeck was one of a small party to sledge from the BAY OF WHALES towards ROOSEVELT ISLAND and achieve a 'farthest south' on the Antarctic continent.

With Bernacchi, Colbeck joined the BRITISH NATIONAL ANTARCTIC EXPEDITION of 1901–04 as commander of the relief vessel *Morning*. When the *Discovery* became ice-bound in MCMURDO SOUND in February 1903, Robert SCOTT ordered the *Morning* north. It joined the TERRA NOVA, in

Below: Stratified layers of coal are visible in the mountains by Taylor Glacier.

Above: The severe cold of the Antarctic winter can cause teeth to crack.

Hobart, Australia, and from there the two ships travelled south in January 1904 and hauled the *Discovery* out of the PACK ICE.

After his return to England, Colbeck rejoined the merchant navy. Bernacchi and Borchgrevink both attended his funeral.

Cold Antarctica is officially the coldest place on Earth: the coldest temperature on record— −89.2°C (−103°F)—was recorded at VOSTOK STATION in 1983. However, on calm summer days, despite low TEMPERATURES, in many parts of Antarctica it may be quite comfortable to be dressed in only one layer of CLOTHING. Once even the slightest breeze springs up, however, the WIND will cut through fabric like a knife, creating a dangerous condition called WINDCHILL.

'The Antarctic cold doesn't feel like air, but like something solid pressed against your skin,' said one visitor describing her first impressions of the SOUTH POLE, and there are many accounts of the extreme cold's effects on humans. On first leaving the relative warmth of an aircraft, nostril hairs freeze, while wet hair will freeze solid instantly. In mid-winter, if a cup of boiling water is thrown up in the air it will snap freeze before hitting the ground, while washing the dishes at field camp more often involves scraping the frozen remains from plates.

The body's response to cold is to increase heat production, initially by raising its rate of metabo-lism, or speeding up body functions, then by increasing muscular activity by shivering. Another instinctive way to combat the cold is by huddling, or curling up, which reduces the skin surface area from which heat is lost.

In very low temperatures human skin 'sticks' to metal. On the 1901–04 NATIONAL ANTARCTIC EXPEDITION, Frank WILD reported, '… one man opened a tin of jam with his sheath knife … and put it in his mouth. The knife immediately froze fast to his lips and tongue and he had to keep it there until it warmed sufficiently for him to remove it without tearing a lot of skin away. As it was his mouth was badly blistered.'

Colvocoresses Bay A cove, 44 km (27 miles) wide at its entrance, near LAW DOME on the WILKES LAND coast. It is named after George Colvocoresses, a midshipman on Charles WILKES's 1838–42 UNITED STATES EXPLORING EXPEDITION. Colvocoresses described the expedition's exploration of the East Antarctic coastline in his book *Four Years in a Government Exploring Expedition*.

Commerce The earliest commercial activity in Antarctica was SEALING, which was at its peak during the first decades of the 19th century. The pattern of resource exploitation established with sealing—a cycle of discovery, boom and bust— was repeated in the 20th century with WHALING and FISHING. Attempts have been made to regulate harvesting so that over-exploitation does not occur again with the establishment of management regimes: the CONVENTION ON THE CONSERVATION OF ANTARCTIC SEALS, the INTERNATIONAL WHALING COMMISSION and the CONVENTION ON THE CONSERVATION OF ANTARCTIC MARINE LIVING RESOURCES. Despite the conventions, fishing remains the major commercial activity in the Antarctic region. Heavy exploitation of fish stocks, mainly by Northern Hemisphere distant-water fleets, began in the 1960s and continued until the early 1990s. It is estimated that in many seasons catches exceeded maximum sustainable yields by considerable amounts.

Interest has been expressed in the possibility of drilling for OIL around Antarctica and mining MINERALS on the CONTINENT. In the 1980s the CONSULTATIVE PARTIES to the ANTARCTIC TREATY attempted to negotiate a regime for managing any future mineral exploitation and signed the CONVENTION ON THE REGULATION OF ANTARCTIC MINERAL RESOURCE ACTIVITIES (CRAMRA); this was abandoned in 1991 and replaced with the MADRID PROTOCOL, which imposed a 50-year moratorium on exploitation of mineral resources in Antarctica.

An important new commercial activity in Antarctica is that of TOURISM. Some airlines carry tourists over the continent, but the main form of tourism is CRUISES around the coastline accompanied by brief landings on the continent. As tourism has grown, so has concern over the potential impact it may have on the Antarctic environment.

Committee for Environmental Protection Created in 1998 as required by the MADRID PROTOCOL, the Committee is intended to provide advice and recommendations to the CONSULTATIVE PARTIES of the ANTARCTIC TREATY; however, it has no decision-making powers of its own. Activities include: maintaining lists of ENVIRONMENTAL IMPACT ASSESSMENTS, audits and reviews; and acting as a forum for discussion on such issues as introducing non-native diseases into Antarctica and protection of specific species.

Common diving petrel (*Pelecanoides urinatrix*) These small DIVING PETRELS, with wing-spans of around 35 cm (14 in), are the most widespread of this group of birds. Although slightly darker, they are very similar in appearance to SOUTH GEORGIAN DIVING PETRELS, making it almost impossible to tell them apart, especially in flight. As a consequence, it is difficult to estimate their at-sea distribution accurately.

They feed on COPEPODS and breed on a large number of sites, including the SNARES, Tasmania and MACQUARIE ISLAND.

Common heritage principle In the 1960s a new approach to the use of global resources—called the common heritage principle—was developed, and articulated by member states of the Non-Aligned Movement and the Group of 77 in the 1970s. In relation to Antarctica, these groups denied the validity of the TERRITORIAL CLAIMS and challenged the international acceptability of the ANTARCTIC TREATY SYSTEM (ATS). They were criti-

Above: The HMNZS Endeavour *prepares to leave Lyttelton harbour for McMurdo Sound as part of the 1955–58 Commonwealth Transantarctic Expedition.*

cal of the exclusive nature of the CONSULTATIVE PARTIES and the secrecy with which ANTARCTIC TREATY CONSULTATIVE MEETINGS were conducted. Advocates of the approach called for a more representative international regime, possibly organized under the UNITED NATIONS.

In response, the ATS changed. Membership of the Antarctic Treaty has expanded, and leading non-aligned states such as INDIA and CHINA became consultative parties. ATS meetings became more open, with the admission of more observers and publication of more documents. Debates in the United Nations' General Assembly on the 'Question of Antarctica' have moderated in their criticism of the ATS, but countries such as Malaysia still insist that the United Nations 'is the most appropriate authority to enforce, administer and monitor the various activities in Antarctica.'

Commonwealth Bay Douglas MAWSON established one of the earliest scientific stations in this bay on the coast of TERRE ADÉLIE in late 1911 as part of the 1911–13 AUSTRALASIAN ANTARCTIC EXPEDITION.

Commonwealth Transantarctic Expedition (1955–58) British Commonwealth expedition, initiated and led by Vivian FUCHS. The expedition was supported by the governments of AUSTRALIA, BRITAIN, NEW ZEALAND and SOUTH AFRICA. The aim was to make the first overland crossing of the Antarctic continent—a 3200 km (1988 mile) journey from the WEDDELL SEA to MCMURDO SOUND in the ROSS SEA, via the SOUTH POLE. There were two parties, one on each side of the continent: the Weddell Sea Party led by Fuchs would make the main crossing, with the Ross Sea Party, led by Edmund HILLARY, acting as support, laying supply depots and scouting out the route from the South Pole. The expedition would make use of modern AIRCRAFT and tracked VEHICLES.

The Ross Sea Party established SCOTT BASE, now the permanent New Zealand research base, then successfully scouted out a route along the Skelton Glacier, laying depots to the South Pole.

On the 3 January 1958 Hillary's party became the first overland crossing to the Pole since Robert SCOTT's in 1912, using vehicles—converted farm TRACTORS—and arriving with just 91 litres (20 gallons) of fuel remaining.

The Weddell Sea Party experienced rough weather in setting up SHACKLETON STATION on the FILCHNER ICE SHELF, losing supplies of coal, timber and oil during a storm. Encountering more difficult terrain, with numerous crevasses in the ice, the party made slow progress and reached the South Pole 16 days later than the Ross Sea Party. Both groups returned to Scott Base, racing to reach it before the winter weather closed in, arriving there on 2 March 1958. The crossing was completed in 99 days.

Communications Communications in Antarctica are important for weather forecasting, and to allow contact between bases, field parties, and home countries and families. The isolated nature of the Antarctic continent and its environment have often made communications difficult. The first use of the telephone was by members of the 1910–13 BRITISH ANTARCTIC EXPEDITION to transmit messages from Cape EVANS to Hut Point, a distance of 24 km (15 miles). The earliest postal services were reliant on ships that can only safely visit the continent for a few months of each year. The introduction of RADIO and telegraph systems to Antarctica by the 1911–14 AUSTRALASIAN ANTARCTIC EXPEDITION, via a relay on MACQUARIE ISLAND, greatly reduced the sense of isolation.

Today, communications in Antarctica are organized by the CONSULTATIVE PARTIES in conjunction with the World Meteorological Organization. A network of low-earth orbit SATELLITES allows voice, data, fax and paging services between Antarctica and the rest of the world, and the largest bases have internet access, and telephone and fax capabilities.

See also GLOBAL COMMUNICATIONS SYSTEM.

Congelation ice Long ICE needles that form in calm conditions at the base of existing ICE FLOES

(and in other places where ice freezes, such as lakes). Energy for their formation is supplied by heat transferred from the surface of the ice layer. Congelation ice stalactites beneath FAST ICE may grow to 6 m (20 ft) long, and sometimes extend to the sea floor.

Conservation Antarctica is seen as a pristine emblem for global conservation issues: an isolated continent with a unique environment that is largely untouched, representing a last chance to reassess attitudes to the planet and environment. For these reasons, conservation issues in Antarctica have a high profile and are the subject of ongoing debate and negotiation. Environmental groups such as the ANTARCTIC AND SOUTHERN OCEAN COALITION and GREENPEACE continue to push for Antarctica to be classified as a WORLD PARK and, under the framework of the ANTARCTIC TREATY SYSTEM, governments have recognized the need to reduce the 'human footprint' on the continent through such measures as the CONVENTION ON THE CONSERVATION OF ANTARCTIC MARINE LIVING RESOURCES, SPECIALLY PROTECTED AREAS and the MADRID PROTOCOL.

The human presence directly threatens the Antarctic environment in two ways: the inevitable impact of constructing and maintaining scientific bases, and the ever-present interest in exploiting the continent's MINERALS and FISHING resources. Since the 1950s, when SEALING and WHALING in the SOUTHERN OCEAN had all but ceased, science has had by far the greatest environmental impact of any activity in Antarctica. Research requires land, sea and air TRANSPORT networks, and increasingly sophisticated scientific bases. On ice-free areas, where most countries choose to establish bases, the mark left by human habitation can be much greater than on snow- and ice-covered areas. These sites are often attractive breeding areas for mammals and BIRDS, and the development of roads and AIRSTRIPS inevitably has a major impact on wildlife.

FUEL, which is imported for aviation and to run large machinery, does not evaporate if it is spilt in the arid Antarctic CLIMATE. Instead it seeps into the PERMAFROST and freezes. Past WASTE DISPOSAL practices at many bases have left a lasting impact on the land and ocean.

The abandonment of the CONVENTION ON THE REGULATION OF ANTARCTIC MINERAL RESOURCE ACTIVITIES and the signing of the MADRID PROTOCOL in 1991 marked an international shift in attitude towards Antarctic conservation. As well as prohibiting MINING for at least 50 years, the protocol set up procedures for ENVIRONMENTAL IMPACT ASSESSMENTS to be made of all activities, and contains provisions for the protection of ecosystems and WASTE DISPOSAL. Despite the Protocol, Roger Wilson, Chair of the Board of Greenpeace in New Zealand, believes that Antarctica is now entering the age of commercial opportunity—a time in which there will be continued pressure for economic exploitation of the continent. 'The question constantly put to us in civilization was, and still is: What is the use? Is there gold? Is there coal? The commercial spirit of the present day can see no good in pure science.

Now unless a man believes that such a view is wrong, he has no business to be down south.' wrote Apsley CHERRY-GARRARD in the first decades of the 20th century.

A future loophole in the protocol may be bioprospecting by the pharmaceutical industry, as MICROORGANISMS and deep-sea organisms, such as SPONGES, are not yet protected. Also, the introduction of exotic species, including pathogens, may threaten Antarctic species. Australian research presented at the 1998 ANTARCTIC TREATY meeting indicated that some PENGUIN colonies had been affected by a poultry pathogen, presumed to have been carried by humans to Antarctica.

Conservation areas Under the ANTARCTIC TREATY, the MADRID PROTOCOL and the AGREED MEASURES FOR THE CONSERVATION OF ANTARCTIC FLORA AND FAUNA, the concept of designating conservation areas in which specific activities are limited or prohibited has been established. The areas are SITES OF SPECIAL SCIENTIFIC INTEREST, protected areas in which research can be carried out; SPECIALLY PROTECTED AREAS, in which all activities are prohibited, and SPECIALLY MANAGED AREAS.

Consultative party A consultative party to the ANTARCTIC TREATY is a state with the right to full participation in activities regulated under the treaty, and is entitled to vote at ANTARCTIC TREATY CONSULTATIVE MEETINGS.

The original 12 signatories of the treaty are consultative parties. Other states have the right to gain consultative party status: to do so they must first accede to the Antarctic Treaty by depositing instruments of ratification and accession with the USA, which is the depositary government for the treaty. The state must also demonstrate its interest in conducting substantial research activity in Antarctica, such as by establishing a scientific station or base, or by sending a scientific expedition.

In 2001 there were 27 consultative parties, and 17 non-consultative parties, to the Antarctic Treaty. In theory, a consultative party that withdrew from activity in Antarctica and that was not one of the original 12 signatories could be downgraded to a non-consultative party, but this has never happened.

Continent See ANTARCTIC CONTINENT.

Continental drift The Earth's rocky outer crust is not a solid shell, but is broken up into thick plates that drift on the soft molten mantle. The theory of continental drift, which had first been proposed in the early 1900s, was developed by the German scientist Alfred Wegener. He assembled geological and palaeontological evidence to show how the continents could have floated across the surface of the globe.

In 1915 Wegener published *On the Origin of Continents and Oceans*, in which he proposed that 200 million years ago there was a gigantic supercontinent, which he named 'Pangaea', meaning 'All-lands', and which began to break up during the Jurassic Period, forming the continents Laurasia and GONDWANA. Antarctica began drifting away from Gondwana about 100 million

years ago, to begin its slow journey to its current polar position.

Continental shelf The submerged edge of the ANTARCTIC CONTINENT. Each year it is fed by an estimated 500 millions tonnes (500 million tons) of material, which is eroded from beneath the Antarctic ICE SHEET and transported out to sea by ICE SHELVES and GLACIERS that fringe the margin of the continent. The deposits are a mixture of gravel, sand and mud called 'diamicton', which may be compacted and bulldozed into MORAINES. In Prydz Bay in EAST ANTARCTICA, glacial moraines formed by the LAMBERT GLACIER are up to 100 m high (320 ft) and several kilometres (miles) wide. The shells and skeletons of marine plants and animals accumulate in shallow basins on the shelf, and millions of microscopic plants called DIATOMS produce layers of green diatom ooze.

Most continental shelves occur at depths of less than 200 m (656 ft). In Antarctica, however, the shelf is at an average depth of 500 m (1640 ft) because the weight of the ice cap pushes it further below sea level. Glacial valleys, up to 1000 to 2000 m (3280 to 6560 ft) deep, are carved on the shelf surface. The largest sea-floor valleys occur where large outlet glaciers cross the shelf in the ROSS and WEDDELL SEAS—the bedrock beneath the Lambert Glacier, for example, is more than 3400 m (11,152 ft) below sea level. Grounded ICEBERGS also plough the continental shelf, particularly in more shallow water. This generates coarse, poorly sorted sediment, called 'iceberg turbate'.

Convention for the Conservation of Antarctic Seals In 1972, after several seasons in which NORWAY had taken 1100 and the USSR 1000 CRABEATER SEALS in Antarctic waters, prompting fears of commercial exploitation of seal stocks, the ANTARCTIC TREATY nations signed the convention prohibiting taking of seals other than for scientific purposes or 'rational use' (that is, 'to provide indispensable food for men and dogs').

Convention on the Conservation of Antarctic Marine Living Resources (CCAMLR) Signed on 20 May 1980 in Canberra, AUSTRALIA, and entered into force on 7 April 1982, CCAMLR was negotiated by the CONSULTATIVE PARTIES of the ANTARCTIC TREATY because of their concern about the prospect of the marine resources of the SOUTHERN OCEAN being over-exploited. At the time there was a lot of interest in FISHING for KRILL—CCAMLR has sometimes been called the 'krill convention'.

The major innovation embodied in CCAMLR is its ecosystem approach to CONSERVATION. As well as aiming to conserve species targeted directly for harvesting, CCAMLR also attempts to take into account the impact on associated and dependent species and the need to restore depleted stocks, and seeks to minimize the risks of actions that are not reversible within two to three decades. (In 1985 the Ecosystem Monitoring Program was established to help provide the information necessary to achieve these goals.) In the convention, conservation is defined as 'rational use', a compromise necessary to gain the agree-

ment of both states interested in fishing and states motivated by conservation.

The area to which CCAMLR applies is different from the Antarctic Treaty, as its jurisdiction follows the ANTARCTIC CONVERGENCE, the point where cold southern waters meet warm northern waters and an ecosystem boundary for many species. This results in a boundary line fluctuating between 45ºS and 60ºS. CCAMLR includes all marine species in the Southern Ocean, excluding WHALES, which are managed by the INTERNATIONAL WHALING COMMISSION, and seals, which were already covered by the CONVENTION FOR THE CONSERVATION OF ANTARCTIC SEALS.

Because of DISPUTED CLAIMS to territory in Antarctica, CCAMLR had to be negotiated to accommodate the different positions held by its members on SOVEREIGNTY. CCAMLR includes Article IV from the Antarctic Treaty, but with the addition of a phrase allowing coastal states to exercise jurisdiction. This compromise has been called the 'bifocal approach' because its ambiguity allows claimants and non-claimants to interpret it differently: the claimants find no challenge to their claims, and the non-claimants can interpret it as applying only to the SUBANTARCTIC ISLANDS, where sovereignty is undisputed.

CCAMLR created an institutional body that operates on a permanent basis in Hobart, Australia. The key policy and regulating body is the Commission for the Conservation of Antarctic Marine Living Resources, which meets for two weeks each year in Hobart. In 2001, there were 24 members of the commission, and another six states had ratified the convention but had not become members.

Decision-making in the commission is made by consensus—a source of criticism as it potentially allows the fishing states to veto conservation measures restricting harvesting activities. The commission is advised by the Scientific Committee, a group of scientific experts who study and exchange information about marine living resources.

In the 1980s CCAMLR struggled to deal with the problems of fish stocks that had been severely depleted by exploitation in the 1970s. The commission was deadlocked as fishing states blocked proposals for new conservation measures. This impasse was overcome when the commission agreed that decisions would be based on the advice given by the Scientific Committee.

The number of conservation measures adopted by CCAMLR has steadily increased since then. The expected interest in fishing for krill has not materialized due to difficulties in processing the fish for human consumption, and the collapse of subsidized fishing by the Eastern Bloc countries following the end of the Cold War.

The greatest challenge to CCAMLR in the 1990s was from illegal, unregulated and unreported fishing in the Southern Ocean, particularly focused on TOOTHFISH FISHING.

In addition, it is estimated that longline fishing techniques have killed tens of thousands of ALBATROSSES, GIANT PETRELS and WHITE-CHINNED PETRELS. Despite efforts by CCAMLR members, this problem is not yet under control.

Above: The concentration of fishing vessels in just this small area of subantarctic waters gives an indication of the extent of exploitation occurring in the world's Southern Ocean.

Convention on the Regulation of Antarctic Mineral Resource Activities (CRAMRA) Anticipating problems that could arise if the exploitation of MINERAL RESOURCES began before any agreements were in place, the CONSULTATIVE PARTIES to the ANTARCTIC TREATY discussed the issue of mining in the 1970s and in 1982 began formal negotiations for a minerals convention. Key issues in the negotiations were the DISPUTED CLAIMS to SOVEREIGNTY in Antarctica, and the need to accommodate the interests of the non-aligned countries and of potential mining states.

CRAMRA was signed on 2 June 1988, but has not entered into force. The convention has a detailed set of rules and procedures for regulating mineral resource activity in Antarctica. Guidelines for establishing liability for accidents and environmental damage are included. It maintains the compromise on sovereignty in the Antarctic Treaty, but decision-making is to be by a three-quarters majority vote rather than consensus.

If mining were to occur in Antarctica then a headquarters would be established in Wellington, NEW ZEALAND, and a commission and advisory committee established.

Environmental groups were unhappy with the prospect of mining in Antarctica and its potential for environmental damage. PROTESTS against CRAMRA and in favour of declaring Antarctica a WORLD PARK occurred. On 22 May 1989, AUSTRALIA announced it would not ratify CRAMRA, and it was followed by FRANCE. This split among the Consultative Parties had the potential to threaten the stability of the ANTARCTIC TREATY SYSTEM. The eventual response was a turning point for environmental protection in Antarctica with

the adoption, in 1991, of the PROTOCOL ON ENVIRONMENTAL PROTECTION TO THE ANTARCTIC TREATY, known as the MADRID PROTOCOL—which essentially prohibits mining in Antarctica for 50 years. It appears unlikely that CRAMRA will enter into force.

Conventions The actions of states and individuals in Antarctica and the SOUTHERN OCEAN are regulated by a variety of international conventions, or legally binding agreements between states. One important group of agreements are those which form the ANTARCTIC TREATY SYSTEM (ATS). International agreements lying outside the ATS also influence activities in Antarctica and the Southern Ocean. Many of these agreements relate to the ocean and the environment. Under international law, the agreements are binding only on those states that are members of the agreement.

The ATS is formed from a core group of treaties and conventions. This includes the 1959 ANTARCTIC TREATY, the 1972 CONVENTION FOR THE CONSERVATION OF ANTARCTIC SEALS, the 1980 CONVENTION ON THE CONSERVATION OF ANTARCTIC MARINE LIVING RESOURCES, and the 1991 MADRID PROTOCOL. The CONVENTION ON THE REGULATION OF ANTARCTIC MINERAL RESOURCE ACTIVITIES was signed in 1988 but, as it has never been ratified, it has been set aside.

A number of other international LAWS, conventions and treaties are of relevance to Antarctica. The Convention for the Regulation of Whaling, negotiated in 1931 and revised in 1946, established the INTERNATIONAL WHALING COMMISSION to regulate WHALING.

The 1972 Convention on the Prevention of

Marine Pollution by Dumping of Wastes and Other Matters entered into force on 30 August 1975; it prohibits noxious wastes being dumped at sea. The 1972 Convention on the International Regulations for Preventing Collisions at Sea entered into force on 15 July 1977, and provides for common standards for light and sound signals on ships.

The 1973 International Convention for the Prevention of Pollution from Ships entered into force on 2 October 1983, and is referenced in one of the Annexes of the Madrid Protocol. The 1982 United Nations Convention on the LAW OF THE SEA, which entered into force on 16 November 1994, regulates activities in the world's oceans.

Several recent international conventions also affect Antarctica and its environments. The 1985 Convention for the Protection of the Ozone Layer (entered into force on 22 September 1988) and the 1987 Protocol on Substances that Deplete the Ozone Layer (entered into force on 1 January 1989) emphasize concerns about the depletion of the ozone layer and the OZONE HOLE above Antarctica. The 1992 UN Framework Convention on Climate Change (which entered into force 21 March 1994) has the goal of stabilizing GREENHOUSE gases in the ATMOSPHERE to limit the effect of climate change. The 1992 Convention on Biological Diversity (entered into force on 29 December 1993) aims for the conservation and sustainable use of biological diversity. Parties to the 1989 Convention on the Control of Transboundary Movements of Hazardous Wastes and their Disposal (entered into force 5 May 1992) agree not to send hazardous wastes south of 60°S.

Cook, Frederick (1865–1940) American doctor and explorer, born in Hortonville, New York. In 1892 he was surgeon and ethnologist on Robert Peary's expedition to North Greenland and on his return began practising medicine in Brooklyn.

Adrien de GERLACHE invited Cook to join the 1897–99 BELGIAN ANTARCTIC EXPEDITION as surgeon. This was the first expedition to overwinter in the Antarctic and members suffered from inadequate diet, lack of preparations, cold and dark. Cook's previous polar experience was invaluable during what he described as a 'hellish existence'. He persuaded de Gerlache of the need for fresh seal and penguin meat to combat scurvy. Cook published his account of the expedition in *Through the First Antarctic Night 1898–1899*.

In 1906 Cook announced he had made the first ascent of Mount McKinley, the highest peak in North America, a claim greeted with scepticism by some and later disregarded. However, the attention he received enabled him to secure adequate funds to finance an expedition to the Arctic. On his return, Cook claimed to have reached the North Pole on 21 April 1908 with two Eskimos, the first to do so. This was discredited five days later when Robert Peary, also just returned from the Arctic, announced that he had reached the Pole on 6 April 1909, and produced navigational records to prove his claim, something Cook was unable to do.

In 1923 Cook was convicted of fraud. Sentencing him, the judge said: 'You have come to a mountain and reached a latitude both of which are beyond you.' After he was released from prison, Cook spent the rest of his life unsuccessfully trying to legitimize his North Pole claim.

Cook, James (1728–79) One of the greatest navigators and explorers, born in Yorkshire, England. Cook was the son of a farm labourer. He spent several years sailing trade vessels, then joined the

Below: Captain James Cook.

Above: Scientists take samples of ice to determine past temperatures, precipitation, atmospheric and environmental conditions.

British Navy in 1755, and spent eight years surveying the St Lawrence River and the coasts of Labrador and Newfoundland.

Appointed lieutenant, he was placed in command of the *ENDEAVOUR* expedition of 1768–71 to the Pacific to observe the transit of Venus, in the course of which he circumnavigated NEW ZEALAND and charted the east coast of AUSTRALIA.

Promoted to the rank of commander, his second voyage of discovery of 1772–75 with the *Resolution* and *Adventure* was to explore Antarctica. On 17 January 1773 Cook's ships became the first to cross the ANTARCTIC CIRCLE and reached 67°15'S the following day. They spent several months sailing around the ice, before turning north to spend the winter in New Zealand waters. In the summer of 1774 Cook headed south once more and on 30 January reached 71°11'S, the highest latitude attained thus far. On turning back again he recorded in his journal not that it was impossible to proceed further south, 'but I will assert that the bare attempting of it would be a very dangerous enterprise and what I believe no man in my situation would have thought of.' The winter and spring were spent in further exploration of the Pacific and a series of island groups in the South Atlantic were charted: SOUTH SHETLANDS, SOUTH ORKNEYS, SOUTH SANDWICH. His experiences convinced Cook that if there were land further south, it would be cold, desolate and inhospitable.

On his return to England he was promoted to captain and seven months later, in December 1776, was sent off on a third voyage, during which, on their way across the southern Indian Ocean, the expedition made landfall at Îles KERGUÉLEN, which Cook called Desolation Island. On 14 February 1779 Cook was killed by natives in Hawaii.

Copepod Copepods are key organisms in the ocean FOOD CHAIN, feeding on DIATOMS and other PHYTOPLANKTON, and in turn being consumed by ZOOPLANKTON, FISH, BIRDS and WHALES. They are considered by some scientists to be the largest ani-

mal biomass on Earth, competing with KRILL for this title.

Most copepods are free living and are found from the surface to great depths. When they hover on the ocean surface, their long, feathered antennae act as parachutes to prevent sinking. They filter food from the water with their legs around their mouth. Their faeces constitute a part of the 'marine snow', whereby nutrients and minerals from the surface waters enrich the deeper sea. Free-living copepods are able to escape predators by folding their antennae close to their body and darting at speeds of between 40 to 200 body lengths per second—making them some of the fastest escapists observed in water.

Some copepods are parasitic. The females attach themselves to a fish host after a free-swimming juvenile stage, then remain stationary for life, concentrating all their energy on reproduction. The males continue to move or swim around to find females and reproduce.

Coral (Cnidaria) Both soft corals (*Alcyonacea*) and true corals (*Scleractinia*) belong to the phylum Cnidaria or 'stinging threads' and are some of the most primitive of true animals to inhabit the ocean BENTHOS.

Several Antarctic soft coral species produce chemicals that they release into the surrounding water to deter predators, such as SEASTARS, and to inhibit bacterial growth. Some Antarctic NUDIBRANCHS that feed on soft corals incorporate these chemical defences into their own deterrent systems with no apparent ill effects.

Colonies of the soft coral *Gersemia antarctica*, which is found in MCMURDO SOUND, SOUTH GEORGIA, SOUTH SANDWICH ISLANDS and BOUVETØYA, can inflate to over 2 m (6 ft) in height and have evolved a particularly specialized grazing behaviour in response to the seasonal availability of PLANKTON as food.

As well as the upright filter-feeding posture normally seen in a coral colony, an entire colony of *G. antarctica* is able to bend down so that the polyps can feed on benthic DIATOMS and particles of organic matter on the sea floor. Sweeping in a full circle, *G. antarctica* searches the immediate area for food, then—even more remarkably—inches forward, pulling itself across the sea floor to find another ungrazed spot. When a colony encounters an area that has already been grazed by another *G. antarctica*, it retracts from it. Individual colonies may move as much as 14 m (46 ft) in a year: impressive behaviour from a primitive organism that lacks a brain.

Core samples The Antarctic ICE SHEET, the ICE SHELVES and ocean sediments are repositories of history. Sealed within their ancient layers are indicators of past TEMPERATURE, PRECIPITATION and atmospheric conditions. By DRILLING deep vertical core samples, scientists obtain an archive of CLIMATE change, from which they hope to find solutions to two linked problems: understanding past climate fluctuations to predict future change, and understanding ice sheet evolution to help predict sea-level change.

Evidence of past environmental conditions is revealed in ice cores. This includes the amount of annual precipitation, which can be measured directly—for example, a thin slice can mark the onslaught of a single storm thousands of years ago. Yearly snowfall cycles may be seen in cross-section in a core streaked with solid particles, such as dust and volcanic material; this material can often be traced back to a particular eruption and used to date the age of the ice.

Impurities, such as methane, lead and radioactive fallout from nuclear weapons testing, are also layered into the ice. This enables scientists to track the extent of global pollution and the effect of industrialization on the climate record. Levels of both lead and mercury, for example, have at least doubled in concentration from 10,000 years ago.

Tiny parcels of air from past ATMOSPHERES, including GREENHOUSE gases, are trapped in isolated bubbles in core samples and can be extracted for isotope analysis. At VOSTOK STATION, scientists have used ice cores to map climate change over the last 420,000 years, revealing that levels of greenhouse gases varied with air temperatures—the first convincing evidence that greenhouse gases play a key role in global warming.

As the glacial environment around Antarctica changed over millions of years, 'packages' of sediment were layered onto the sea floor. Whereas analysis of ice cores gives scientists information about the last half a million years, to obtain a record of earlier ice, they need to date the stacked sequence of sediment from a deep-ocean core. GLACIERS and the Ice Sheet churn and grind sediment into angular shapes; ice shelves deposit much finer material on the seabed. Differences in this ice-rafted debris on the seabed are believed to indicate the onset of the Antarctic Ice Sheet: as the ice expands, so does the quantity of debris and the distance it is spread beyond the coast. In this way expansions and contractions of the Ice Sheet are recorded in sea-floor sediment.

Microfossils in sediment cores mark changes in species and ecosystems over time. If scientists know the preferred living conditions of species—such as water temperature, SALINITY and depth—

Above: *Crabeater seals (*Lobodon carcinophagus*) spend most of their lives in and around the Antarctic pack ice and are rarely seen ashore.*

the environments necessary to support them can be reconstructed. This type of analysis, apart from providing information about the evolution of marine species, can 'fine tune' knowledge about larger-scale climate changes that are recorded in the sediment record.

Cormorant (Phalacrocoricidae) Of the 29 species of cormorants, sometimes called shags, five are found in the Antarctic and subantarctic regions: ANTARCTIC SHAGS, which breed on the ANTARCTIC PENINSULA; BLUE-EYED CORMORANTS and several subspecies, including KERGUELEN CORMORANTS, SOUTH GEORGIA CORMORANTS, CAMPBELL ISLAND CORMORANTS and AUCKLAND ISLANDS CORMORANTS. The name 'cormorant' originated from the Latin 'corvus marinus', meaning 'sea raven'; although many cormorant species are darkly coloured like the ominous, glossy black ravens, all the Antarctic and subantarctic species have extensive white underparts. Their bodies vary from duck to goose size, with short legs set well back, their wings are short and they have long, 'S'-shaped necks.

Cormorants are experts at fishing, feeding on an extensive array of marine life, including FISH, CEPHALOPODS and CRUSTACEANS. They have been caught at depths of well over 50 m (164 ft). Their feet act as powerful propellers underwater and they press their short wings close to their bodies to reduce drag. Long necks allow them to thrust their heads out quickly and secure fish in their serrated bills. For centuries, people used them as hunting birds, placing rings around their necks so that they could catch but not swallow the fish.

Cormorants can live near both salt- and freshwater sources. They have small throat pouches which, as well as allowing them to swallow large fish, are used as a cooling mechanism in place of sweat glands. Cormorants have permeable

plumage—that is, they must perch and spread their wings to dry after swimming. Because of this, they rarely stray far from land.

Although they do not start breeding until they are about two or three years old, cormorants are unusual in that their clutches contain between one and six eggs, and they can also lay twice in a season. They nest in colonies and are seasonally monogamous: the males find the nesting sites and gather nesting materials for the females, who then build the nests.

Crabeater seal (*Lobodon carcinophagus*) Despite their name, these phocid seals feed mainly on KRILL. They use specially adapted cheek teeth to sieve the krill from the water. Like ROSS SEALS, crabeater seals spend most of their life in and around the Antarctic PACK ICE and are rarely seen ashore.

They have a circumpolar distribution, but have been known to wander far to the north: they have been seen in southern NEW ZEALAND and South Australia. Although populations are hard to measure, estimates range from 11 to 14 million. Edward WILSON, zoologist on the 1901–04 NATIONAL ANTARCTIC EXPEDITION, found a mummified crabeater carcase on the FERRAR GLACIER at an altitude of 1100 m (3608 ft)

The crabeaters grow about 2.6 m (9 ft) long and weigh up to 230 kg (500 lb). Their slim bodies are usually mottled a brownish-white colour, with some variability between individuals. Older animals are often paler. Extremely agile in water, they can also cross ice at speeds of up to 25 km (15 miles) per hour. They swim in groups, usually of about 20 or so, but sometimes much larger schools of several hundred.

Breeding takes place from late September to early November on the pack ice near food. The mother nurses the single pup for three to four weeks, during which time a male is often present, trying to mate with the female, and fighting off other males.

Crean, Thomas (d. 1938) Irish sailor and explorer. This tall, powerfully built Irishman was a crew member of HMS *Ringarooma* when he was selected to join the TERRA NOVA at Port Chalmers, NEW ZEALAND, for the 1901–04 NATIONAL ANTARCTIC EXPEDITION. Crean returned to Antarctica with the 1910–13 BRITISH ANTARCTIC EXPEDITION, and was a member of the polar support party, but he was not chosen for the final assault on the SOUTH POLE. Fellow expedition member Trygge Gran later wrote that Crean was 'a man who wouldn't have cared if he'd got to the Pole and God Almighty was standing there, or the Devil. He called himself "The Wild Man from Borneo" ... And he was!' On the descent from the POLAR PLATEAU, Crean was largely responsible for saving the life of Edward EVANS who had collapsed with scurvy.

A member of the shipwrecked IMPERIAL TRANS-ANTARCTIC EXPEDITION of 1914–17, Crean was chosen to make the journey in the *James Caird* from ELEPHANT ISLAND to SOUTH GEORGIA, during which he helped to sustain the others with albatross stew prepared on the tiny primus stove and with his singing. 'One of the memories that comes to me of those days,' Ernest SHACKLETON wrote, 'is of Crean singing at the tiller. He always sang while he was steering, and nobody ever discovered what the song was. It was devoid of tune and as monotonous as the chanting of a Buddhist monk at his prayers; yet somehow it was cheerful.

Below: Irish sailor Petty Officer Crean made the trans-alpine journey across South Georgia with Ernest Shackleton in 1916.

Above: Despite the advantages gained by modern technology, crevasses remain a threat to any transportation over ice.

In moments of inspiration [he] would attempt "The Wearing of the Green".' On arrival at SOUTH GEORGIA Crean and Frank WORSLEY accompanied Shackleton over the mountains from Haakon Bay to Stromness Bay.

Crean retired to a hotel at Annaschaul, County Kerry, Ireland, which he named 'South Pole Inn'.

Crevasses Cracks or splits on the surface of a GLACIER. They are formed by stresses created when ICE is stretched over ridges and flows through valleys. When SEA ICE spreads out over the surface of the ocean similar stresses cause cracks, which are called LEADS. They range in length from a few centimetres (inches) to several kilometres (miles). Some are only a hair's breadth wide; others open up with a large crack, like a rifle shot. Sailors on the 1901–04 NATIONAL ANTARCTIC EXPEDITION reported seeing crevasses on the BEARDMORE GLACIER that were wide enough to 'swallow their ship twice over'.

Depending on the stresses in the ice, crevasses can be longitudinal (aligned along the length of the ice) or transverse (aligned across the ice). Deep crevasses can have spectacular vertical blue ice walls. Near the glacier terminal or in steep places, crevasses may become jumbled, forming 'seracs', or pinnacles of ice.

Crevasses can become covered by blown SNOW and expeditions venturing across 'snow bridges' must be careful. The safest way to travel over crevassed terrain is to distribute weight as widely as possible. Nowadays, most parties in Antarctica travel on skis or on VEHICLES with caterpillar tracks. DOGS used to be used, harnessed in pairs or in fan formations.

On 14 December 1912, during Douglas MAWSON's epic sledging journey across GEORGE V LAND,

one of his two companions, Belgrave Ninnis, broke through a snow bridge with his sledge and dogs and disappeared—the first death caused by an Antarctic crevasse.

Crozet, Îles A French national park since 1938, the islands are situated in the Indian Ocean midway between the PRINCE EDWARD ISLANDS and Îles KERGUÉLEN. Divided into two main groups, L'Occidental, or western islands, and 100 km (62 miles) east, L' Oriental (Île de l'Est and Île de la Possession or POSSESSION ISLAND). Volcanic in origin, they have a combined land area of 325 sq km (125 sq miles). On average it rains 300 days per year, and westerly winds exceed 100 km (62 miles) an hour on 100 days annually. Summer temperatures are cool, rarely exceeding 18°C (64°F), and in winter it seldom drops below 5°C (41°F).

French explorer Marion DU FRESNE, in his ship *Mascarin*, discovered Îles Crozet in 1772 and claimed them for FRANCE. The first sealers arrived in 1804 and, when that resource was decimated, were replaced by profitable WHALING ventures.

A scientific base, Alfred-Faure Station, was established in 1964 on Possession Island, the largest in the Crozet group and home to several million KING PENGUINS.

Crozier, Cape Named by James Clark ROSS after Francis CROZIER, captain of the *Terror*, Cape Crozier lies on the eastern tip of ROSS ISLAND at the junction of the ROSS SEA and the ROSS ICE SHELF.

In 1902, a party from the NATIONAL ANTARCTIC EXPEDITION landed at Cape Crozier to leave a record of the expedition's progress—in a cylinder nailed to a post. In the winter of 1911, three members of the BRITISH ANTARCTIC EXPEDITION made a five-and-a-half week round trip from

Cape EVANS to Cape Crozier in order to collect eggs from the EMPEROR PENGUIN nesting site located there: the rock shelter built by the party was rediscovered in 1957. The Cape has been a focus for ornithological research (on emperor and ADÉLIE PENGUINS) and OCEANOGRAPHIC RESEARCH.

Crozier, Francis (*c.* 1796–1848) Irish naval officer and explorer. Crozier joined the British Navy in 1810 and between 1821 and 1827 undertook three voyages of exploration to the Arctic with Sir William Parry, and a fourth with James Clark ROSS in 1835.

On Ross's 1839–43 expedition to Antarctica in HMS *Erebus*, Crozier commanded HMS *Terror*. During their winter sojourn at Hobart, AUSTRALIA, he proposed to Sophia Cracroft, the niece of governor Sir John Franklin, who turned him down because 'she thought him a horrid radical and bad speller'. The expedition explored the ROSS SEA area and VICTORIA LAND. Ross was full of praise for his second-in-command's seamanship.

After a short time back in England, Crozier in the *Terror* set out with Sir John Franklin in *Erebus* to search for the north-west passage.

Franklin perished in 1846 and Crozier took over command, but the ships became ice-bound and the entire crew died while trying to walk to safety.

Cruises Cruise liners began operating in the SOUTHERN OCEAN in 1958, and have visited the Antarctic continent every year since 1966. In 1969, Lars-Eric Lindblad built the *Lindblad Explorer*, which was specially designed to carry tourists to Antarctica. SHIPS usually travel from South America, but sometimes depart from AUSTRALIA, NEW ZEALAND or SOUTH AFRICA.

Most cruise expeditions visit the ANTARCTIC PENINSULA and nearby islands, and some venture into the ROSS SEA and around the EAST ANTARCTICA coast. These voyages are the main form of TOURISM in the Antarctic, with around 20 vessels, carrying between 45 and 280 passengers, visiting each year, and occasionally, larger ships with almost 1000 passengers. In the 1999–2000 season over 14,000 cruise-ship passengers visited Antarctica. Landings at RESEARCH STATIONS, HISTORIC SITES and PENGUIN rookeries are popular with tourists.

Cruises to Antarctica can take anywhere from a week to two months. Activities have diversified in recent years to include KAYAKING and scuba diving from vessels. Tours are often accompanied by Antarctic experts, who give shipboard lectures.

Cruise travel to Antarctica is not without risks. Safety problems are formidable: safe landings require careful handling in icy waters, and changing weather conditions can trap a ship in ice. In January 1968, two ships ran aground: the *Magga Dan* at MCMURDO SOUND and the *Lindblad Explorer* in the Antarctic Peninsula. Other ships have been damaged by ice, and the Argentinian supply ship *Bahia Paraiso* was carrying tourists when it sank in 1989, spilling 681,900 litres (150,000 gallons) of OIL into the sea.

See also, INTERNATIONAL ASSOCIATION OF ANTARCTIC TOUR OPERATORS.

Above: *The* Marco Polo *regularly cruises in Antarctic waters.*

Crustaceans Antarctic ZOOPLANKTON is primarily composed of species belonging to the crustacean class of organisms, particularly KRILL and COPEPODS, making crustaceans a critically important component of the Antarctic FOOD WEB.

Crustaceans are a large class of mainly aquatic organisms, defined by having two pairs of antennae, a jointed external skeleton, gills and a body divided into head, thorax and abdomen, of which the head and thorax are often fused. They usually go through several larval stages.

The Antarctic BENTHOS (sea floor) community includes a number of crustacean species—including the GIANT ISOPOD (*Glyptonotus antarcticus*), which is a predator of a variety of other benthic animals. Another 40 isopod species are present along with about 100 species of amphipods, of which *Orchomene plebs*, a scavenger that can swim actively in search of carrion, is the most numerous.

Small, shrimp-like mysid crustaceans are also found, with 19 of the 37 species present endemic to the Antarctic region. Mysids can form large swarms and are an important part of the diet of many FISH, including ICEFISH, ANTARCTIC TOOTHFISH and spiny PLUNDER FISH. Other predators include BRITTLE STARS, BIRDS and CRABEATER SEALS.

Cryptobiosis An adaptation that allows organisms to tolerate severe dehydration; also known as anabiosis or anhydrobiosis. Several Antarctic terrestrial INVERTEBRATES—TARDIGRADES, NEMATODES and ROTIFERS—are able to undergo cryptobiosis, thus greatly extending their lifespans in harsh environmental conditions.

The loss of 20 percent of body moisture is fatal for humans, but cryptobiotic organisms can survive losing 99 percent of their water. When dehydrated, the organisms show no signs of life—it is not known whether metabolism stops completely or occurs extremely slowly. This differentiates cryptobiosis from dormancy or hibernation: in those states, metabolism continues, albeit at a much reduced rate. The drying process must occur slowly, however, to stop cells becoming disrupted and torn apart. Chemical changes may also occur.

In their dehydrated state, cryptobiotic organisms can typically withstand extreme conditions such as temperatures from –270°C (–454°F) up to temperatures of around 200°C (392°F). They can survive immersion in alcohol, and desiccated tardigrades have been shown to be approximately 1000 times more resistant to X-rays than humans. Cryptobiosis also extends the lifespan of organisms: for instance, a tardigrade may survive around 60 years.

Cryptobiosis, first described by the Dutch scientist Anton van Leeuwenhoek in 1720, is also found in non-Antarctic organisms such as brine shrimp larvae and in many plants.

Currents See OCEAN CURRENTS.

Cuverville Island Located in the PALMER ARCHIPELAGO, off the west coast of the ANTARCTIC PENINSULA, Cuverville Island has one of the largest populations of GENTOO PENGUINS in Antarctica. These birds have been the subject of long-term studies to determine the impact of TOURISM on their breeding patterns and behaviour.

Cyanobacteria Photosynthetic BACTERIA that contain photophyll; also known as blue-green ALGAE. Until recently, they were classified as algae, but have been shown to be more similar to bacteria than to plants. Microscopic cyanobacteria fossils have been extracted from Precambrian rocks.

d

Dakshin Gangotri Base Established by INDIA in 1983 on the coast of DRONNING MAUD LAND. The base was converted into a supply and transit base once MAITRI BASE opened in 1989.

Dallman, Eduard (1830–96) German sealer. Dallman led the first German expedition to Antarctica. He had been involved in Arctic sealing and whaling and in 1873 the Society for Polar Navigation of Hamburg placed him in command of the *Grönland* for a WHALING voyage to the SOUTHERN OCEAN. Sailing south from the SOUTH SHETLANDS, the ship became the first steamer in Antarctic waters. The BISMARCK STRAIT and the adjacent ANTARCTIC PENINSULA coast were charted for the first time, but no right whales were found.

In 1877 Dallman was involved in trading across the Barents and Kara Seas to northern Siberia, and in 1884 he played a part in the annexation of the Bismarck Archipelago, Papua New Guinea. He retired in 1893.

Danco Island Charted by Adrien de GERLACHE in 1897–99, Danco Island is 1.5 km (1 mile) long and is flanked by a wide stony beach, a nesting ground for GENTOO PENGUINS. It lies in the PALMER ARCHIPELAGO, off the western coast of the ANTARCTIC PENINSULA. A hut called 'Base O', constructed by the Falkland Islands Dependencies Survey (now BRITISH ANTARCTIC SURVEY) in 1955–56, has a large anthracite dump nearby. The building is maintained as a refuge, and is occasionally used by researchers.

Dark Sector An area about a kilometre (½ mile) from other buildings at the SOUTH POLE used for ASTRONOMICAL RESEARCH. Human-produced light and electromagnetic radiation are prohibited in the Sector in order to preserve the unique atmospheric qualities available to scientists.

Darwin Mountains Discovered by the 1901–04 NATIONAL ANTARCTIC EXPEDITION and named for Major Leonard Darwin of the ROYAL GEOGRAPHICAL SOCIETY, this mountain group lies west of the ROSS SEA between the Cook Mountains and the BRITANNIA RANGE in southern VICTORIA LAND due southwest of MCMURDO SOUND.

Dating Samples of ICE, ROCKS and marine and glacial sediment from Antarctica can reveal past environmental conditions. In order to piece together the history of the continent, the samples must be dated accurately. There are two types of dating: relative and absolute. Relative dating involves finding out what sequence of events occurred in an environment; absolute dating places a fixed date on a sample.

Within ice CORE SAMPLES, annual layers of snowfall can be detected. If the accumulation of SNOW is high, samples can be taken from each layer and dated accurately. If snow accumulation is low or if a core penetrates deep into the ice, the layers will be compressed into fine sections. Although this makes dating more difficult, it may be possible to model the ice flow characteristics in these deep layers, then deduce the amount of time it would have taken for the flow patterns to develop.

Like snow, hydrogen peroxide has an annual accumulation pattern. It is formed in the ATMOSPHERE by a chemical reaction involving ULTRAVIOLET RADIATION, and is dissolved in ice. Sulphate also appears in the record, either dissolved in ice, or in visible particles that have been blasted from a VOLCANO, possibly on the other side of the globe, and eventually layered onto the Antarctic ICE SHEET. If this material can be matched to a particular volcanic eruption, the age of the ice can be estimated.

Elements found in rocks or volcanic material or dissolved in ice decay naturally and can be dated radiometrically. This process involves using sensitive instruments to find the proportion of the original 'parent' element and comparing this with the proportion of the newly formed 'daughter' element to measure the decay rate.

The Carbon-14 method, commonly known as carbon dating, is a widely used radiometric method for dating samples less than 70,000 years old. Carbon-14 is absorbed by plants and animals; when the organism dies, the carbon begins to decay into nitrogen at a constant rate. Radiometric dating of carbon has been conducted on Antarctic samples, such as tree FOSSILS, bone and marine shells, and is accurate to about 40,000 years.

David, Tannant William Edgeworth (1858–1934) Australian geologist, born in Wales. After graduating from Oxford University in classics, David worked as a geological surveyor in Sydney, AUSTRALIA. In 1891 he became professor of geo-logy at Sydney University, where his students included Douglas MAWSON.

David took part in the 1907–09 BRITISH ANTARCTIC EXPEDITION. Together with Mawson and Dr Alistair Mackay he made the first ascent of Mount EREBUS; in January 1909, the three were also first to reach the SOUTH MAGNETIC POLE, making geological surveys as they went. When visited by PENGUINS on this trip, David wrote: 'they evidently took us for penguins of an inferior type and the tent for our nest.' The journey, which took 122 days and covered 2028 km (1260 miles) is the longest unsupported man-hauling SLEDGE journey attempted.

David gave much support to Nobu SHIRASE and his JAPANESE ANTARCTIC EXPEDITION when they were stranded in Sydney with few resources. During World War I, he was consultant geologist to the Australian Tunnelling Battalion in France. He resigned from Sydney University in 1924.

Davis, John (c. 1550–1605) British sailor and explorer, born Devonshire. Discovered the FALKLAND ISLANDS in 1592. The quadrant he invented was used by navigators throughout the 17th and 18th centuries. His treatise on navigation, *The Seaman's Secret*, was published in 1594.

Davis, John (c. early 18th century) American sealer. According to logbooks discovered in 1952, Davis may have been the first to step onto the Antarctic continent. The *Huron* from New Haven under Davis and the *Huntress* from Nantucket under Christopher Burdick were among 30 American sealers working from Yankee Bay, Livingston Island, in the vicinity of PALMER LAND in the summer of 1820–21. On 9 December 1820 Davis took the *Cecilia*, a shallop from the *Huron*, in search of new seal colonies. By the end of the month he had collected only about 2500 skins; a rival fleet had collected 21,000 over the same period. Seeing no profit around the SOUTH SHETLANDS, Davis looked further south.

On 7 February 1821, at 64°01'S, he came ashore for an hour at what was 'a Large Body of Land ... I think this Southern Land to be a Continent'. This is the earliest known reference to Antarctica as a continent, and it predated Carsten BORCHGREVINK's 'first-footing' by nearly three-quarters of a century. From the records in his logbook it is assumed that Davis's landing was made at what is now called Hughes Bay on the ANTARCTIC PENINSULA.

On 10 February he was back at Yankee Bay with a further 1670 seal skins. In March the *Huron* and the *Huntress* headed back to the USA.

Davis, John King (1884–1967) Irish-Australian navigator, born in London, England. He went to sea at the age of 16. On the 1907–09 BRITISH ANTARCTIC EXPEDITION, he was chief officer of the

Below: An ice profile showing ice stratification.

NIMROD, and took over as captain on the return voyage to England. He had applied 'on the spur of the moment' when led by chance to the expedition offices.

Davis was captain of the *AURORA* on the 1911–14 AUSTRALASIAN ANTARCTIC EXPEDITION, led by Douglas MAWSON. Although nicknamed 'Gloomy' because he rarely smiled, Mawson wrote that, 'he entered upon the enterprise with enthusiasm tempered with prudence and sound good sense.'

Ernest SHACKLETON asked him to captain the *ENDURANCE* on the 1914–17 IMPERIAL TRANS-ANTARCTIC EXPEDITION. Davis refused, saying he thought the expedition doomed, but in 1916 he organized the rescue of the surviving members of the expedition's Ross Sea Party in the *Aurora*. He returned to Antarctica in command of the *DIS-COVERY* on the first summer of Mawson's 1929–31 BRITISH-AUSTRALIAN-NEW ZEALAND ANT-ARCTIC RESEARCH EXPEDITION.

Davis was Australia's director of navigation from 1920 until 1949.

Davis Sea Lies between the SHACKLETON ICE SHELF and the WEST ICE SHELF off the coasts of QUEEN MARY LAND and WILHELM II LAND. It was discovered by the crew of the *AURORA* during the 1911–14 AUSTRALASIAN ANTARCTIC EXPEDITION and is named after John King DAVIS.

Davis Station AUSTRALIA's Davis Station was established in January 1957 during INTERNATION-AL GEOPHYSICAL YEAR and named after John King DAVIS. It is located on the eastern side of Prydz Bay in PRINCESS ELIZABETH LAND. Closed temporarily in 1965, it was reopened four years later and has been occupied continuously since. Research is conducted into atmospheric sciences, biology, glaciology, geosciences, human impacts and medicine.

Day length See SUMMER and DAYLIGHT HOURS.

Daylight hours Most areas of Antarctica experience continual daylight over SUMMER and perpetual darkness over WINTER (March to mid-August). The SUNRISE at the end of winter occurs in mid-August. From 70°S, from late September to the end of February, the sun circles overhead in the sky, illuminating the landscape 24 hours a day. On 21 December at the SOUTH POLE, the sun completes a perfect 24-hour circle without changing its altitude at all.

DDT The PESTICIDE, dichlorodiphenyltrichloro-ethane, which was widely used in malaria-eradication programmes and as an agricultural pesticide in the 1950s and 1960s, has been carried to Antarctica via OCEAN CURRENTS and airborne particles. As long ago as 1962 Rachel Carson in her book *Silent Spring* recorded that DDT residues had been found in the body fat of a WEDDELL SEAL. DDT does not break down easily, but persists in the environment for decades.

Although DDT has been banned in many countries, it is still used as an agricultural pesticide in some parts of the world.

Above: Frank Debenham, Antarctic explorer and scientist.

Deaths Mortality rates in Antarctica are the lowest in the world because the population is temporary, mainly young and medically fit. Nearly 600 deaths have been recorded on the continent; almost half of these were caused by the *San Telmo* SHIPWRECK and the Air New Zealand DC10 AIR-CRAFT CRASH on Mount EREBUS in 1979. Other deaths have been the result of SHIPWRECKS, pre-existing medical conditions such as heart failure, accidents, trauma; and exposure (see HYPOTHER-MIA and FROSTBITE).

Debenham, Frank (1883–1965) Australian geologist, born in New South Wales, AUSTRALIA. He studied geology with Edgeworth DAVID at Sydney University and took part in the 1910–13 BRITISH ANTARCTIC EXPEDITION, in which he explored the geology of VICTORIA LAND together with Griffith Taylor, Charles Wright and Edgar EVANS. After being wounded in World War I, Debenham worked at Cambridge University, England. He was the first director of the SCOTT POLAR RESEARCH INSTITUTE, which he co-founded in 1926. He was also the university's first professor of geography from 1931 until his retirement in 1946.

At an address on Antarctic issues to the Geographical Section of the British Association in 1935, Debenham recommended that Antarctica should become a WORLD PARK. He argued that a park would be a resource for people the world over and could provide food, tourist potential and a site for scientific studies.

Deception Island A collapsed volcanic cone in the SOUTH SHETLANDS that forms an almost complete ring, with one windy 230 m (750 ft) wide gap in the coastline, Neptune's Bellows, which encloses a natural harbour.

The volcano was active throughout last century, and in 1923 the harbour water boiled and

Above: Frozen foods at Mizuho are one food item easily stored in Antarctica.
Left: Fresh fruits and vegetables, however, are valuable commodities. Arçtowski Station houses Antarctica's first greenhouse.

and young fish. Some demersal fish have young that are pelagic, or live in the surface waters of the SOUTHERN OCEAN, and are thus able to take advantage of summer PHYTOPLANKTON abundance; once they mature they inhabit the benthos.

Denison, Cape Located on the coast of TERRE ADÉLIE and dubbed the 'windiest place on Earth' by Douglas MAWSON, the Cape is battered by fierce KATABATIC WINDS. It was the site of the 1911–14 AUSTRALASIAN ANTARCTIC EXPEDITION's base, where Mawson and other expedition members spent two winters. Several of the expedition's HUTS are still standing.

Deserts See POLAR DESERTS.

Diamond dust Fine crystals of ICE that float over the interior of Antarctica in clear weather. They are formed from water vapour, which in warm areas would descend to the ground as fog or dew, but in polar conditions remains suspended in the air.

In the sunlight, diamond dust reflects light, like prisms, giving the air a sparkle. This phenomenon was described by Robert SCOTT: 'From these drifting crystals above, the sun's rays were reflected in such an extraordinary manner that the whole arch of the heavens was traced with circles and lines of brilliant prismatic or white light. The coloured circles of a bright double halo were touched or intersected by one which ran round about us parallel to the horizon; above this again, a gorgeous prismatic ring encircled the zenith ...'.

Diatoms Single-celled photosynthesizing ALGAE that have cell walls strengthened by silica

deposits. The silica shells, or frustules, of diatoms often have intricate designs typical of each species. Nearly all (99 percent) Antarctic PHYTO-PLANKTON is composed of diatoms: there are around 100 diatom species in the SOUTHERN OCEAN which, when blooming in summer, can give the waters a greenish tinge and stain ICE FLOES red-brown.

Joseph HOOKER, geologist on the 1839–43 expedition led by James Clark ROSS, was the first to describe this phenomenon: 'they occurred in such countless myriads, as to stain the Berg and the Pack-Ice ... they imparted to the Brash and the Pancake-Ice a pale ochreous colour.'

As well as phytoplankton, diatoms are found in a variety of wet and sunlit environments. Some diatoms are also able to live and grow while frozen inside winter PACK ICE.

Under favourable laboratory conditions diatoms can replicate by dividing once every day, although in the extreme cold of the Southern Ocean the generation time is likely to be slower.

Diet In bitterly cold Antarctic conditions, humans need a high calorie intake to maintain energy levels. This is provided by food high in carbohydrates and fats. Where there is air access, fresh vegetables and fruit are flown in every few weeks; fresh bread is cooked daily, and many bases grow hydroponic produce. Dehydration is a serious threat in the arid environment and a high fluid intake is essential.

Modern packaging and freeze-drying techniques allow field parties to have a balanced diet and still travel light. A person camping out in the field will need about 3500 kilocalories per day, about double the amount needed in warmer regions. Field-party food will include basics such as freeze-dried meat, dried soup and vegetables, rice, pasta, crackers and cheese, drinks, sugar, porridge and milk powder. Little fresh food can be taken on long trips, but parties take plenty of snacks and chocolate for energy.

On early expeditions, rations were seldom sufficient to keep members healthy for long periods. Lack of Vitamins C and B in Robert SCOTT's party's rations may have contributed to their deaths on the return from the SOUTH POLE. A year later, in January 1913, Dr Xavier Mertz died after being slowly poisoned by Vitamin A from the livers of dogs that he and Douglas MAWSON were consuming on a harsh trek in GEORGE V LAND.

Combined with long, hard physical work, dehydration could cause a rapid deterioration in health, and the explorers often did not have enough FUEL to melt sufficient drinking water.

Pemmican, originally used by Native Americans, was a staple food on sledging journeys. Made from dried and crushed meat, it was mixed to a paste with fat and berries. Hot meals were usually a 'hoosh', a porridge made of reconstituted pemmican and dried biscuits.

To supplement dried rations, parties killed their own animals or lived off wildlife. Roald AMUNDSEN ate his sledging dogs on his journey to the Pole on a planned schedule, and Scott's Northern Party survived WINTER in an ICE CAVE, eating mainly PENGUIN and SEAL meat.

stripped paint off ships' hulls. In 1930 the floor of the harbour dropped by 3 m (10 ft) in an earthquake, and an eruption in December 1967 caused the evacuation of the Argentinian, British and Chilean bases sited there.

The southernmost WHALING station operated out of Whalers' Bay from 1907. Jean-Baptiste CHARCOT, who visited the next year, wrote that, 'The smell is unbearable. Pieces of whale float about on all sides, and bodies in the process of being cut up or waiting their turn lie alongside the various boats.'

The remains of an AIRCRAFT hangar in Whalers' Bay mark the site where Hubert WILKINS took off in November 1928 on the first powered flight over Antarctica.

Demersal fish The coastal bottom-dwelling or demersal fish of Antarctica are a varied and highly endemic group of about 15 families, dominated by members of the suborder NOTOTHENIOIDEI, which includes ANTARCTIC COD and PLUNDER FISH. About 127 notothenioid species are known, most of them demersal. Other bottom-dwelling fish include eelpouts, sea snails and rat-tailed fishes.

Demersal fish feed mainly on INVERTEBRATES such as amphipods, ISOPODS and MOLLUSCS commonly found in the BENTHOS, as well as fish larvae

Above: The Discovery *(centre) with relief ships* Morning *and* Terra Nova *in 1904.*

Dinosaurs FOSSILS of carnivorous dinosaurs from the Jurassic period (208 to 144 million years ago) were found in 1990–91 in the sandstones of the TRANSANTARCTIC MOUNTAINS, along with evidence of a diverse ecosystem needed to support such large animals. Cretaceous (144 to 65 million years ago) plant-eating dinosaur fossils have also been discovered on JAMES ROSS ISLAND. Scientists originally thought that dinosaurs were wiped out by one event: a massive meteor colliding with Earth about 65 million years ago. However, fossil evidence shows that they began to struggle as long as eight million years before the impact, when the Earth began to cool.

Discovery British ship, constructed specifically for the 1901–04 NATIONAL ANTARCTIC EXPEDITION, led by Robert SCOTT. The *Discovery* was a wooden sailing ship, 52 m (172 ft) long with 63 cm (25 in) thick sides, a steel-plated prow, 10 m (33 ft) beam and auxiliary power, built in Dundee, Scotland in 1900. It cost £50,000.

The *Discovery* left NEW ZEALAND for Antarctica in November 1901 and wintered at Hut Point, MCMURDO SOUND. In January 1903 it was still frozen into the ice and a relief ship, the *Morning*, took supplies and some crew back to England. A year later, the *Morning* returned with the *TERRA NOVA*. Using explosives and the swell, the *Discovery* was freed from ice on 16 February 1904, just as plans for abandoning the vessel were being prepared.

After the expedition, the *Discovery* was used in the Arctic fur trade. It was involved in the search for Ernest SHACKLETON's shipwrecked expedition, then Russian ventures. It returned to Antarctic waters in 1925 as part of a British programme investigating WHALING. In 1929, the *Discovery* sailed with the BRITISH-AUSTRALIAN-NEW ZEALAND ANTARCTIC RESEARCH EXPEDITION. The *Discovery* was restored in the 1980s and is now on display at Discovery Point in Dundee, Scotland.

Discovery Expeditions (1924–51) Series of British scientific expeditions organized by the Discovery Committee. The committee was established by the British government to investigate the potential of a WHALING industry in the SOUTHERN OCEAN area. Sustainability as well as commerce was an object of study. The first expedition in 1924–26, undertaken on Robert SCOTT's old ship DISCOVERY, was based in SOUTH GEORGIA. Later, a second research vessel, the *William Scoresby*, joined the expedition, and in 1929 the *Discovery II* replaced the original ship. Numerous voyages were undertaken in the course of the quarter-century long Discovery programme, including two CIRCUMNAVIGATIONS of the Antarctic continent, and it marked the beginning of efforts to conserve the whale.

Disputed claims A number of TERRITORIAL CLAIMS in Antarctica are disputed by rival claimants and non-claimants, making it impossible to clearly establish SOVEREIGNTY.

Of the claimant states, only AUSTRALIA, BRITAIN, FRANCE, NEW ZEALAND and NORWAY mutually recognize each other's claims.

The claims of ARGENTINA, Britain and CHILE overlap in the ANTARCTIC PENINSULA. Argentina and Britain also have a sovereignty dispute over the FALKLAND ISLANDS and associated areas in the BRITISH ANTARCTIC TERRITORY. Argentina and Chile have had disputes in the past over islands near the DRAKE PASSAGE. During World War II, Britain mounted OPERATION TABARIN to support its claim. After the war, the presence of naval vessels in the region threatened militarization of disputes. This was in part averted by a 1949 agreement among Argentina, Britain and Chile not to send warships below 60°S. In 1952 there was an incident at HOPE BAY, when an Argentinian party fired over the heads of a group of British scientists.

The USA and the USSR (now represented in Antarctic matters by RUSSIA) have reserved the right to make claims based on past discoveries and EXPLORATION activities in Antarctica.

In 1959 the disputed claims were 'frozen' by the ANTARCTIC TREATY. Despite this, claim-strengthening activities continue: these include the establishment and/or maintenance of bases for substantial SCIENTIFIC RESEARCH activities, as required by the treaty.

One problematic area is that the United Nations Convention on the Law of the Sea (UNCLOS) has changed since the time the Antarctic Treaty was signed, and it is not clear how this change affects Antarctic territorial claims. Economic Exclusion Zones (EEZs), established since the 1970s, can be expanded by states under UNCLOS. Debate has begun on how this expansion should be handled for Antarctica and the SOUTHERN OCEAN. The ongoing disputes add an underlying current of tension to Antarctic POLITICS.

See also COMMON HERITAGE PRINCIPLE and LAWS, INTERNATIONAL, REGARDING ANTARCTICA.

Above: Above the water's surface, icebergs are shaped by sun, wind, rain and wave action. Their submerged flanks, sculpted by melting and ocean currents, provide an ethereal world for filmmakers to explore.

Diving The first Americans dived in Antarctica in January 1947 as part of OPERATION HIGHJUMP. Diving is now a routine activity for many scientists and some construction workers in Antarctica, but it necessitates precautions over and above those involved with diving in most other parts of the world.

At –1.8°C (28°F), the sea is just above freezing point. Divers must wear several layers of warm clothing underneath drysuits which have sealed, three-fingered gloves; these reduce dexterity but allow heat to radiate between adjacent fingers and so keep the hands warmer than would otherwise be possible. Generally only small parts of their faces are exposed to the water, but even so divers frequently suffer from severe 'ice cream' headaches due to exposure to extreme cold.

The regulators used to supply air to the diver from SCUBA tanks are specially modified to prevent freezing. Each diver also invariably has a back-up breathing system, usually in the form of a 'bail-out bottle', which has a separate regulator attached.

Most Antarctic diving is done in and around the SEA ICE. Holes may be drilled to gain an entry and exit point and a line, usually marked with flags and/or lights, is dropped down to ensure divers can readily locate this. Diving around ICE-BERGS also occurs. As 70 percent of the mass of an iceberg is underwater, this diving is often spectacular, while in open water the potential for icebergs to roll over adds an exciting and sometimes dangerous edge. Even in sea ice there is a certain amount of movement within the ice and it is advisable to have an alternate exit route planned.

CURRENTS and cold WINDs present the biggest danger to dive activities. The sea ice makes for an overhead environment similar to cave diving, in that the only exit point is where the diver entered the water, and strong currents can create problems for divers trying to get back to their exit point. For this reason much diving involves tethers. These allow each diver to follow their own tether back to the exit hole if necessary, and also enable a diver to communicate with those on the surface by sending signals along the line. WIND-CHILL is a problem for those people holding the other end of the tether line; as long as divers are under water the surface support team must tend the lines and cannot leave to seek shelter.

Early in the season the visibility in the ROSS SEA can be as good as 1 km (½ mile). Later, as the plankton blooms occur, the visibility plummets to just metres (several feet).

Above: A team prepares to make a dive under the sea ice.

The cold water and strenuous workload are both predisposing factors to an increased risk of decompression illness (DCI). As a consequence, strict time and depth limits (with extra margins of safety) are imposed on divers using decompression tables designed for heavy work in cold conditions.

Diving petrels (Pelecanoidae) Very small, chunky, black-and-white birds, diving PETRELS have short, rounded wings. Although they are easy to recognize as a group, it is difficult to distinguish between individual species. Their style of propulsion has been compared to that of bumblebees—they speed over the water and disappear into swells like bullets, emerging unscathed seconds later. Their wings are equally powerful underwater, and allow the birds to wing-row after their prey: small FISH, PLANKTON and various CRUSTACEANS.

They gather for breeding in large colonies and nest either in burrows or crevices to lay their single egg. There are four species of diving petrels and a number of subspecies, two of which nest in the SUBANTARCTIC ISLANDS: the SOUTH GEORGIAN DIVING PETREL and the SUBANTARCTIC DIVING PETREL.

Dogs An historical and often nostalgic symbol of the human presence in Antarctica, dogs helped expeditions explore, map and study the continent. On their successful journey to the SOUTH POLE, Roald AMUNDSEN's party took four teams of 13 dogs, killing the animals for food as they went; in contrast, Robert SCOTT sent his dogs back to base and his team man-hauled their SLEDGES over the horrors of the POLAR PLATEAU.

With their thick pelts, the Artic-bred sledge dogs, or huskies, are well suited to Antarctic conditions. They can haul great loads through strong WINDS and BLIZZARDS, travelling about twice as fast as a team man-hauling a similar load. After World War II, huskies became quite common at scientific bases. From the INTERNATIONAL GEOPHYSICAL YEAR onwards, they were used on research and MAPPING expeditions. A team of dogs travelled with the COMMONWEALTH TRANS-ANTARCTIC EXPEDITION to scout routes up steep GLACIERS for the TRACTOR team led by Edmund HILLARY. However, from the mid-1960s, motorized VEHICLES began replacing dogs, and the remaining huskies were kept on bases mainly for sentimental reasons. Eventually ANTARCTIC TREATY nations decided that all dogs should be removed from the continent by 1994 for environmental reasons.

Driving a team of dogs was a skilled job. Four to six pairs of dogs were harnessed together, with one lead dog in front. If the sledge was travelling over crevassed terrain, the dogs were harnessed in fan formation, to brace the load if some of the dogs or equipment broke through the ICE. The driver would stand on the back and yell instructions to the dogs—an experienced team would instinctively weave around obstacles and negotiate dips on the snow surface.

Huskies were known for vicious fighting, and a driver would sometimes have to untangle overturned sledges with a team of barking, fighting dogs attached.

Dolphin (Delphinidae) Three members of the family Delphinidae—the KILLER WHALE, SOUTHERN RIGHT WHALE DOLPHIN and HOURGLASS DOLPHIN—are found in the SOUTHERN OCEAN.

Dome Argus Also known as Dome A. One of the three highest points on the POLAR PLATEAU, Dome Argus lies in the middle of EAST ANTARCTICA, at about 81°S 77°E. It is the highest point in the AUSTRALIAN ANTARCTIC TERRITORY with an elevation of 4039 m (13,248 ft) above sea level and contains the thickest ice—nearly 5 km (3 miles)—in the Plateau.

Below: Indispensible on early expeditions, dogs have been banned from Antarctica since 1994.

Dome Circe Also known as Dome Charlie or Dome C. It is one of the three highest points on the POLAR PLATEAU at approximately 3500 m (11,480 ft) above sea level. Dome Circe lies far from any major geographical landmark in the middle of WILKES LAND at 74°39'S, 124°10'E. In 1983, the USA established a weather station at Dome Circe and in 1997, Concordia Base, a joint French and Italian research station, was opened nearby.

Dominican gull (*Larus dominicanus*) See KELP GULL.

Don Juan Pond A SALINE LAKE in the DRY VALLEYS. For every kilogram (2.2 lb) of water in Don Juan Pond, there is 500 grams (1.1 lb) of dissolved salt. Concentrated levels of calcium chloride in the lake make the water so dense that it does not even ripple in light winds. Calcium chloride evaporates into a mineral called 'antarcticite', which crusts the lake edges. Antarcticite is very unstable in warmer environments, and is unique to Don Juan Pond.

Donoso-La Rosa Declaration In March 1948 ARGENTINA and CHILE affirmed the existence of a South American Antarctic, and declared that only they possessed rights of SOVEREIGNTY. This was later reaffirmed with the Act of Puerto Montt in February 1978.

Drake, Sir Francis (*c.* 1540–96) English navigator and pirate, born near Tavistock, Devon. He went to sea at the age of 13, and after suffering some losses to the Spanish in the West Indies and becoming the first Englishman to see the Pacific Ocean, undertook a voyage to explore MAGELLAN STRAIT. His 1577–80 expedition comprised the 100 tonne (100 ton) *Pelican* (which he renamed *Golden Hind*), the 80 tonne (80 ton) *Elizabeth* and three smaller ships, of which two were abandoned. Like Ferdinand MAGELLAN, he sailed south along the coast of South America and through the Magellan Strait. On the Pacific side of the strait he encountered a storm and was blown back to 57°S where 'there is no maine nor Iland to be seen to the Southwards, but that the Atlanticke Ocean and the South Sea meete in a most large free scope.' This area between TIERRA DEL FUEGO and the Antarctic continent is called DRAKE PASSAGE.

Drake returned to England a hero, the first Englishman to circumnavigate the globe, and was knighted by Elizabeth I. He made a number of other voyages to the Americas, and played a leading role in the English Navy's defeat of the Spanish Armada in 1588. He died on an expedition to the West Indies.

Drake Passage A stretch of open water, approximately 1000 km (620 miles) wide, that separates TIERRA DEL FUEGO and the South American continent from the SOUTH SHETLAND ISLANDS and the Antarctic mainland, and which marks the junction of the Pacific and Atlantic Oceans. It is named after Sir Francis DRAKE, who discovered it in September 1578. Drake Passage is one of the wildest stretches of open water in the world.

Above: The water in Don Juan Pond in the Dry Valleys contains concentrated layers of salt left behind by evaporation.

Dredging The earliest solid evidence of the existence of the Antarctic continent were continental rocks, such as granite, quartz, sandstone and limestone, strewn on the seabed and obtained by deep-sea dredges working in the southern Atlantic and Pacific Oceans during the CHALLENGER EXPEDITION of 1872–76. John MURRAY concluded that the rocks originated in a southern continent and had been carried north by ICEBERGS. The expedition also identified DIATOMS in the PHYTOPLANKTON and a varied benthic FAUNA. Dredging continues to be an important technique in OCEANOGRAPHIC RESEARCH.

Drescher Station Established by GERMANY in 1986 on the Riiser Larsen Ice Shelf in the WEDDELL SEA area. It is occupied in summer only and research is conducted into geophysics, glaciology and meteorology.

Drilling Large drill rigs, usually internationally operated, extract CORE SAMPLES from Antarctic ICE and sediment. By analysing these diagnostic tubes of sediment and ice, scientists can piece together the glacial, tectonic and geological history of an area, and the timeframe over which environmental changes occurred.

Drilling technology has developed rapidly since engineers at BYRD STATION in EAST ANTARCTICA drilled right through the depth of the ice sheet—2164 m (7098 ft) in 1968. The VOSTOK STATION sits on the EAST ANTARCTIC ICE SHEET and has become synonymous with drilling programmes. RUSSIA began drilling ice cores in the 1970s, and was joined by a French team in 1984. On the POLAR PLATEAU near Vostok Station, a drilling programme reached a depth of 3350 m (11,000 ft), stopping just short of the surface of LAKE VOSTOK, an enormous subglacial lake.

As part of the USA's West Antarctic Ice Sheet programme, about 1000 m (3280 ft) of ice cores were drilled at SIPLE DOME in 1997–98; these samples have been compared with similar cores drilled in Greenland to gauge differences between climate fluctuations in Antarctica and the Arctic. Siple Dome is located between two ICE STREAMS, and the dynamic processes that occur at the ice stream beds were also investigated.

In a recent project, at Cape ROBERTS, a multinational group of scientists drilled through SEA ICE and about 170 m (558 ft) of water and into the sea floor in order to learn more about the tectonic and climatic history of western MCMURDO SOUND between 25 and 70 million years ago. The project, which produced about 1500 m (4920 ft) of core samples, has been an environmental test case, proving the ability of a large drilling project to meet stringent clean-up procedures.

Cape Roberts has set the standards for the next major international drilling programme: ANDRILL. This project is still in the planning and development phase, but is likely to involve eight ANTARCTIC TREATY nations, and will attempt to drill beneath twice as much water as the Cape Roberts project.

Dronning Maud Land The area of EAST ANTARCTICA between 20°W and 40°E, also known as Queen Maud Land. Dronning Maud Land makes up one-sixth of Antarctica's land area and is covered by an ice cap averaging 2000 m (6560 ft) in thickness. It includes several mountain ranges, which are popular mountain-climbing destinations, and one of the largest bird colonies on the Antarctic continent—of ANTARCTIC PETRELS at Svathamaren, 200 km (124 miles) inland from the Princess Martha Coast.

The coast of Dronning Maud Land was possibly spotted as early as the 1820s, but it was not until the mid-20th century that its coastline and interior were extensively explored and mapped. NORWAY made a TERRITORIAL CLAIM for Dronning Maud Land on 14 January 1939—partly in order to protect Norwegian whaling interests in the area and partly to forestall German claims.

Druzhnaya Station There have been four stations named Druzhnaya. The first was built by the USSR in the WEDDELL SEA area in 1975. Druzhnaya I was dedicated to prospecting for MINERALS and operated each summer from 1975 to 1986. In the winter of 1986, the ice on which the station was located calved off, carrying Druzhnaya I with it. Druzhnaya II operated on the Lassiter Coast on the ANTARCTIC PENINSULA in the summer seasons from 1980 to 1982. Druzhnaya III operated as a summer season airstrip on the Princess Astrid Kyst on the DRONNING MAUD LAND coast in the late 1980s. Druzhnaya IV in WILKES LAND has been operated as a summer season base since 1986–87.

Dry Valleys *See following pages.*

Drygalski, Erich von (1865–1949) German scientist and explorer, born at Königsberg. Drygalski led expeditions to Greenland in 1891 and 1893 to study glacial ice, and published his two-volume *Grönland-Expedition* in 1897. He was professor of geography at Berlin University when chosen to lead the GERMAN ANTARCTIC EXPEDITION of 1901–03 in the *Gauss*. Research, not exploration, was the main aim of the expedition, which spent a month surveying and gathering oceanological data around Îles KERGUÉLEN before sailing on towards Antarctica.

When the ship became trapped in PACK ICE in the DAVIS SEA, scientific investigations continued, particularly in glaciology. At one point Drygalski surveyed the land from a BALLOON at a height of 488 m (1600 ft). Drygalski's practical approach was evident when he devised the scheme of spreading dark-coloured rubbish on the ice in front of the *Gauss* to hasten melting. The ship broke free in February 1903, by which time significant research had been accomplished. Drygalski published *Zum Kontinent des eisigen Südens* and the 20-volume *Deutsche Südpolar-Expedition 1901–03* about the expedition and its findings.

In 1910 he travelled with Count Zeppelin on a dirigible flight to Spitsbergen. From 1906 to 1934 he held the chair of geography at Munich University and in 1942 he co-authored a comprehensive textbook on glaciology.

Dufek, George (1903–77) American admiral, born in Rockford, Illinois. He joined the US Navy and was navigator on the 1939–41 UNITED STATES ANTARCTIC EXPEDITION, on which he discovered the mountains of THURSTON ISLAND. He returned to the Antarctic as leader of the eastern group on OPERATION HIGHJUMP, which was also under the overall leadership of BYRD. The next year he led a naval expedition to the Arctic.

Dufek was leader of the second stage of OPERATION DEEPFREEZE, when the AMUNDSEN-SCOTT SOUTH POLE STATION and four other bases were built. On a preliminary site inspection, on 31 October 1956, he became the first person to stand at the SOUTH POLE since Robert SCOTT in 1912. 'It was like stepping out into a new world,' he wrote. 'We stood in the centre of a sea of snow and ice that extended beyond our vision. ... Bleak and desolate, it was a dead world, devoid of every vestige of life except us.'

After the death of Byrd in 1957, Dufek became head of the US Antarctic Programmes.

Dumont d'Urville, Jules-Sébastien (1790–1842) French explorer, born in Condé, Calvados. He joined the French Navy in 1807, and during hydrographic surveys of the Mediterranean in 1819–20, saw the recently unearthed statue of Venus de Milo and arranged its purchase for France. He visited the Pacific in 1822–25 as a member of the round-the-world biological expedition led by Louis-Isidore Duprey and published a book on the flora and fauna he had researched in the FALKLAND ISLANDS. During his 1826–29 circumnavigation, he surveyed the northern coast of the South Island of NEW ZEALAND.

After a period of poverty and illness, although he had proposed an ethnological exploration of the Pacific, he was instead sent on a government-supported expedition in search of the SOUTH MAGNETIC POLE and the Antarctic continent in 1837–40.

D'Urville headed south from TIERRA DEL FUEGO for the WEDDELL SEA in early 1838. Just below 62°S, the ships *Astrolabe* and *Zélée* were trapped in PACK ICE for five days, breaking free on 9 February and turning west to a coastline D'Urville named 'Louis-Philippe Land' (now the ANTARCTIC PENINSULA) and JOINVILLE ISLAND. The expedition charted and mapped the northern extremity of what is now known as GRAHAM LAND, then returned to CHILE.

After 18 months exploring the Pacific, in January 1840 the ships sailed south from Hobart, AUSTRALIA, to the Antarctic coast D'Urville named TERRE ADÉLIE after his wife (he also named the ADÉLIE PENGUIN). He went on to determine the approximate position of the South Magnetic Pole and reached 64°S before being driven back by severe storms. D'Urville's narrative of the expedition, *Voyage au Pole Sud et dans l'Océanie* was published in 10 parts in Paris in 1844 and accompanied by *Atlas Pittoresque*, followed by other volumes of scientific records.

On 8 May 1842 he was killed with his wife and second son in a train accident when travelling to Versailles.

Dumont d'Urville Base Dumont d'Urville Base, on Pétrels Island near the coast of TERRE ADÉLIE, was built by FRANCE in 1956 to replace PORT MARTIN BASE.

The base, which has operated continuously since then, houses 30 personnel during winter and up to 120 in summer. Dumont d'Urville Base has limited access for ships as it is only ice-free for two months a year.

In the 1980s, there were widespread protests when France began construction of an airstrip that caused considerable environmental damage to the island and to its plant and animal life, including an ADÉLIE PENGUIN colony.

Dumont d'Urville Sea A marginal sea in EAST ANTARCTICA named after the French explorer, who sailed into these seas in 1840. Inland is TERRE ADÉLIE, which Dumont d'Urville named after his wife.

Dundee Island Five km (3 miles) northwest of PAULET ISLAND, near the ANTARCTIC PENINSULA, Dundee Island was discovered by British whaler Captain Thomas Robertson in 1893. He named the island after his homeport in Scotland. Dundee Island was the departure point for Lincoln ELLSWORTH's 1935 trans-Antarctic flight.

Dry Valleys

An ancient POLAR DESERT located in South VICTORIA LAND on the west side of MCMURDO SOUND, at approximately 78°S. Known for strong winds, extremes of cold and dryness, the landscape is a mosaic of ROCK, gravel and ICE. Some soil deposits have been dated at over 2 million years old and FOSSILS found in the area indicate the CLIMATE here was much warmer in the past.

Three valleys—the Victoria, Wright and Taylor— run from and through the TRANSANTARCTIC MOUNTAINS to the ROSS SEA coast and are separated by ranges up to 2500 m (8200 ft) high. The Dry Valleys have a combined area of around 2500 sq km (965 sq miles); these OASES represent a significant proportion of the exposed rock on the Antarctic continent, of which 98 percent is covered by ice. Mean annual valley floor temperatures are around –20°C (–4°F). The region is dotted with large rocks called VENTIFACTS, stone that has been carved into strange sculptural shapes by winds, and includes several frozen saline LAKES and ponds, and Antarctica's longest non-glacial RIVER.

The Transantarctic Mountains shelter the valleys from the prevailing southerly STORMS, causing a PRECIPITATION shadow. Although little rain falls here, very light showers have been recorded. The air in the Dry Valleys could hold the moisture of 30 times more snow than actually falls. Snow that does fall is invariably sublimated, evaporated by KATABATIC WINDS. Consequently, the area receives less moisture than the Sahara Desert, and is the closest landscape on Earth to that of Mars. For this reason, the United States space programme has trained astronauts in the Dry Valleys, and carried out tests there before launching the space probes to Mars and Venus.

Carved by 'outlet' GLACIERS, which drained ice from the POLAR PLATEAU to the sea, the valleys became 'dry' when the land was raised faster than the glaciers were able to carve through the valleys, effectively cutting the glaciers off. The glaciers have now receded, leaving the lower part of the valleys ice-free apart from lakes and small alpine glaciers that drape down the valley walls. Many of these end abruptly in steep ice cliffs. Broad, low, gently sloping glaciers on low coastal hills extend into the ROSS SEA and separate the area from McMurdo Sound.

The Dry Valleys ecosystem clings to existence in the rock, lakes and small meltwater pools and streams. It represents life at environmental limits: the MICROORGANISMS, MOSSES and LICHENS found here are among the oldest life forms on Earth. One of the most fascinating species discovered is a microscopic worm, NEMATODE, which can survive being freeze-dried for years at a time. Some microorganisms—called endoliths—live inside the rocks in the Dry Valleys and survive on warmth, moisture and the small amount of light that seeps between rock crystals. The microscopic life forms in the Dry Valleys give scientists clues about how life began not only on Earth but also on other planets.

The only bird life here is the SKUA, which visits in search of food and sometimes is able to scavenge on the CRABEATER SEALS and PENGUINS that occasionally migrate in the wrong direction and travel inland. Their mummified, wind-blasted bodies are preserved by the aridity of the climate and are eventually eroded into skeletons by the persistent winds.

The Dry Valleys were 'discovered' in November 1903, during the 1901–04 NATIONAL ANTARCTIC EXPEDITION. Returning from the East Antarctic Ice Sheet, a three-man team led by Robert SCOTT descended into the Taylor Valley from the FERRAR GLACIER. A field party from the 1907–09 BRITISH ANTARCTIC EXPEDITION undertook a geological survey of the Ferrar and Upper Taylor Glaciers in 1909. The Wright and Victoria Valleys remained unseen until 1955 when the US Navy and the COMMONWEALTH TRANSANTARCTIC EXPEDITION began aerial mapping of the area. This began an era of intensive scientific investigation, which has raised concerns that the 'area has

evolved from one of pristine desolation to one of possible scientific despoliation,' as the New Zealand glaciologist T J H Chinn warned in 1990. Such concerns have prompted the development of environmental monitoring and a scientific 'code of conduct' for the Dry Valleys.

Above left and right: *The barren Dry Valleys landscape is thought to be similar to that of Mars, and has been used as a test site for NASA space programmes.*

Right: *The lack of humidity and extreme cold act as preserving agents for any corpses. This crabeater seal evidently wandered too far inshore.*

e

Earthquakes See SEISMIC ACTIVITY.

East Antarctica Also known as Greater Antarctica, East Antarctica makes up about two-thirds of the area of the Antarctic continent and faces the Atlantic and Indian Oceans. Its foundation is a single ancient landmass that geologists call the Precambrian Shield. More than 570 million years old, the BEDROCK is similar to that found on other continents to which Antarctica was once joined as GONDWANA. The ICE SHEET that covers the bedrock is stable and slow-moving. It is also very thick: the average height of the EAST ANTARCTIC ICE SHEET above sea level is greater than any other continent. The central part of East Antarctica—known as the POLAR PLATEAU—rises to about 3050 m (10,000 ft) above sea level near the SOUTH POLE, and the high elevation causes the CLIMATE to be much colder than that of WEST ANTARCTICA. Strong KATABATIC WINDS blow from the Plateau to the coast.

East Antarctic Ice Sheet East Antarctica contains 88 percent of the massive Antarctic ICE SHEET. Land-based, or 'terrestrial', most of the sheet's base is located above sea level. It is stable and slow-moving.

The thickest ice lies in a subglacial trench only 400 km (248 miles) from the coast at TERRE ADÉLIE. At this point it is 4776 m (15,665 ft) deep; about half the height of Mount Everest.

East Base (Stonington) The USA's oldest remaining Antarctic RESEARCH STATION. Built on STONINGTON ISLAND on the western coast of the ANTARCTIC PENINSULA in 1940 by the UNITED STATES ANTARCTIC SERVICE EXPEDITION, it operated until 1941, and again in 1947–48 by the US RONNE ANTARCTIC RESEARCH EXPEDITION. The first WOMEN to live in Antarctica, Edith RONNE and Jennie Darlington, stayed here in 1947. East Base was used and visited infrequently. In 1989 it was declared an HISTORIC SITE.

East Wind Drift Strong airflow off the Antarctic continent creates polar easterly WINDS that drive this east-to-west flowing surface current, also referred to as the Antarctic Coastal Current. The East Wind Drift is shallower and slower moving than the WEST WIND DRIFT; the two are separated by the ANTARCTIC DIVERGENCE.

Echinoderms Echinoderms or 'spiny skins' are organisms with a unique five-point (usually star-like) body symmetry and a skeleton made up of many ossicles of calcium carbonate. The Echinodermata phylum contains five distinct classes of organisms, all found in the Antarctic BENTHOS: SEASTARS, feather stars, BRITTLESTARS, SEA URCHINS and SEA CUCUMBERS.

There are at least 45 species of seastars, or starfish, in Antarctica, which prey on SPONGES, MOLLUSCS, other echinoderms and various zoophytes. Seastars can be very long-lived—one species, *Anasterias rupicola*, is known to reach an age of at least 39 years.

At least 10 species of feather stars—the most primitive of echinoderms—are found in Antarctica. Brittlestars are the most mobile echinoderm and get their name from their ability to avoid capture by shedding arms, then growing replacements. Sea urchins show a high level of brooding of young, with 39 out of the 60 Antarctic and subantarctic species rearing offspring internally within their test (shell) rather than spawning eggs into the ocean. Sea cucumbers are deposit feeders and show an extreme adaptation of the basic five-fold echinoderm symmetry, with the two poles of the spherical urchin shape drawn out into a cylinder, with five lines of tube feet along its length.

Below: Much of the continent's ice is contained in the East Antarctic ice sheet.

Echo sounder Widely used in OCEANOGRAPHIC RESEARCH in the SOUTHERN OCEAN, echo sounders send out signals that bounce off the sea floor to indicate the water depth: the longer a signal takes to return to the surface, the deeper the water. The first oceanographic expedition to use an echo sounder was the 1925–26 *METEOR* EXPEDITION. Echo sounders can also be mounted under planes, and airborne instruments have been used to show the way in which ICEBERGS move with large WAVES. They are also used to estimate KRILL populations.

Ecological research Antarctica's pristine environment offers ecologists a unique opportunity to study the relationship between FAUNA and FLORA and the environment. Many of the first ecological studies were carried out in the SUBANTARCTIC ISLANDS. The development of sophisticated, automatic instruments for measuring temperature, wind speed, humidity and other environmental conditions has since enabled ecologists to gather long-term data. Because of the relative simplicity of the Antarctic ECOSYSTEMS, they are suited to mathematical modelling, which allows ecologists to predict the behaviour of communities under different environmental conditions, such as global warming.

Economic Exclusion Zones (EEZs) The development of EEZs in the 1970s, which allow a state to have jurisdiction over marine living resources within 200 nautical miles of its shoreline, affected Antarctica in several ways. EEZs were established around the SUBANTARCTIC ISLANDS to which ownership was undisputed. And as EEZs were declared around the world, FISHING activities were shifted into the high seas region of the SOUTHERN OCEAN, leading to the over-exploitation of some species, particularly KRILL and more recently toothfish.

EEZs also complicate DISPUTED CLAIMS, because the zones can only be enforced where a coastal state has SOVEREIGNTY. It is uncertain whether or not an EEZ can be declared under the freeze on extending or making new TERRITORIAL CLAIMS in the ANTARCTIC TREATY. Drawing boundaries is also complicated by the permanent floating ICE SHELVES surrounding parts of the Antarctic continent, which may not suffice as a baseline.

Ecosystem A system involving the interactions between a community of PLANTS and ANIMALS and their non-living environment. In Antarctica, there are several distinct environments—marine, fresh water and terrestrial—with their own FOOD CHAINS. They, in turn, differ from south to north, affected by variations in TEMPERATURES, nutrient levels and SEASONS. Further north, the MARITIME and SUBANTARCTIC ISLAND zones have their own distinctive ecosystems.

Edward VII Land Bordered on the west by MARIE BYRD LAND and on the east by the edge of the ROSS ICE SHELF.

Eights, James (1798–1882) American scientist and explorer, born in Albany, New York. Eights studied medicine before turning to geology, botany and

Above: Lincoln Ellsworth (centre front) with members of his expedition.

zoology. As a member of the *Seraph* expedition to the SOUTHERN OCEAN and southern Chile in 1830, led by Benjamin Pendleton, he made an outstanding contribution to Antarctic SCIENCE, discovering new PLANT and LICHEN species, identifying 16 different BIRD species (including five PENGUINS), illustrating three new species of INVERTEBRATES, and making a study of tidal flows. His theory on the transportation of rock and animals by drifting ICE pre-dated Charles Darwin's and his speculations on the origins of ICEBERGS have been proved correct. On his return, Eights published five papers on his Antarctic findings. The *Seraph*'s logbook, containing much valuable early meteorological and oceanographical data, is in the Library of Congress.

Eights Base Established by the USA on the BELLINGSHAUSEN SEA coast of the ANTARCTIC PENINSULA and named after James EIGHTS, who in 1830 became the first American scientist to visit the Antarctic. It operated from January 1963 to January 1965, mainly in support of upper atmosphere physics. In 1963 the plasmapause, a distinctive region of the magnetosphere, was discovered here.

Ekström Ice Shelf On the coast of DRONNING MAUD LAND and first mapped by the NORWEGIAN-BRITISH-SWEDISH ANTARCTIC EXPEDITION of 1949–52. It was named after a member of this team, Bertil Ekstrom, who died when his vehicle plunged through the PACK ICE.

Elephant Island At the northeastern end of the SOUTH SHETLANDS, Elephant Island is the site of several CHINSTRAP PENGUIN rookeries and ancient beds of MOSS more than 2000 years old. Twenty-two men from Ernest SHACKLETON's 1914–17 expedition were stranded here in 1915 after their ship was crushed in PACK ICE in the WEDDELL SEA. They sheltered under upturned boats at Point Wild for 135 days before being rescued.

Elephant seal See SOUTHERN ELEPHANT SEAL.

Ellsworth, Lincoln (1880–1951) Wealthy American adventurer. Ellsworth was a sportsman, surveyor, prospector, engineer, World War I pilot, and, in 1924, a member of a geological expedition to Peru. In 1926, along with Roald AMUNDSEN, he made the first trans-Arctic flight across the North Pole.

With the intention of flying over the SOUTH POLE, he purchased the herring boat *Fanefjord*, which he renamed *Wyatt Earp*, and a twin-engined Northrop monoplane. In January 1934, just after it was unloaded at the BAY OF WHALES, the plane was seriously damaged and was returned to the USA for repairs.

After several abandoned attempts at a transcontinental Antarctic flight from DECEPTION ISLAND, Ellsworth returned south with a new pilot, Herbert Hollick-Kenyon, and a new starting base, DUNDEE ISLAND. They took off on 23 November 1935, crossed the continent in three stages, and ran out of fuel about 25 km (16 miles) short of LITTLE AMERICA, which they reached on foot on 15 December. They had flown 3360 km (2100 miles).

In 1938 Ellsworth returned to Antarctica and on 11 January 1939 he flew with his pilot J H Lymburner from ENDERBY LAND over WILKES LAND to 70°S, where he dropped a canister to make a TERRITORIAL CLAIM for the USA, despite it being part of the AUSTRALIAN ANTARCTIC TERRITORY.

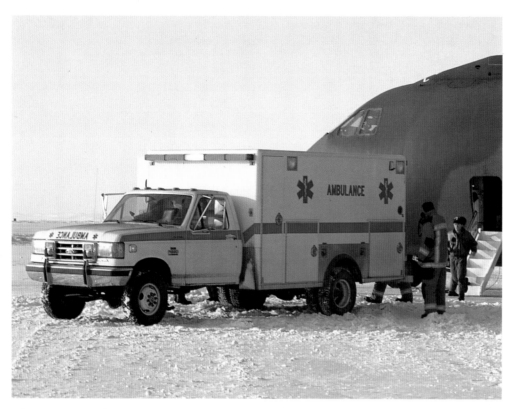

Above: A USA ambulance delivers a patient to a waiting plane.

Ellsworth Land In WEST ANTARCTICA, the area that stretches westwards from the BELLINGSHAUSEN SEA across the base of the ANTARCTIC PENINSULA to the ELLSWORTH MOUNTAINS.

Ellsworth Mountains First sighted by Lincoln ELLSWORTH on his 1935 flight across Antarctica, the Ellsworth Mountains overlook the RONNE ICE SHELF, on the west of the Foundation Ice Stream. They consist predominantly of sedimentary rocks; although of the same age, they are more deformed and their structural grain lies at right angles to that of the TRANSANTARCTIC MOUNTAINS. It is believed the Ellsworth Mountains may have been torn from EAST ANTARCTICA and rotated along the east-west fracture line. The entire range was mapped by the UNITED STATES GEOLOGICAL SURVEY between 1958 and 1961. The highest mountain in Antarctica, the VINSON MASSIF, is located in the SENTINEL RANGE in the northern half of the Ellsworth Mountains.

Ellsworth Station Established by the USA on the FILCHNER ICE SHELF as part of INTERNATIONAL GEO-PHYSICAL YEAR. From 1958 to 1963 Ellsworth Station was managed by ARGENTINA. In 1986 it floated off to sea when an ICEBERG calved off the Filchner Ice Shelf.

Emerald Island A non-existent island or PHANTOM ISLAND, reported by early 19th-century whalers to lie in the region of 57°S and 162°E, due south of MACQUARIE ISLAND. In January 1840 Charles WILKES made a fruitless search for it during his UNITED STATES EXPLORING EXPEDITION. Half a century later, the 1893–95 ANTARCTIC EXPEDITION also attempted to find the island. On 16 November 1894 they set course for its supposed location and spotted what looked like a large land mass some 80 km (50 miles) across. It was named Svend Foyn Island, but turned out to be a massive iceberg. SATELLITE surveying conclusively disproved the existence of Emerald Island.

Emergencies Routine problems can quickly become severe at isolated Antarctic field camps and bases. With medical emergencies, evacuations are logistically difficult over winter, and flying in medical supplies is expensive and risky (see RES-CUES). In 1999, American doctor Jerri Nielsen OVERWINTERING at the AMUNDSEN-SCOTT SOUTH POLE BASE found a lump, possibly cancerous, in her breast. Medical experts in America assessed her condition through SATELLITE imaging and decided she urgently needed drugs. On 11 July, just after MIDWINTER, an US Air Force Starlifter cargo jet swept over the Pole at 320 kph (200 mph) in pitch darkness and driving snow and dropped emergency supplies—six pallets of equipment, including an ultrasound, a digital microscope and drugs—aimed at a blazing arc of barrels. Base staff then had seven minutes to retrieve the cargo before it was damaged in the –55°C (–67°F) COLD.

In April 2001, a Royal New Zealand Air Force plane flew to Antarctica to evacuate four ill USA personnel from MCMURDO STATION, and two Canadian planes flew to the SOUTH POLE to airlift out American doctor Ronald Shemenski, who was suffering from pancreatitis.

Before researchers and support personnel go to remote sites or on the SEA ICE, they must complete a field training course to learn how to deal with and operate effectively during emergencies. The most common emergencies are accidents with heavy equipment and machinery, but groups must also learn about RESCUE procedures. This may include training in CREVASSE rescue for travel on glaciated terrain, and procedures for deploying a search and rescue team if members of the party are lost in severe WEATHER. They must also learn about FIRE prevention and control, as this is an ever-present danger in Antarctica.

Emperor penguin (*Aptenodytes forsteri*) The world's largest living PENGUIN, it stands about 1 m (39 in) tall, and weighs in at between 30 and 40 kg (66 and 88 lb). Emperors are similar in appearance to, but three times the mass of, KING PENGUINS. The adult emperor has a black head and back, orange ear patches and a white breast that is pale yellow near the top. A powerful diver, it feeds on seafood such as SQUID, KRILL and Antarctic silver fish.

These are truly Antarctic birds. They are found only in this zone and, astonishingly, they incubate their eggs through the darkness of the Antarctic winter. They do this to ensure their fledglings leave the ice at a time when food is more plentiful. In April and May, as darkness begins to settle on the Antarctic continent, courting and mating take place in breeding sites located on flat, featureless PACK ICE. By June, the female emperor lays a single green-shelled egg, transfers it to the feet of the male, then returns to the ocean to feed. Unlike other penguins, emperors do not build nests—the male carries the large egg in a warm, highly vascularized abdominal pouch between his feet and belly for 60 to 70 days (the longest incubation period of any bird), keeping the egg at about 35°C (95°F). This prolonged incubation in the appalling Antarctic winter is a feat unmatched in the animal kingdom. 'One very interesting fact we saw was that these birds are so anxious to incubate an egg that they will incubate a rounded piece of ice instead if they cannot obtain an egg,' Edward WILSON noted in his 1911 journal recording the grim midwinter journey across ROSS ISLAND to Cape CROZIER he made with Henry BOWERS and Aspley George CHERRY-GARRARD to collect emperor penguins' eggs. 'The weirdest bird-nesting expedition that has ever been made,' Wilson commented. Cherry-Garrard went on to describe the trek in *The Worst Journey in the World*. It was made in darkness and at temperatures of around –57°C (–70°F) and even lower: conditions to which the emperors had long since adapted.

During incubation, the driving winter cold forces the male bird colonies into large, densely packed 'huddles' of up to 5000 birds that move continuously downwind over the ice, the birds shuffling and changing places so no member of the huddle spends too much time on the bitterly cold outer edge. During this time the males do not eat and are completely non-aggressive to avoid disrupting the life-saving warmth of the huddle.

The birds survive on fat reserves—in the process their body weight may drop by half. The females return when the egg is due to hatch. They

Opposite: Once the female emperors return to the colony, the parents alternate care for their chicks.

roads, a battery of four guns and (on nearby Shoe Island) a jail—the only occupant was the settlement's surgeon (Dr Rudd), following a drunken binge. At its peak the settlement had a population of 306. However, because insufficient whales were caught, less than two years after its establishment the settlement was abandoned.

Enderby Land Lying between DRONNING MAUD LAND to the west and KEMP LAND to the east. Discovered by John BISCOE in February 1831, Enderby Land was the first area of the Antarctic mainland sighted in the INDIAN OCEAN sector. Biscoe named his discovery after his employers. Between 1929 and 1931 Douglas MAWSON's BRITISH-AUSTRALIAN-NEW ZEALAND ANTARCTIC RESEARCH EXPEDITION charted and explored much of Enderby Land.

Endurance Norwegian-built wooden ship used on the IMPERIAL TRANSANTARCTIC EXPEDITION (1914–17), originally named *Polaris* and built for Lars CHRISTENSEN and Adrien de GERLACHE to carry polar-bear hunting parties to the Arctic. The

Above: Male emperor penguins (Aptenodytes forsteri) *protect their newborn chicks by enveloping them in a pouch of abdominal skin.*

Right: Shackleton's ship Endurance *trapped in ice in 1915.*

locate their mates through voice recognition, and from then on the parents take turns caring for the chick and going back and forth to the ocean to gather food. If the female does not return in time, the male will finally desert the chick. Successful young penguins are brooded for a total of 40 days but remain dependent on their parents for the next six months.

About 40 nesting sites are known to be dotted around the periphery of the Antarctic mainland. The known world population of adult emperors is just under 200,000 breeding pairs.

Enderby Brothers Charles, Henry and George Enderby took over the London-based whaling and sealing firm, the British Southern Whale Fishing Company, following the death of their father Samuel in 1829 and carried on his interest in exploration. Charles Enderby, in particular, encouraged his captains to explore for new lands in the course of their sealing and whaling trips, even though this greatly reduced their profitability. Enderby Brothers' captains included John BISCOE, John BALLENY and possibly Peter KEMP.

Enderby Island Settlement On 18 August 1849 the *Samuel Enderby*, belonging to the British Southern Whale Fishing Company, left Plymouth for the AUCKLAND ISLANDS, accompanied by the *Brisk* and *Fancy*. On board was Lieutenant-Governor Charles Enderby, who planned to find a southern headquarters for the company's WHAL-ING enterprise. At Erebus Cove, Port Ross, a town called Hardwicke was established with cottages, barracks, a colonial-style Government House,

ship was 43 m (144 ft) long with a 7.5 m (25 ft) beam, three masts and a coal-fired steam engine.

Ernest SHACKLETON bought the ship for £67,000 and rechristened it in keeping with his family's motto: *Fortitudine vincimus*—'By endurance we conquer'. It sailed to Antarctica in 1914 under the command of Frank WORSLEY, became trapped in PACK ICE in the WEDDELL SEA and was eventually crushed by ice floes. The party was forced to abandon ship on 27 October 1915. The sinking of the *Endurance* on 21 November was captured on FILM by Frank HURLEY.

Environmental Impact Assessments (EIA) The 1991 MADRID PROTOCOL recognizes that all human activities in Antarctica cause some impact. Annex 1 of the Protocol provides for Environmental Impact Assessments (EIAs) to be made of any activity undertaken in Antarctica. This involves planning to ensure that activities are carried out in the most environmentally sound way possible, and that both short- and long-term impacts are minimized and reported. Even simple activities, such as setting up a small field camp, are subject to a Preliminary Assessment, and if environmental questions arise, then an Initial Environmental Evaluation is undertaken.

For projects that have potentially large impacts, such as AIRSTRIP and base construction or DRILLING, a Comprehensive Environmental Evaluation is required; this must include consideration of ways that the site can be restored after the activity has ceased.

Equipment On the 1819–20 expedition led by Thaddeus BELLINGSHAUSEN, the first Antarctic voyage devoted to scientific investigations, ships were equipped with instruments purchased in London to carry out geodesic, astronomical and meteorological research—instruments that today would seem clumsy and inefficient, as would the TRANSPORT equipment used by early explorers. Part of Roald AMUNDSEN's 1912 success in reaching the SOUTH POLE can be attributed to his obsession with obtaining equipment best suited to Antarctic conditions, and the meticulous preparation over the previous winter getting items such as tents and skis ready for the journey to the Pole. Among the 1907–09 BRITISH ANTARCTIC EXPEDITION's equipment were a printing press, type, plate-making equipment and paper, which were used in the production of the book AURORA AUSTRALIS over the winter months.

Now scientific parties take a range of sophisticated equipment into the field. Along with specialized scientific instruments, parties must also carry first aid equipment, radios, tents, sleeping bags, CLOTHING, stoves and FUEL. Equipment may range from EMERGENCY equipment for a day trip or several helicopter loads of field gear, to the large-scale deployment of equipment needed for DRILLING projects. The logistics of organizing equipment for projects requires the expertise of full-time staff at bases.

Erebus, Mount The southernmost active VOLCANO in the world, Mount Erebus on ROSS ISLAND

Above: *All equipment destined for use in Antarctica, particularly fragile electronic, mechanical or scientific gear, must be rated for or protected against extreme cold and snow that can be as fine as talcum powder. Padded covers are used, and batteries are kept warm inside clothing.*

is also Antarctica's highest at 3795 m (12,450 ft). It was discovered by James Clark ROSS in 1841, who named the volcano after his ship. He noted in his journal that Erebus was '... emitting flame and smoke in great profusion ... some of the officers believed that they could see streams of lava pouring down its sides until lost beneath the snow.' A party of six from the 1907–09 BRITISH ANTARCTIC EXPEDITION were the first to climb Erebus. In November 1979 an Air New Zealand DC10 on a tourist flight crashed into the side of the mountain, killing all 257 people aboard.

Very large eruptions from Erebus have occurred in the past, and ash from the volcano has been found in ice as far as 300 km (186 miles) away. The volcano still erupts about 10 times daily, but it is rare for lava to be thrown beyond the 600 m (2000 ft) diameter crater. In 1984, however, gas trapped under a solid lava crust was

released, and a series of cannon-like explosions showered lava 'bombs', some as big as cars, 3 km (2 miles) beyond the crater rim. They whistled as they fell to the ground and crackled as they cooled on the surface. The eruptions lasted for three months, and scientists had to abandon a research station near the crater rim. The most unusual feature of Erebus is a permanent magma-filled lake, which reaches temperatures of around 1000°C (1830°F). The magma, called phonolite, is very rich in sodium and potassium. The crater rim is crusted with feldspar crystals that grow in the neck of the volcano and are broken up when they are thrown out; they are up to 10 cm (4 in) long and are among the largest and most perfectly formed volcanic crystals on Earth. Towering around the crater are bizarre formations, known

Continued page 78

Exploration

Above: *Left to right, Roald Amundsen, Helmer Hanssen, Sverre Hassel and Oscar Wisting take formal leave of their flag at the South Pole in December, 1911.*

The ancient Greeks believed that a great south land, which they named TERRA AUSTRALIS INCOGNITA, existed to balance the Northern Hemisphere. To prove or disprove this theory, James COOK was sent south by the British Admiralty and, on 17 January 1773, made the first crossing of the ANTARCTIC CIRCLE. The honour of first sighting Antarctica perhaps belongs to Thaddeus BELLINGSHAUSEN who, on 27 January 1820, may have sited the POLAR PLATEAU.

Sealers were soon active in the area and on one voyage, led by John DAVIS, the first landing may have been made on the ANTARCTIC PENINSULA, on 7 February 1821. Further sightings included one by Nathaniel PALMER, on 17 November the same year. Subsequently, James WEDDELL's voyage in 1823 achieved the record for reaching the point furthest south and voyages by the SEALING and WHALING ships owned by the ENDERBY BROTHERS, and others, explored unknown waters.

In the 1830s three scientific expeditions were launched within two years—by Frenchman Jules-Sébastien DUMONT D'URVILLE (1837–40), the American Charles WILKES (1838–42) and British explorer James Clark ROSS (1839–43). They were focused on taking magnetic observations and increasing geographical knowledge. Using newly developed scientific equipment, the British naval voyage of HMS *CHALLENGER* between 1872–76 began the science of oceanography. Between 1874–76 and 1882–83 (the FIRST INTERNATIONAL POLAR YEAR), British, French, German and American expeditions travelled south to observe transits of Venus.

Once the sealing boom was over, sights turned to WHALING. On one Norwegian expedition led by Henryk BULL, a landing was made on 24 January 1895 at Cape ADARE on the Antarctic continent.

Adrien de GERLACHE's 1897–99 BELGIAN ANTARCTIC EXPEDITION revived interest in Antarctica when, in 1898, the ship *BELGICA* was beset for 12 months, making this the first expedition to winter south of the Antarctic Circle. A year later, the 1898–1900 BRITISH ANTARCTIC EXPEDITION led by Carsten BORCHGREVINK, who had been a member of Bull's party, spent the first winter on land, at Cape Adare, and returned with a valuable record of meteorology, magnetism and marine biology.

As the 20th century began, several expeditions were ready for sea. The first away was Britain's NATIONAL ANTARCTIC EXPEDITION led by Robert SCOTT. While the expedition's ship, the *DISCOVERY*, was frozen in at MCMURDO SOUND, a major scientific programme and the first extensive exploration on land were accomplished.

Research, not exploration, was the purpose of Erich von DRYGALSKI's 1901–03 GERMAN SOUTH POLAR EXPEDITION and important scientific observations were made. Drygalski's ship, the *Gauss*, also became trapped in ice for a time. Otto NORDENSKJÖLD's 1901–04 SWEDISH SOUTH POLAR EXPEDITION situation was more serious: in February 1903 the *Antarctic* sank and the entire party endured a second polar winter. Scottish and French expeditions, led by William BRUCE and Jean-Baptiste CHARCOT, followed, completing comprehensive scientific programmes and making major discoveries in oceanography, botany and geography.

The achievements of Ernest SHACKLETON's 1907–09 BRITISH ANTARCTIC EXPEDITION included the introduction of motorized transport, the discovery of the SOUTH MAGNETIC POLE and sledging to within 180 km (112 miles) of the SOUTH POLE.

Further expeditions to the ROSS SEA in the 1910–13 period were Roald AMUNDSEN's, which successfully reached the South Pole; Nobu SHIRASE's, which undertook some science; and Robert Scott's BRITISH ANTARCTIC EXPEDITION which added much to knowledge of glaciology, geology, biology, meteorology and magnetism, but was marred by the deaths of Scott and his party, when returning from the Pole.

During this period Charcot's second FRENCH ANTARCTIC EXPEDITION (1908–10) and Wilheim FILCHNER's GERMAN SOUTH POLAR EXPEDITION (1911–12) both wintered on the ice and Douglas MAWSON's AUSTRALASIAN ANTARCTIC EXPEDITION based at COMMONWEALTH BAY completed a detailed science programme, including a visit to the region of the South Magnetic Pole.

At the close of this 'HEROIC' ERA, one major goal remained—a crossing of the continent. Shackleton's IMPERIAL TRANSANTARCTIC EXPEDI-

Above: Scott's party, defeated, at the South Pole in January, 1912. 'Great God!' Scott wrote, 'This is an awful place and terrible enough for us to have laboured to it without the reward of priority.'

TION involved two parties: a Weddell Sea party (1914–16) led by himself, and a depot-laying Ross Sea party (1914–17) led by Aeneas Mackintosh. However, the grandiose scheme failed. After losing the ENDURANCE, Shackleton rescued his entire party, but three of the Ross Sea party, including Mackintosh, died.

Although World War I curtailed much Antarctic exploration, in late 1920 John Cope led his BRITISH GRAHAM LAND EXPEDITION to the Antarctic Peninsula, which involved two members overwintering. A year later, during the QUEST EXPEDITION of 1921–22, Shackleton died of a heart attack at SOUTH GEORGIA.

It was 1928 that shaped the future of Antarctic exploration. On 16 November, Hubert WILKINS made reconnaissance flights along the Antarctic Peninsula, then a year later Richard BYRD established a base named LITTLE AMERICA, on the Ross Ice Shelf, with advanced communications. On 28 December 1929, Byrd flew to the vicinity of the South Pole.

AIRCRAFTS were also used on Mawson's BRITISH AUSTRALIAN AND NEW ZEALAND ANTARCTIC RESEARCH EXPEDITION (1929–31) and the period 1929 to 1933 saw further OCEANOGRAPHIC RESEARCH, during the DISCOVERY EXPEDITIONS. Two years later Lincoln ELLSWORTH completed the first trans-Antarctic flight and in 1934–37 John RYMILL made flights along the Antarctic Peninsula. Byrd led further expeditions in 1933, 1939 and 1946, again utilizing aircraft, land vehicles and ships.

Extensive aerial photography and geological programmes were features of these expeditions.

After World War II many countries began to show serious interest in Antarctica and a number of permanent bases were established during the INTERNATIONAL GEOPHYSICAL YEAR of 1957–58, heralding the beginning of a new era of research. It was appropriate that Shackleton's dream was realized—by Vivian FUCHS's COMMONWEALTH TRANSANTARCTIC EXPEDITION—in 1958.

Today, there are few of the overland traverses that were a feature of the 1950s and 1960s. Remote sensing by SATELLITES has revolutionized exploration in Antarctica and the quest for knowledge has extended to beyond the ATMOSPHERE and beneath the polar ice sheet and sea floor.

Perhaps with increased TOURISM to the continent and remote private expeditions a new phase of exploration has already begun.

Below: Ernest Shackleton's team in 1909, at the camp nearest to the South Magnetic Pole.

Above: *Mount Erebus is the southernmost active volcano in the world.*

as fumarolic ice towers, which form when hot gas, rich in water vapour, seeps out of fractures around the vent and freezes; they are layered into top-heavy shapes and are often undercut by tunnels and ICE CAVES.

Visible from both MCMURDO STATION and SCOTT BASE, Erebus is the subject of ongoing research and monitoring. As well as its volcanic activity and geological interest, it supports the only Antarctic ALGAE that need high temperatures to survive; known as thermophilic algae, they will not grow at temperatures below 25°C (77°F) and thrive at 45°C (113°F).

Escudero Declaration In July 1948 CHILE responded to USA proposals for the internationalization of Antarctica with the Escudero Declaration. This proposed a five-year suspension of discussions on SOVEREIGNTY to allow SCIENTIFIC RESEARCH in Antarctica and political neutrality for expeditions sent there. These ideas were eventually adopted and incorporated into the 1959 ANTARCTIC TREATY.

Esperanza Station Built by ARGENTINA in March 1952, Esperanza Station is located on an island off the northern tip of the ANTARCTIC PENINSULA near a GENTOO PENGUIN colony. The station, which is operated year-round by the Argentinian army, is a colonial settlement with a school, chapel and radio station, and several families are stationed here to reinforce Argentina's TERRITORIAL CLAIM. The first BIRTH recorded in Antarctica was at Esperanza, on 7 January 1978.

Evans, Cape Site of the winter HUT of Robert SCOTT's 1910–13 BRITISH ANTARCTIC EXPEDITION, located on the western side of ROSS ISLAND at the foot of Mount EREBUS and named after Scott's second-in-command, Edward EVANS. In 1915, the second party of the IMPERIAL TRANSANTARCTIC EXPEDITION spent the winter stranded in Scott's

hut at Cape Evans. Today, the restored hut is a much-visited HISTORICAL SITE.

Evans, Edgar (1876–1912) British seaman and explorer, born in Rhossili, South Wales; known as 'Taff' or 'Seaman Evans'. He joined the British Navy in 1891. He volunteered for the 1901–04 NATIONAL ANTARCTIC EXPEDITION, during which he sledged over 1100 km (680 miles) across VICTORIA LAND with Robert SCOTT. When he returned to England, Evans became a naval physical training and gun instructor.

He joined the 1910–13 BRITISH ANTARCTIC EXPEDITION and was known as the 'strong man' of the expedition. As part of Scott's party, he reached the SOUTH POLE on 17 January 1912. On the return journey to the base camp he suffered from frostbite, malnutrition and depression, and rapidly deteriorated. On 4 February 1912 he fell into a crevasse, and it was suspected he suffered minor concussion. He collapsed and died 13 days later and was buried at the foot of BEARDMORE GLACIER.

Evans, Edward (1881–1957) British naval lieutenant, born in London; known as 'Teddy'. He joined the British Navy in 1896. As a sub-lieutenant on the *Morning*, he took part in the relief of the NATIONAL ANTARCTIC EXPEDITION in 1902. He was then accepted on the 1910–13 BRITISH ANTARCTIC EXPEDITION as Robert SCOTT's second-in-command and captain of the expedition ship, the TERRA NOVA.

As preparation for the attempt to reach the SOUTH POLE, Evans set out ahead of Scott, laying supply depots, marking out the route and selecting camp sites. He was the leader of the last support party to leave Scott and his four companions, 270 km (150 miles) from the Pole: 'I had a narrow squeak, thank God I was not included in the advance party,' he wrote to friends later. Evans, badly affected by scurvy, and his two companions

barely made it back to base, and he was invalided home in 1912. He returned to Antarctica in command of the *Terra Nova* at the beginning of the next summer to collect the surviving expedition members. His book, *South with Scott*, was published in 1921. Evans served in World War I as commander, sinking two German destroyers. He commanded the Royal Australian Squadron from 1929, and later the Africa Station, and was created Lord Mountevans in 1929.

Expeditions Polaires Français A state-backed French organization established in 1947 by Paul-Emile Victor, who worked with Jean-Baptiste CHARCOT. The organization gives logistical support to FRANCE's Antarctic expeditions.

Exploration *See previous pages.*

Exploratory Age (1772–1894) From James COOK's 1772–75 southern voyage of discovery through Thaddeus BELLINGSHAUSEN, Charles WILKES and Jean-Sebastien DUMONT D'URVILLE to James Clark ROSS in the 1840s, the great journeys south were for the purposes of discovery—geographical and scientific. Sailing into the unknown must have been frightening; even the remarkable Captain Cook admitted as much in his Journal, when PACK ICE forced him to turn north: 'I, who had ambition not only to go farther than anyone had been before, but as far as it was possible for man to go, was not sorry at meeting with this interruption; as it, in some measure, relieved us; at least, shortened the dangers and hardships inseparable from the navigation of the southern polar region.' The sealers and whalers who ventured into the treacherous SOUTHERN OCEAN were also searching for the unknown. By 1895, the emphasis had moved from science to heroic expeditions of adventure. These expeditions were often state-sponsored for reasons of territorial expansions.

f

Fairy prion (*Pachyptila turtur*) Small bluish-grey PRIONS similar to FULMAR PRIONS, they are 25 cm (10 in) long, have conspicuous black-tipped tails and bold wing patterns, short blue bills and blue feet with cream webbing. They breed in peat burrows on SUBANTARCTIC ISLANDS, including SOUTH GEORGIA, CROZETS, PRINCE EDWARD and MACQUARIE. Common in subantarctic and subtropical waters of the central INDIAN OCEAN, eastern Australasian region, and east of South America outside the breeding season, they are surface feeders, taking KRILL and other small CRUSTACEANS as well as small FISH by surface-seizing and dipping into the water.

Falkland Islands Two main islands, East and West Falkland, along with about 700 islets, that lie 490 km (300 miles) east of Patagonia in the South Atlantic Ocean. About 2200 people currently inhabit the islands; most in the capital, Stanley, on East Falkland, and the rest in 'Camp', the local term for the countryside. The Falklands are known by the Argentinians, who have a long-standing claim to the islands, as the Malvinas. In 1992 BRITAIN and ARGENTINA went to WAR over possession of the Falklands.

Lying outside the ANTARCTIC CONVERGENCE, the Falklands are not strictly subantarctic. However, the CLIMATE is influenced by cold surface currents seeping north from the SOUTHERN OCEAN and the islands' proximity to the ANTARCTIC PENINSULA make them at least geographically subantarctic. They are also a stopover on many Antarctic voyages.

The terrain on most islands is mountainous, apart from the southern half of East Falkland, Lafonia, which is low-lying. Ribs of quartzite boulders splay out from the peaks and ridges of both islands, and most uplands are bare peat and scree slopes. There are few trees, and most of the vegetated land is covered with shrubs and heath. The Falklands are home to large colonies of birds and marine mammals, including the world's largest colony of BLACK-BROWED MOLLYMAWKS and five species of PENGUIN.

Faraday Station BRITAIN's Argentine Islands Base was moved to Marina Point, Galindez Island in 1954, and in 1977 renamed Faraday Station, after Michael Faraday, the discoverer of electromagnetism. The station was operated by the BRITISH ANTARCTIC SURVEY until 1996, when control was transferred to the UKRAINE and it was renamed VERNADSKY. The oldest operational station in the ANTARCTIC PENINSULA area, this is where British scientists first identified the OZONE HOLE in 1985.

Fast ice A region of SEA ICE around the margin of the ANTARCTIC CONTINENT within the PACK ICE zone. Fast ice is multi-year sea ice (which remains for two or more years) and may reach a thickness of several metres (yards). It is distinct from the floating ICE SHELVES, which are formed from glacial ice.

Fauna All the ANIMAL life of a given time or place is known as the fauna of that period or region. With the exception of tiny terrestrial INVERTEBRATES living in Antarctica's MOSS and ALGAE habitats, most naturally occurring Antarctic ANIMALS are part of the rich marine ECOSYSTEM. Antarctic mammals include SEALS, SEA ELEPHANTS, DOLPHINS and WHALES. Cats, rats and mice are among the INTRODUCED SPECIES on SUBANTARCTIC ISLANDS. Numerous BIRD species also live and breed in Antarctica and on subantarctic islands, including a number of PENGUIN species as well as coastal birds such as SKUAS, TERNS and GULLS and deepwater seabirds such as ALBATROSSES and PETRELS.

A large proportion of the FISH species occurring in Antarctic waters are endemic to the region. Antarctica's rich marine invertebrate fauna includes such organisms as BRYOZOANS, CORALS, SPONGES, SEA ANEMONES, CRUSTACEANS, MOLLUSCS, SEA SPIDERS and ECHINODERMS.

Above: *Despite the harshness of the environment, Antarctic fauna is hugely varied, from marine invertebrates on the ocean floor to marine mammals such as the Weddell seal* (Leptonychotes weddellii).

Below: *The oldest operational station in the Antarctic Peninsula area, Faraday Station was handed over to the Ukraine by Great Britain in 1996 and renamed Vernadsky.*

Fauna

FOSSIL records show DINOSAURS, MARSUPIALS and reptiles such as turtles and crocodiles lived in a more temperate Antarctic CLIMATE in prehistoric times.

Fellfield Communities of MOSS and LICHEN that cling to exposed, weathered rocks in Antarctica.

Ferrar Glacier Flows from VICTORIA LAND, west of the Royal Society Range, and joins the TAYLOR GLACIER before flowing into MCMURDO SOUND. The glacier was discovered by the 1901–04 NATIONAL ANTARCTIC EXPEDITION, and named after Hartley Ferrar, the expedition's geologist.

Filchner, Wilhelm (1877–1957) German surveyor and army officer, born in Munich. Filchner joined the army and crossed the Pamir Mountains in 1900 and led a 1903–05 expedition to Tibet, carrying out cartography work and taking magnetic observations.

He was chosen to lead the 1910–12 SECOND GERMAN SOUTH POLAR EXPEDITION to cross the ANTARCTIC CONTINENT from the WEDDELL SEA to the ROSS SEA on SLEDGES, using a prefabricated hut as a base camp. The ship, *Deutschland*, was trapped in PACK ICE from March to November 1912 and, when storms finally broke up the ice, the hut and men were carried northward for a considerable distance before they could be reached by the ship. Filchner carried out important OCEANOGRAPHICAL RESEARCH on the movement of the pack ice. On his return, Filchner wrote of his experiences in *Zum sechsten Erdteil*, published in 1923.

During an expedition to Nepal in 1939–40, he carried out further magnetic surveys of the Himalayas region. A book about his various travels, *Ein Forscherleben*, was published in 1950.

Filchner Ice Shelf Over 320 km (198 miles) long and 160 km (99 miles) wide, the Filchner Ice Shelf lies at the head of the WEDDELL SEA and extends inland on the east side of BERKNER ISLAND (opposite the RONNE ICE SHELF) for more than 400 km (248 miles). It is largely sustained by the Slessor and Recovery Glaciers. The Ice Shelf was discovered by Wilhelm FILCHNER in 1912, and was claimed first by BRITAIN and subsequently by ARGENTINA. It often marks the start of 'transantarctic' ski expeditions.

In early 1986, 13,000 sq km (5019 sq miles) of ICE calved off from the Ice Shelf into the Weddell Sea, carrying Argentina's Belgrano I Base and the USSR's summer base, DRUZHNAYA, with it. A Soviet expedition found their base buried deep below snowdrifts on an ICEBERG a few weeks later.

Film The first moving images of Antarctica were captured by William BRUCE's 1902–4 SCOTTISH NATIONAL ANTARCTIC EXPEDITION, but the most famous movie images of the frozen continent from the HEROIC AGE are probably those of Herbert PONTING, who joined the 1910–13 BRITISH ANTARCTIC EXPEDITION and took with him a movie camera as well as equipment for still PHOTOGRAPHY. The resulting film *90° South with Scott to the Antarctic* is regarded as a classic.

Roald AMUNSDEN took a movie camera to Antarctica on his 1910–12 expedition, although he did not take it to the SOUTH POLE. An early fragment of 35 mm film entitled 'Roald Amundsen has planted the Norwegian flag on the South Pole' and showing the FRAM arriving back in Norway captured the imagination of his home audience, and marked the beginnings of the European travel film.

Frank HURLEY's footage of the IMPERIAL TRANSANTARCTIC EXPEDITION, most particularly of the *ENDURANCE* disintegrating in PACK ICE in 1915, was released as *South*, and has been restored. Other notable early films include *With Byrd at the South Pole: The Story of Little America* made by Joseph Rucker and Willand van der Veer.

Most notable among the few Antarctic dramas are the romanticized 1948 *Scott of the Antarctic* (filmed in Norway), starring John Mills and directed by Charles Frend; Harushi Kadokama's *Virus*, about survival in a post-nuclear world; Koreyoshi Kurahara's *Antarctica*, which recounts the fate of dogs accidentally abandoned at a base and is perhaps most notable for its MUSIC soundtrack and the fact it was filmed on the ice; and horror movie director John Carpenter's feature *The Thing*. Many newsreels were shot during the INTERNATIONAL GEOPHYSICAL YEAR (1957–58), among them a documentary film about the COMMONWEALTH TRANSANTARCTIC EXPEDITION.

In the latter part of the 20th century the region became a popular destination for commercial natural history film producers, working mainly on 16 mm film, with the first series probably being *The Big Ice*, by New Zealand's National Film Unit in 1983. Subsequent significant programming included Natural History New

Above: Herbert Ponting's record of the 1910–1913 British Antarctic Expedition has become a classic.

Below: Aside from technical and logistical considerations, commercial filming in Antarctica today requires strict adherence to guidelines laid down by the country from which the film crew originates. These guidelines are consistent with the protocols of the Antarctic Treaty, and are aimed at minimizing the impact of human activities on the environment.

Zealand's *Antarctic Trilogy* and 1999 film *The Crystal Ocean*, and the BBC series *Life in the Freezer*.

The advent of videotape technology and, subsequently, digital cameras made it easier to capture moving images of Antarctica without needing to deal with the tendency of film to become brittle in the cold. However, the extreme conditions still necessitate specialized preparation. Cameras and camera equipment may need to be 'winterized', a process in which grease or oil from moving parts in items such as tripods and zoom lenses is replaced by special lubricants designed for use in low temperatures. Large-capacity lithium batteries are best, as the cold diminishes a battery's charge, and these need to be kept warm on the body. The PVC insulation on power cables is also prone to cracking in extreme cold, so cables that are coated in silicone rubber compounds are used.

The previously unseen realm of the Antarctic seabed was revealed to the world in the early 1980s when remote-operated vehicles (ROVs) were used to film at depths inaccessible to divers. More recently, American researchers have mounted small video cameras on WEDDELL SEALS and EMPEROR PENGUINS to investigate the diving and feeding habits of these marine creatures.

Fin whale (*Balaenoptera physalus*) The second largest of the whales, fins are BALEEN WHALES. They can grow 26 m (85 ft) long and weigh up to 90 tonnes (90 tons). Distinctive in appearance, they have predominantly black dorsals, white ventrals and flat heads split into darker and lighter zones. The lower lip on the right side is almost white and it is thought that fins use this light skin to frighten schooling fish into a tighter formation, thus making them easier to catch. Their dorsal fins are pronounced and slightly hooked. They were called 'Razor-backs' by whalers because of their distinctly ridged tailstocks.

Initially, whalers based on the SUBANTARCTIC ISLANDS left them alone: the fins' ability to swim at speeds of up to 30 km (19 miles) an hour meant that they were difficult to catch. But once boats became faster and whaling methods more efficient, the fins were one of the important catches.

Fin whales reach Antarctic waters during October and November, the adult males and pregnant females arriving first. They feed on KRILL, and live off their blubber in winter. They breed in warmer northern waters during June and July and gestation lasts 12 months. They become sexually mature between six and 12 years of age. The females have a two- to three-year reproductive cycle, and generally give birth to a single calf.

When diving, fin whales can reach depths of over 230 m (754 ft) and stay submerged for around 15 minutes. They have a circumpolar distribution and are found in the oceans of the Northern and Southern Hemispheres. Although they usually travel singly or in pairs, they have been seen in groups of 100 or more in feeding areas.

There is no reliable information about current population sizes, but the Southern Hemisphere population is estimated at about 15,000.

Above: Low humidity and strong winds make fire in Antarctica a major risk. Most people who go to the continent undergo specific fire training.

Fire Fire is one of the biggest risks at Antarctic camps and bases, and base personnel are usually given fire-fighting training before leaving for the continent. The low humidity causes materials to shrink and dry, and strong winds can quickly blow a fire out of control. In 1960, a fire killed eight men at the MIRNYY STATION. Help is usually a long way away. According to Mike Masterman, winter site manager at the SCOTT-AMUNDSEN SOUTH POLE STATION in 1998–99, 'When the fire alarm goes off or the power goes out, one of the things running through your mind is that we're on our own, there is no way out.'

First International Polar Year (1881–82) Coordinated by the newly established INTERNATIONAL POLAR COMMISSION, the International Polar Year involved 12 nations establishing 14 bases in polar regions to study the Earth's climate and magnetism. Most work was carried out in the Arctic. Of the four geomagnetic and meteorological stations planned for Antarctic regions, only the German base on SOUTH GEORGIA was established. The second International Polar Year was held 50 years later, in 1932–33, with a similar agenda, but the focus was again on the Arctic. In 1950, the American scientist Dr Lloyd Berkner suggested another international collaboration: this became the INTERNATIONAL GEOPHYSICAL YEAR, which had a greater emphasis on the Antarctic.

Fish The first Antarctic fish specimens were collected at Îles KERGUÉLEN during the 1839–42 expedition led by James Clark ROSS. Until then, it was believed the Antarctic environment was too harsh to support life. Fish are important in the diet of SEALS, BIRDS and other fish. It is estimated that 15 million tonnes (15 million tons) of fish are consumed annually by birds and seals in the SOUTHERN OCEAN. Over 200 fish species are known to live in Antarctica, equating to about 1 percent of all known species. Approximately 88 percent of these are found only in Antarctica. Most are DEMERSAL FISH, which live near the sea bottom but can swim freely, and the most abundant species are those belonging to the suborder NOTOTHENIOIDEI, a group of bony, perch-like fish that are mostly bottom-dwelling and lack swim-bladders. Over half the demersal species in Antarctica are notothenioids, most of them ANTARCTIC COD and ICEFISH. Other bottom-dwelling fish include eelpouts, sea snails and rat-tailed fishes. Hagfish, lantern fish, skates and barracuda are also found in Antarctic waters.

Only one PELAGIC FISH—the Antarctic silverfish (*Pleurogramma antarcticum*)—lives permanently in the upper waters of the ocean. But a number of species are adapted to live in pelagic waters temporarily (usually during a larval stage), so that they can use food resources such as KRILL and summer PHYTOPLANKTON blooms.

Antarctic fish are uniquely adapted to survive their harsh environment. They tend to have large, bulging eyes, designed to capture as much light as possible in the darkness below the ice. Some species have GLYCOPROTEIN in their blood, a form of ANTIFREEZE, blocking the formation of ice crystals. At very low temperatures, blood becomes thicker and harder to pump, and to reduce blood viscosity many Antarctic fish have fewer red blood cells than normal, and can survive because freezing water contains high levels of dissolved oxygen. Although they can tolerate very cold temperatures, Antarctic fishes can only cope with a narrow temperature range—a rise of about 8°C (14.4°F) can kill them.

Fishing The SEALING and WHALING industries represented the earliest COMMERCE in the SOUTHERN OCEAN. Today, the most important economic activity is fishing. The two main fisheries in the Southern Ocean have focused on various finfish species and KRILL. Smaller fisheries for crabs and SQUID may have some future potential.

Large-scale harvesting of finfish species began in the 1960s, when trawl fisheries were established in the South Atlantic sector of the Southern Ocean. The main species targeted were marbled rockcod, mackerel, ICEFISH, Patagonian rockcod, subantarctic lanternfish and Wilson's icefish. Most were fished for human food, but some smaller species were turned into fishmeal. After initial good yields, stocks declined as they were over-exploited. Despite CONSERVATION measures implemented by the Commission for the CONVENTION ON THE CONSERVATION OF ANTARCTIC MARINE LIVING RESOURCES (CCAMLR), these stocks have not yet recovered. In 2001, fishing for many finfish species was prohibited, and the Total Allowable Catch (TAC) for others was limited.

In the mid-1980s longlines were introduced to catch the valuable PATAGONIAN TOOTHFISH. The fishery has suffered from illegal, unregulated and unreported fishing (known as IUU fishing), and some stocks are already commercially extinct. Through the incidental BY-CATCH mortality of BIRDS on longline hooks, several bird species are now endangered. Because of this, organizations such as GREENPEACE have called for a moratorium on TOOTHFISH FISHING.

Commercial fishing for krill began in the early 1970s. Most harvesting has taken place around SOUTH GEORGIA, SOUTH SANDWICH and SOUTH ORKNEY ISLANDS. The annual catch peaked at 500,000 tonnes (500,000 tons) in the early 1980s, and has since dropped fivefold. Because of difficulties in processing high fluoride levels in krill shells, early catches were used mainly for animal feed. However, these problems have been solved, and interest in krill may increase for aquaculture and biochemical products. Krill play an important role in the ECOSYSTEM of the Southern Ocean; if the stocks are over-exploited

Above: A variety of lichens, algae and mosses grow on the Antarctic Peninsula.

the impact on associated and dependent species could be catastrophic.

Until the 1990s most fishing for finfish was conducted by Eastern Bloc states, but now ARGENTINA, AUSTRALIA, CHILE, FRANCE, NEW ZEALAND, SOUTH AFRICA and UKRAINE are involved. RUSSIA, Ukraine and JAPAN have been responsible for the bulk of krill harvesting. Any illegal fishing usually takes place using vessels registered under a flag of convenience, making enforcement of regulations difficult.

Flora Cryptogamous PLANTS (those that do not produce seed) dominate the terrestrial flora of

Below: In the milder climate of the subantarctic, flowering plants such as this Pleurophyllum speciosum *are common.*

Antarctica. Often encrusting, they are small or microscopic and include ALGAE, LICHENS, FUNGI, MOSSES and LIVERWORTS. Two species of FLOWERING PLANTS—ANTARCTIC HAIRGRASS and ANTARCTIC PEARLWORT—are found on the ANTARCTIC PENINSULA but do not survive further south.

The first list of terrestrial plants in Antarctica was compiled in 1841 by Joseph HOOKER, who catalogued 18 species and recognized the essentially cryptogamous nature of the flora.

Today, about 100 species of mosses, 25 liverworts, 300 to 400 lichens and about 20 species of fungi are known to exist in the region. Species diversity increases the further northward the latitude, as the CLIMATE becomes warmer and wetter.

The cold TEMPERATURES and virtual absence of WATER, SOIL and nutrients on the continent limit the terrestrial plants. Lichens grow on rock faces warmed by the summer sun, which melts ice crystals and snow. Moss and algae grow in areas receiving water from melting snowbanks and glaciers. Lichens, algae and fungi survive in rocks in the very dry, very cold DRY VALLEYS region. The most vigorous growth is often near BIRD colonies, which provide much-needed nutrients.

Further north, in the SUBANTARCTIC ISLANDS, the dominant vegetation is TUSSOCK with TUNDRA-like vegetation that includes flowering plants and ferns.

Flowering plants Only two flowering plants grow below 60°S. They are the ANTARCTIC HAIRGRASS (*Deschampsia antarctica*) and the ANTARCTIC PEARLWORT (*Colobanthos quitensis*). Both are found on the western side of the ANTARCTIC PENINSULA, growing in small clumps near the shore. They have tiny, inconspicuous blooms.

Antarctic hairgrass is the more common and needs the soil created by LICHENS and MOSSES in which to flourish. Antarctic pearlwort grows in compact cushions about 25 cm (10 in) across. As well as moisture and soil, both need nutrients that they can obtain near BIRD breeding grounds.

Fogbow An OPTICAL PHENOMENON, a type of HALO. Fogbows occur when light is refracted through the water vapour that forms halos in the upper atmosphere.

Food chain A series of organisms in a community, each of which feeds on another in the chain and is eaten in turn. A number of interconnecting linear food chains go together to form the network of relationships that make up the Antarctic FOOD WEB.

A food chain always begins with a primary producer, a PLANT or MICROORGANISM that is able to fix light energy into food. PHYTOPLANKTON is the first component of all Antarctica's marine food chains and is, therefore, central to the marine food web. The primary producers are then eaten by primary consumers in the food chain: the HERBIVORES, a group that includes ZOOPLANKTON in the marine system, and INVERTEBRATES (such as TARDIGRADES, ROTIFERS and NEMATODES), feed on MOSS, CYANBACTERIA and ALGAE in Antarctica's terrestrial food chains.

Secondary consumers may be carnivores or omnivores and include FISH, SEALS, WHALES and BIRDS. In most oceans, fish are among the main consumers and a prey for animals higher in the chain, but the Antarctic marine system is unusual in that birds and MAMMALS replace fish as the main consumers.

The oft-quoted Antarctic food chain example—phytoplankton-KRILL-whales—is significant because it is so short and thus illustrates the extreme fragility of the marine food web; however, since krill are believed to be omnivorous and also feed on small zooplankton, it might be more correctly represented as: phytoplankton-zooplankton-krill-whales. FISHING of krill could thus have profound and far-reaching effects on mammal and bird species in the Antarctic, which depend heavily on krill as their main, sometimes only, food source.

Food web Although Antarctica's terrestrial and freshwater food webs are fairly simple, the marine environment is extremely rich, forming a diverse web of interacting PLANT and ANIMAL species. The interdependent relationships between plants and animals in an ECOSYSTEM and the ways in which energy moves through that system make up a network of pathways known as the food web. Food webs are developed from FOOD CHAINS. Most organisms have more than one food source and/or are eaten by more than one predator, so that a food web is composed of a number of food chains interrelating together.

In the marine food web, all other organisms depend on the primary production of the annual summer PHYTOPLANKTON bloom. The SOUTHERN OCEAN food web differs from northern ones in that the single most important ZOOPLANKTON is the KRILL species *Euphausia superba*. Without krill, the Southern Ocean web would collapse. Krill, in turn, are a food source, both directly and indirectly, for FISH, BIRDS, SQUID, SEALS and WHALES. When a shortage of krill occurs in any year, the breeding success of birds and FUR SEALS is adversely affected.

Freshwater and terrestrial environments in Antarctica vary in their conditions and suitability for life from area to area. FRESHWATER LAKES contain relatively few species of small INVERTEBRATE herbivores, such as PROTOZOA, ROTIFERS and TARDIGRADES, and there is one carnivorous COPEPOD. Primary production is by freshwater phytoplankton and benthic ALGAE and depends on ice melting in summer and the subsequent availability of light.

Primary production in the terrestrial environment by MOSSES, algae, CYANBACTERIA and LICHENS is limited by the availability of nutrients and water. Subsequent grazing of moss by invertebrate herbivores is minimal compared with the NUTRIENT CYCLING that occurs through the activity of two types of decomposers: BACTERIA and FUNGI. Even these organisms, however, are limited by cold temperatures: in Antarctica, peat formed by moss banks may accumulate for up to 5000 years, but in some subantarctic areas, such peat deposits may be 11,000 years old.

Forests Antarctica—the coldest, windiest, driest and highest continent on Earth—was once covered in forests of *GLOSSOPTERIS* and other associated deciduous conifers: their remains have been fossilized in COAL seams of Permian age (about 245 to 286 million years ago) in the TRANS-ANTARCTIC MOUNTAINS. FOSSIL evidence also shows that southern beech (*Nothofagus*) grew there. Fossilized beech leaves collected only 400 km (248 miles) from the SOUTH POLE have been revealed by carbon DATING to be only 3 million years old. Such trees would have required summers of 5°C (41°F) or more to survive, suggesting Antarctica was at least 20°C (68°F) warmer then. Similar fossil finds elsewhere indicate that WEST ANTARCTICA may have been completely ice-free as recently as 100,000 years ago.

Forster, Johann Rheinhold (1729–98) English naturalist, born in Prussia. Forster was teaching natural history when James COOK offered him a position on his 1772–75 voyage as naturalist (to replace Joseph BANKS). Forster brought with him his 16-year-old son, Johann Georg Adam Forster, and Andreas Sparrman, a pupil of Swedish naturalist Linnaeus. Although Forster was difficult to get along with, he was an outstanding scientist. He recorded 31 BIRDS, 14 of which were new species, including *Aptenodytes forsteri*, the EMPEROR PENGUIN. On their return to England, Forster became bitter that he did not gain the recognition that Banks had after Cook's previous voyage.

Fossils Fossil finds in Antarctica have provided scientists with evidence for the existence of GONDWANA; they have also given an insight into the evolution of Antarctic species and the climatic history of the continent.

Over 800 fossils have been found on SEYMOUR ISLAND, off the northeast tip of the ANTARCTIC PENINSULA. In ROCKS ranging in age from 40 to 120 million years, some weird and wonderful species have been discovered, including a 2 m (6.5 ft) tall penguin that weighed up to 135 kg (300 lb) and a turtle the size and shape of a Volkswagen Beetle car. A fossil of the small mar-

Below: Petrified wood in Allan Hills is an ancient remnant of the forests that once grew in Antarctica.

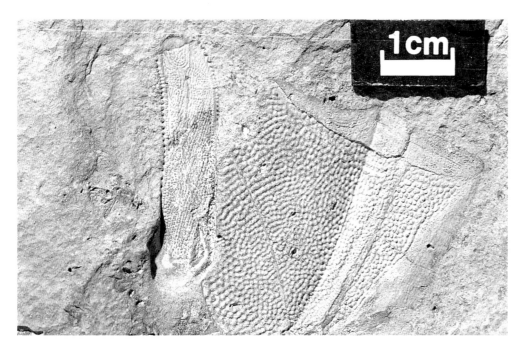

Above: A fossilized bony plate from a fish found in Devonian epoch rocks.

supial *Polydolops* unearthed on SEYMOUR ISLAND was the first land mammal find in Antarctica. This was a crucial piece in the puzzle of how early marsupials migrated across the world: it proved that they travelled to AUSTRALIA, via Antarctica, on a land bridge.

Fossils have shown that Antarctica once had a much warmer CLIMATE. The SHACKLETON and BEARDMORE GLACIERS regions, for example, harbour abundant plant fossils, including remains of ancient FORESTS, mineralized peat deposits and pollen. The remains of forests of *GLOSSOPTERIS* from an earlier ICE AGE about 250 million years ago have been fossilized in the TRANSANTARCTIC MOUNTAINS.

Antarctica was believed to be at a latitude of about 65°S when TEMPERATURES rose again, about 200 million years ago. The first fossils of carnivorous DINOSAURS from this period were found in the Transantarctic Mountains, along with evidence of a diverse ecosystem needed to support such large animals. A fern called *Dicroidium* was the most common plant, and it is believed that a vast range of amphibians and reptiles roamed the plains and waters of Antarctica during the Jurassic period (208 to 144 million years ago).

One of the warmest periods, however, was the Cretaceous, 144 to 65 million years ago. Fossils from exposed sedimentary rocks formed under the sea during the Cretaceous period have now become exposed on ALEXANDER and JAMES ROSS ISLANDS. They show that for much of the era, rainforests grew around the Antarctic coast, representing the highest latitude rainforests known to have existed in the Southern Hemisphere. The most abundant species from this time was the southern beech (*Nothofagus*). Growth rings in the fossilized wood suggest the trees may have existed in a seasonal light-dark cycle, similar to the present pattern in Antarctica. Incomplete skeletons of Cretaceous plant-eating dinosaurs were also discovered on James Ross Island.

Right: The Fram, *in a photo presented by Roald Amundsen to Ernest Shackleton for his ship, the* Aurora.

Foyn, Svend (1809–94) Norwegian sealing magnate. Foyn went to sea at 14 and became wealthy through his involvement in Arctic SEALING during the 1840s. Frustrated by the difficulties of hunting fast-moving FIN WHALES, he developed a grenade-headed harpoon gun and small steam-powered chaser capable of matching the whale's speed. Persuaded by Henryk BULL to fund his 1893–95 expedition to Antarctica to search for RIGHT WHALES, Foyn died before the expedition returned.

Fram Norwegian ship, commissioned by Fridtjof NANSEN to withstand the pressures of polar ICE. The vessel, which Nansen specified must be 'as round and slippery as an eel', was 34.5 m (113 ft) long and 11 m (36 ft) wide and weighed 400 tonnes (400 tons). The *Fram*'s hull was designed so the pressure of ice would push the ship upwards rather than crush it. Nansen took the *Fram* to the Arctic on his 1893–96 attempt to reach the North Pole, and Roald AMUNDSEN borrowed it for what became the NORWEGIAN ANTARCTIC EXPEDITION. The *Fram*, captained by Lieutenant Thorvald Nilsen, delivered expedition members to FRAMHEIM in the ROSS ICE SHELF area, spent the 1911 winter in Buenos Aires, then returned to collect the party the following summer. On 7 March 1912, the ship reached Hobart, AUSTRALIA, from where Amundsen sent a cable around the world announcing they had reached the SOUTH POLE.

The *Fram* is now in the Fram-museet in Oslo.

Framheim A base built by Roald AMUNDSEN in the BAY OF WHALES to support his SOUTH POLE journey in 1911–12. The base, named after the expedition's ship, was buried under snow by the time Richard BYRD established LITTLE AMERICA in the same area in 1929. Framheim was carried out to sea sometime in the 1960s, when an ICEBERG calved from the ROSS ICE SHELF.

France In 1738 a French expedition under the command of Jean-Baptiste BOUVET DE LOUZIER ventured into the SOUTHERN OCEAN. Seeking a 'New Europe' to act as a staging post to the Indies, they discovered BOUVETØYA in 1739. Subsequent expeditions between 1771 and 1910 led by Yves KERGUÉLEN-TRÉMAREC, Jules-Sébastian DUMONT D'URVILLE and Jean-Baptiste CHARCOT penetrated further south, crossing the ANTARCTIC CIRCLE, discovering Îles KERGUÉLEN and TERRE ADÉLIE, and charting thousands of kilometres (miles) of coastline.

In 1924 France made a TERRITORIAL CLAIM for the territories of Îles CROZET, St Paul and Amsterdam Islands and the Îles Kerguélen, along with Terre Adélie. The boundary was extended to the SOUTH POLE in 1938, after negotiations with

BRITAIN. The territories were under jurisdiction of the governor of Madagascar until 1955, when the French created the autonomous TERRES AUSTRALES ET ANTARCTIQUES FRANÇAISES.

France participated in the INTERNATIONAL GEO-PHYSICAL YEAR and was one of the original CONSULTATIVE PARTIES to the ANTARCTIC TREATY. The French maintain one station, DUMONT D'URVILLE BASE in Terre Adélie.

Although France had expressed interest in MINING in Antarctica, in 1989 it was one of the first countries to refuse to ratify the CONVENTION ON ANTARCTIC MINERAL RESOURCE ACTIVITIES. It has been involved in FISHING, especially around its SUBANTARCTIC ISLANDS. France is a member of the CONVENTION ON THE CONSERVATION OF ANTARCTIC MARINE LIVING RESOURCES (CCAMLA), and significantly influenced its contents by insisting that subantarctic islands, where SOVEREIGNTY is uncontested, be excluded from the area of application for CCAMLR, thus allowing the respective governments to impose tighter regulations in waters over which they have jurisdiction.

Frazil ice Formed from SEA ICE crystals that are disturbed by ocean turbulence. They look like slush and are distributed through a surface layer of water several metres (around 10 ft) thick, in a very disorganized manner. In calm conditions frazil ice may consolidate into flexible and thin sheets—up to 10 cm (4 in) thick—called 'nilas'. Frazil ice can also form below the surface, giving rise to ice crystals and larger platelets (about 10 cm/4 in in diameter); these float up and gather on the undersurface of the sea ice, becoming what is known as 'platelet ice'.

French Antarctic Expedition (1903–05) Privately funded French scientific expedition, led by Jean-Baptiste CHARCOT. Originally intending to explore the northern polar regions, the wealthy Charcot had the 250 tonne (250 ton) vessel *FRANÇAIS* purpose-built for polar conditions. He altered his plans when he learnt that Otto NORDENSKJÖLD's *ANTARCTIC* EXPEDITION was missing, and set sail on a rescue attempt.

The *Français* explored the ANTARCTIC PENINSULA and spent the winter of 1904 at Booth Island off GRAHAM LAND, before running aground at ALEXANDER ISLAND and from there struggling to TIERRA DEL FUEGO. Charcot charted the northwestern side of the Antarctic Peninsula and carried out extensive scientific research. When the *Français* reached Patagonia in March 1905, 75 packing cases of collected materials and scientific notes were offloaded.

French Antarctic Expedition (1908–10) Government-sponsored French expedition led by Jean-Baptiste CHARCOT in the *POURQUOI PAS?* After calling at DECEPTION ISLAND in 1908, Charcot found a new base at PETERMANN ISLAND and surveyed uncharted coast on the western side of the ANTARCTIC PENINSULA. He discovered what he named Charcot Land (after his father) and reached 124°W before turning back. Charcot's charts were used exclusively before the area was systematically explored by the British in 1935.

Above: Freshwater lakes in the Dry Valleys are fed by glacial meltwater over summer.

French Institute for Polar Research and Technology Also known as the French Polar Institute, formed in 1992 as a result of the merger between the Paul-Emile Victor French Polar Expeditions Association and the French Austral and Antarctic Territories, the Institute is responsible for the development of FRANCE's polar and subpolar scientific programmes.

Freshwater lakes The smallest Antarctic lakes with through-flow containing fresh WATER are short-lived pools and depressions on GLACIERS and ice fields. Glacial retreat leaves larger depressions, which are fed by glacial meltwater over summer.

The lakes are supplied with sufficient minerals and nutrients by surrounding ROCK to form the basis of an ECOSYSTEM. Many fresh water lakes support abundant populations of ALGAE, luxuriant stands of aquatic mosses and even fairy shrimp (*Branchinecta gainii*), the largest freshwater INVERTEBRATE in Antarctica. Lakes near PENGUIN rookeries or SEAL communities become eutrophic, or enriched with nutrients from excreta, to the point where they may become toxic to life.

du Fresne, Marion (1724–72) French explorer. On an exploratory voyage to the South Seas, two small flutes, the *Mascarin* under du Fresne and the *Marquis de Castries* under Julien Crozet, headed southeast from the Cape of Good Hope. On 13 January 1772 they discovered the Austral Islands, which they named 'Land of Hope' (now MARION ISLAND), and 'The Island of the Cavern' (now PRINCE EDWARD ISLAND). Further to the east, in the southern INDIAN OCEAN, on 23 January, they discovered a group of smaller islands, Îles CROZET. They then sailed to NEW ZEALAND, where du Fresne was killed by Maori.

Friends of the Earth International (FoEI) A federation of over 60 environmental groups from around the world founded in 1971 by four organizations from BRITAIN, FRANCE, SWEDEN and the USA. FoEI campaigns at the national and international level on environmental issues. In 1978 FoEI was one of the founding members of the ANTARCTIC AND SOUTHERN OCEAN COALITION.

Frostbite A high risk in polar regions, frostbite occurs when human skin is exposed to the COLD for a long period of time and freezes. The first signs are white patches on extremities, such as ears, noses, cheeks and feet, caused by blood vessels constricting as they try to slow heat loss. This degree of frostbite can be fixed by rubbing or slight warming. However, if it develops further, the sufferer will lose all sensation and may no longer be able to feel the cold. If the frozen skin tissue dies, gangrene may set in, and the affected area usually has to be amputated.

Fuchs, Vivian (1908–99) British archaeologist and explorer; known as 'Bunny'. Born on the Isle of Wight, England, he studied geology at Cambridge, where he was tutored by Sir James Wordie, a member of the IMPERIAL TRANSANTARCTIC EXPEDITION.

Fuchs went to Greenland in 1929, and on several expeditions to Africa during the 1930s. During World War II, he served with the British Army in North Africa and Europe.

In 1947 Fuchs was appointed leader of the Falklands Islands Dependencies Survey (FIDS). Severe conditions kept his FIDS party in Antarctica without relief for two seasons in 1949–50, during which Fuchs carried out a series of long surveying journeys with R J Adie and first conceived of the journey—to follow Ernest SHACKLETON's plan to cross the ANTARCTIC CONTINENT via the SOUTH POLE—that became the COMMONWEALTH TRANSANTARCTIC EXPEDITION of 1955–58.

Above: Vivian Fuchs, leader of the 1955–58 Commonwealth Transantarctic Expedition.

Above: Fur seal populations were once decimated by sealers. Despite being protected by the CCAS, they still fall victim to the fishing industry through entanglement in nets and debris.

Fuchs was overall leader, and directly responsible for the WEDDELL SEA party, of the Transantarctic expedition; Edmund HILLARY led the ROSS SEA party, which was charged with laying supply depots. Fuchs's party left SHACKLETON STATION on the FILCHNER ICE SHELF and reached the South Pole on 19 January 1958. Leaving the Pole five days later, they reached SCOTT BASE on 2 March 1958, having traversed 3472 km (2152½ miles) in 99 days. As well as being the first overland crossing of the continent, the expedition carried out a series of seismic soundings and gravity determinations.

From 1958 to 1973 Fuchs was director of the BRITISH ANTARCTIC SURVEY (BAS) and developed BAS's long-term biological research programme. He was also president of the International Glaciological Society from 1963 to 1966, the British Association for the Advancement of Science in 1971, and the ROYAL GEOGRAPHIC SOCIETY from 1982 to 1984. In 1985 he published *Of Ice and Men* about his work with BAS, which was followed by an autobiography in 1985.

Fuel Warmth in the early HUTS was provided by stoves, usually burning coal or anthracite. SEAL blubber was used when fuel was low, providing quick heat but a pungent smell. Cooking was done on small primus stoves. These are sturdy and effective and are still used, burning kerosene. In some huts, stoves use a mixture of butane and propane, which works well in cold conditions. WASTE DISPOSAL of fuel has caused problems but, since the 1980s, pressure from environmental groups and the ratification of the MADRID PROTOCOL have led to most bases 'repatriating' and recycling fuel drums, and to the appointment of environmental officers to many national programmes.

Fulmars (Procellariidae) None of these PETRELS nest in burrows or under any sort of cover, and their body design suits gliding more than the erratic flight patterns of some petrels, such as the GADFLYS or PRIONS. When feeding, rather than diving, they alight on the water and peck at food on the surface. Their diet consists of SQUID and CRUSTACEANS. Fulmars lay a single egg on nests of either GRASSES and mud in the subantarctic, or stones and rock chips in the far south.

Fulmars have five genera and there are six Antarctic or subantarctic species: SOUTHERN GIANT PETREL, NORTHERN GIANT PETREL, CAPE PETREL, ANTARCTIC FULMAR, ANTARCTIC PETREL and SNOW PETREL.

Fulmar prion (*Pachyptila crassirostris*) With their small bluish-grey bodies, 28 cm (11 in) long, black-tipped tails and bold wing patterns, fulmar prions are similar in appearance to FAIRY PRIONS, but have slightly paler heads and shorter, stouter bills. They have a unique looping flight pattern. Fulmar prions are sedentary and are rarely seen away from their breeding grounds on the SNARES, AUCKLAND, BOUNTY, CHATHAM, FALKLANDS and HEARD ISLANDS. They feed in subantarctic waters, mostly on CRUSTACEANS.

Fungi The Antarctic FLORA includes less than 30 species of macrofungi (toadstools), but filamentous microfungi occur widely in organic and inorganic soils. Few species are endemic.

Macrofungi are found only on the northern end of the ANTARCTIC PENINSULA, which is generally warmer and moister than the rest of the continent. Microfungi, however, have been found surviving even in the extreme conditions of the DRY VALLEYS where temperatures may drop to –80°C

(–112°F) in winter. Here, in 1978, biologists found widespread MICROBIAL ACTIVITY—including fungi, ALGAE and LICHENS—within light-coloured, semi-translucent ROCKS. The MICROORGANISMS live in a microclimate inside cracks and pores in the rocks, and some are believed to have survived here for 200,000 years.

Fur seals (*Arctocephalus*) The ANTARCTIC FUR SEAL and its close relative the KERGUELEN FUR SEAL belong to the family Otariidae, and are the only Antarctic SEALS that have pinnae or visible ear flaps. Fur seals are more agile on land than phocid seals: their hind flippers point forward, allowing them to stand with all four limbs (pairs of hind and fore flippers) under their body. Fur seals have thick coats, which made them a favourite prey for SEALERS from the 18th century on. Although hunting is no longer allowed, some fur seals are still accidentally killed through becoming caught up in marine debris and entangled in fishing lines.

Furious Fifties See ROARING FORTIES.

Furneaux, Tobias (1735–81) British naval officer, born near Portsmouth, England. Furneaux was the first person to circumnavigate the globe in both directions: westerly with Samuel Wallis on the HMS *Dolphin* in 1766–68, and easterly with James COOK.

Furneaux joined the British Navy at 13 and on Cook's 1772–75 voyage to the Pacific, he was given command of the *Adventure*. It was during this voyage that the ANTARCTIC CIRCLE was crossed for the first time. Invalided from the navy in 1778, after serving in the American War of Independence, he died aged 46.

g

Gadfly petrel These small to medium-sized, widely distributed PETRELS are named for their swift, dashing flight. Of the 23 gadfly species, there are three Antarctic species: MOTTLED PETREL, GREAT-WINGED PETREL and WHITE-HEADED PETREL. The gadfly petrels have a large and complex taxonomic distribution, and include some of the rarest and least understood seabirds. All have short, chunky black bills, but their appearance varies between species, with a dark patch around the eye and a dark 'M' shape across the wings being the most common markings. Gadfly petrels have adopted both solitary and colonial habits, but nearly all are solitary at sea.

Garbage disposal See WASTE DISPOSAL.

Garcia de Nodal, Bartoleme and Gonzalo (c. 17th century) Spanish navigators. On instructions from the Spanish India Office and with a Portuguese crew, these two Spaniards set off from Lisbon in 1619 in two 80 tonne (80 ton) caravels. Blown off-course while attempting to sail around Cape Horn, they discovered the tiny islands about 90 km (56 miles) to the south-southwest, which they named Islas Diego Ramirez after the expedition's geographer.

Gateways A number of Southern Hemisphere cities have close Antarctic connections, many of them long-standing, and are starting points for trips to the polar region. They are: Christchurch in NEW ZEALAND, the base for ANTARCTIC NEW ZEALAND and support centre for the Antarctic programmes of New Zealand, the USA and ITALY; Hobart in Tasmania, AUSTRALIA, an old WHALING port where the AUSTRALIAN ANTARCTIC DIVISION is headquartered; Punta Arenas in CHILE, a major port for research vessels; Ushuaia in ARGENTINA, the main tourist departure point; Stanley in the FALKLAND ISLANDS, and Cape Town in SOUTH AFRICA.

Gauss A purpose-built polar vessel constructed for the 1901–03 GERMAN SOUTH POLAR EXPEDITION, named after 18th-century mathematician Carl-Friedrich Gauss, and modelled on the design of Fridtjof NANSEN's ice-resistant FRAM. A three-masted, 1440 tonne (1440 ton) schooner, it was 46 m (150 ft) long and 11 m (37 ft) wide and was equipped with an auxiliary engine. Commanded by Hans Ruser, it had a complement of five officers and 22 crew in addition to scientists. The ship became trapped in PACK ICE in the DAVIS SEA in February 1902 and was frozen in for 11 months. The Gauss reached Germany on 24 November 1903.

Gauss Expedition (1901–03) See GERMAN SOUTH POLAR EXPEDITION

Gentoo penguin (Pygoscelis papua) Clocked at speeds of up to 27 km (17 miles) per hour, the gentoo is the fastest swimmer of all birds, and can dive to depths of 100 m (328 ft). It has northern and southern subspecies. The northern gentoos—which at 6 kg (13 lb) are slightly larger than their 5.5 kg (12 lb) southern counterparts—breed on the northern SUBANTARCTIC ISLANDS, particularly SOUTH GEORGIA, Îles KERGUÉLEN and the FALK-LANDS. They lay two eggs as early as July. The southern gentoos breed on the ANTARCTIC PENINSULA as well as the more southerly SUBANTARCTIC ISLANDS, often beside ADÉLIE and CHINSTRAP colonies, and lay their two eggs from October to December. Gentoo nests are made of moss,

Below: A southern gentoo penguin (Pygoscelis papua) colony on the Antarctic Peninsula.

Above: The origin of the name 'gentoo' is unclear. In hindustani the word means 'pagan', but it is thought the term may be of Anglo-Indian/Portuguese derivation, or Spanish, perhaps having evolved from the name 'Juanito'.

pebbles and sometimes seaweed, and sites change from year to year. Eggs are spherical, with rough blue-white shells. Incubation lasts about 36 days, with chicks being brooded for 28 days. Fledging is a long process, taking 70–90 days. Gentoos are about 76 cm (30 in) tall, have orange bills and white triangles above their eyes, and are the least aggressive of the penguins. They feed mainly on CRUSTACEANS, SQUID and FISH. The gentoo population is estimated at about 300,000 breeding pairs.

Geographical research Early geographical research focused on exploring the coastline and physical geography of Antarctica. New regions were discovered and explored—often on foot—and the map changed continually. James COOK's circumnavigation in 1772–74 was the first attempt to chart the outline of Antarctica, continued in the 1820s by Thaddeus BELLINGSHAUSEN. In the 1830s the commanders of WHALING expeditions financed by the ENDERBY BROTHERS firm were encouraged to explore the unknown southern regions and they contributed much to geographical knowledge.

Aerial surveying began with Richard BYRD in 1928–29 and by the 1940s most of the coastline had been accurately charted. This work was consolidated by the US Navy's massive OPERATION HIGHJUMP in 1946–47 and during INTERNATIONAL GEOPHYSICAL YEAR. In the modern era, geography has become much more specialized: the majority of geographers in Antarctica are glaciologists and climatologists, studying physical processes on the continent.

Geological research Antarctica has little exposed ROCK. The geological record is mainly hidden beneath the vast ice cover. Because of severe working conditions and the expense of conduct-

ing field-work in this hostile environment, geological research in Antarctica has lagged behind that occurring in other continents. However, modern techniques, such as DRILLING to take CORE SAMPLES from the seabed and the base of ICE STREAMS, and remote sensing techniques, including seismic and RADAR sounding, allow geologists to delve beneath the ICE SHEET into Antarctica's past.

The first geological investigations were made on early expeditions, many of which included geologists to undertake specific research. Notable research was carried out by Hartley Ferrar, on the 1901–04 NATIONAL ANTARCTIC EXPEDITION, who mapped the MOUNTAINS of VICTORIA LAND. Geologists on the 1910–13 BRITISH ANTARCTIC EXPEDITION—Raymond PRIESTLEY, Frank DEBENHAM and Thomas Griffith Taylor—expanded on this work, and made significant FOSSIL discoveries along the MCMURDO SOUND coastline. On their ill-fated return journey from the SOUTH POLE, Robert SCOTT's party stopped at Buckley Island, where Edward WILSON '... with his sharp eyes has picked several impressions the last a piece of coal with beautifully traced leaves in layers also some excellently preserved impressions, of thick stems showing cellular structure,' Scott recorded in his journal. These were the first *GLOSSOPTERIS* fossils found in Antarctica. Such samples were to provide important links to similar fossils from other continents—evidence of GONDWANA.

A new phase of geological science began after World War II, with a series of long sledging journeys by Vivian FUCHS and R J Adie. They confirmed that EAST ANTARCTICA and WEST ANTARCTICA are geologically separate landmasses, and that West Antarctica has experienced more active PLATE TECTONIC and VOLCANIC activity than East Antarctica.

Geological studies gained momentum after the INTERNATIONAL GEOPHYSICAL YEAR. Virtually all exposed rock areas were mapped. The Dry Valleys Drilling Project carried out in 1971–75—a collaborative project conducted by the USA, NEW ZEALAND and JAPAN and the first rock drilling project in Antarctica—provided 2000 m (6560 ft)

Below: New Zealand geologist Professor Bryan Storey researches continental shift on Charcot Island.

of material and 10 million years of the DRY VAL-LEYS' geological history.

Geomagnetic Latitude System A method of pin-pointing the position of magnetic phenomena that occur in the ionosphere (see ATMOSPHERE), such as the AURORA AUSTRALIS. It involves determining the relationship between the direction of the magnetic field in the ionosphere and the geographic co-ordinates of the phenomenon.

Geophysical research Many modern techniques to study the physical properties and structure of the Earth and ATMOSPHERE have been developed in Antarctica. Edmond HALLEY was the first to measure magnetic variations in southern waters, in 1699–1700, and to realize the magnetic field controlled the AURORA AUSTRALIS. This work was continued by James Clark ROSS, who collected data that allowed geophysicists to calculate the exact position of the SOUTH MAGNETIC POLE. Robert SCOTT and Douglas MAWSON were also both interested in geophysics and attempted, rel-atively unsuccessfully, to photograph the aurora australis.

These early investigations demonstrated the unique characteristics of the magnetic field and ATMOSPHERE around Antarctica. But it was not until the INTERNATIONAL GEOPHYSICAL YEAR that geophysics began to focus on Antarctica. Data gathered were collated with information gathered simultaneously around the world to produce a new understanding of the Earth's environment. Over this period, a number of unmanned stations were established. With direct links to SATELLITES that orbit around the South Pole, geophysical data can now be transmitted year-round from Antarctica back to laboratories on the other side of the world.

George V Land Between OATES LAND and TERRE ADÉLIE on the west of the ROSS SEA, it was explored and named by Douglas MAWSON in 1912–13. It is regularly buffeted by vicious gales, which led Mawson to nickname it the 'Home of the Blizzard'. Mawson's base hut at COMMONWEALTH BAY can still be visited.

George VI Ice Shelf The shelf covers George VI Sound and lies between ALEXANDER ISLAND and PALMER LAND on the southern portion of the ANTARCTIC PENINSULA.

Gerlache, Adrien de (1866–1934) Belgian explor-er, born in Hasselt. In 1892 he volunteered to join a Swedish expedition which was abandoned before it started. He then set about organizing the 1897–99 BELGIAN ANTARCTIC EXPEDITION. Members included Roald AMUNDSEN, geologist Henryk ARÇTOWSKI and surgeon Dr Frederick COOK.

De Gerlache purchased the 250 tonne (250 ton) *Patric*, which he renamed the BELGICA. The expedition sailed from Belgium on 16 August 1897, and arrived in Antarctic waters on 20 January 1898, late in the season. After exploring islands off GRAHAM LAND in the waters later called GERLACHE STRAIT, de Gerlache headed into PACK ICE. The ship soon became trapped. Some crew members suspected he had done this intentionally so they would be the first to winter in Antarctica. The *Belgica* was freed from the ice on 14 March 1899. The expedition had been poorly prepared to spend a winter in Antarctica and de Gerlache himself did not cope well.

De Gerlache joined Jean-Baptiste CHARCOT's 1903–05 expedition, and advised Charcot on the construction of the ship, the *Français*. However, once the *Français* had reached Buenos Aires, de Gerlache resigned, claiming he was unhappy away from his fiancée. He made a number of jour-neys to the Arctic, including expeditions to Greenland in 1905 and 1909, and acted as advi-sor to Ernest SHACKLETON on the 1914–17 IMPERI-AL TRANSANTARCTIC EXPEDITION.

Gerlache Strait Running between the western coast of the ANTARCTIC PENINSULA and the string of islands to the west, including ANVERS ISLANDS, the strait is enclosed by mountains, glaciers and rock cliffs. Gerlache Strait is one of the major waterways of the Antarctic, for shipping and for the HUMPBACK WHALES that cruise the waters searching for KRILL swarms. The Gerlache Strait was discovered on 23 January 1898 by Adrien de GERLACHE, who named it the Belgica Strait after his ship.

German Antarctic Expedition (1938–39) German exploratory expedition, led by Captain Alfred Ritscher. The expedition surveyed the coast of DRONNING MAUD LAND in January and February 1939, via aerial photographs taken from hydro AIRCRAFT that were catapult-launched from the ship *Schwabenland*. Photographs—none of which survived World War II—were taken of a 250,000 sq km (96,525 sq mile) area. The expedition's purpose was to prepare for a German TERRITORI-AL CLAIM. To this end, claim markers made from 1.5 m (5 ft) long aluminium darts engraved with swastikas were thrown from the planes.

German Deep Sea Expedition (1898–99) German government-sponsored scientific expedition. The *Valdivia* expedition was led by Karl Chun, pro-fessor of zoology at Leipzig University, and included botanists Andreas Schimper and Emil Werth.

Apart from its valuable contribution to marine biology, the expedition rediscovered BOUVETØYA. From Îles KERGUÉLEN, where a research station was set up, the *Valdivia* headed south towards ENDERBY LAND, reaching 64°15′S. It returned to GERMANY in 1899.

In his book *Aus den Tiefen des Weltmeeres*, published in 1903, Chun expressed his surprise that 'in the icy water of the Antarctic, the temper-ature of which is below 0°C, we find an astonish-ingly rich ANIMAL and PLANT life,' and around Kerguélen he observed 'the ocean is alive with transparent jelly-fish.'

PENGUIN colonies and the vertical distribution of pelagic organisms were studied, and many deep-sea specimens collected. Twenty-four vol-umes of marine biological data were eventually published.

German International Polar Year Expedition (1882–83) German scientific expedition. During the FIRST INTERNATIONAL POLAR YEAR, only two expeditions travelled south; of these, only one sailed into the Antarctic region. The *Moltke* left Hamburg on 3 June 1882 and was the first pow-ered vessel to reach SOUTH GEORGIA.

There, under the leadership of Dr K Schrader, a research station was constructed; it included four OBSERVATORIES, a zoological laboratory, sta-bles and experimental gardens. Hourly meteoro-logical observations were recorded, and magnetic, glacial, astronomical and tidal records kept. These included observations of the tidal effects of the eruption of Krakatoa on 27 August 1882 and of a transit of Venus on 6 December. Many new species of FAUNA and FLORA were discovered in the course of the year. The scientists also monitored INTRODUCED SPECIES—three oxen, 17 sheep, six goats with three kids (which provided fresh meat and milk) and two geese (which flew off to a near-by cliff where they died). Although fodder was taken for the animals, they preferred local GRASS-ES. Of the plants, only cress was successful; the potatoes yielded half-a-dozen pea-sized tubers, and the rye produced one ear and was shortly afterwards destroyed in a storm.

The expedition was relieved after a full year. It had provided the first full set of meteorological data obtained for the Antarctic, tidal observations and glacier movements had been recorded, and comprehensive botanical and zoological investi-gations had been undertaken.

German Magnetic Association The leading 19th-century organization concerned with the magnet-ic crusade to promote research into magnetism. It recommended a system of regular, synchronous observations made around the globe, used by James Clark ROSS on the 1839–43 voyage to the SOUTHERN OCEAN.

German South Polar Commission The govern-ment-funded organization that initiated and sponsored the GERMAN SOUTH POLAR EXPEDITION of 1901–03 led by Erich von DRYGALSKI.

German South Polar Expedition (1901–03) German government-funded scientific expedition. Proposed by the GERMAN SOUTH POLAR COMMIS-SION, the expedition was led by Erich von DRYGALSKI, professor of geography at Berlin University. The aim was to winter over in the ANTARCTIC CONTINENT, to explore the region and make magnetic observations.

The expedition left Kiel on the GAUSS in 1901 and, after calling in at Cape Town, SOUTH AFRICA, headed for Îles KERGUÉLEN where an OBSERVATORY had been established by the 1898–99 GERMAN DEEP SEA EXPEDITION. They then sailed south to the DAVIS SEA where, on 21 February 1902, the ship became ice-bound. Magnetic observation posts were set up on the ice some distance from the *Gauss* and on one occasion 'High Land' was spot-ted; this turned out to be what is now called Drygalski Island. Temperatures reached as low as −28°C (−18°F), which cracked instruments, reduced petrol to an oily liquid and burst dozens of bottles of beer. Meteorological KITES were flown to take weather records, observations were made from a tethered BALLOON, holes were drilled to take ICE CORE SAMPLES, and a Berliner phonograph was used to record PENGUIN sounds. Sledging parties to the south returned with geological samples and the discovery of land 80 km (50 miles) away, which was then named WILHELM II LAND.

Efforts to free the *Gauss* from the ice by blasting and sawing proved useless. Drygalski hastened melting by spreading dark-coloured rubbish on the ice in front of the vessel, and on 3 February 1903 it was freed after having been trapped for 11 months. When the ship reached South Africa on 9 June, the expedition sent a request back to GERMANY to extend the voyage by another year. This was refused, and they returned home in November 1903. Drygalski published scientific results in *Zum Kontinent des eisigen Südens* and the 20-volume *Deutsche Südpolar-Expedition 1901–03*.

German South Polar Expedition (1911–12) German scientific expedition, led by Wilhelm FILCHNER. The original intention was to use two ships, one sailing to the WEDDELL SEA, the other to the ROSS SEA, to land shore parties who would attempt to cross the continent to determine if the two seas were linked by an iced-over channel or separated by continental land mass. In the event, there was only enough money to purchase one ship. The modified aim included an attempt to reach the SOUTH POLE from the WEDDELL SEA.

A Norwegian polar vessel, the *Bjorn*, was obtained and refitted, partly under the supervision of Ernest SHACKLETON. Renamed *Deutschland*, it sailed from Bremerhaven on 4 May 1911 to SOUTH GEORGIA and the SOUTH SANDWICH ISLANDS and thence to the Weddell Sea. The expedition sailed along part of COATS LAND to Vahsel Bay on the edge of the FILCHNER ICE SHELF. Here they began establishing their base; however, on 18 February 1912 the ice broke up, carrying away the newly erected HUT, all but one DOG, and PONIES, all of which had to be rescued. On 6 March the ship became ice-bound and began an aimless drift that lasted eight months. During this time much valuable research was carried out, in particular observation of the behaviour of PACK ICE. On 23 June two sledging parties set off to discover Benjamin MORRELL's 'New Greenland', but after reaching 70°32'S concluded it did not exist. After breaking free from the ice, the *Deutschland* reached SOUTH GEORGIA on 19 December, then sailed home.

Germany The first German expedition to Antarctica, the 1882–83 GERMAN INTERNATIONAL POLAR YEAR EXPEDITION, established a station on SOUTH GEORGIA. Two further expeditions in the first decades of the 20th century explored the INDIAN OCEAN sector of the ANTARCTIC CONTINENT

*Right: Antarctic isopods (*Glyptonotus antarcticus*) are giants, and fill the niche usually occupied by crabs in other marine environments.*

around 90°E, and discovered the FILCHNER ICE SHELF. Under the Third Reich, German involvement in WHALING increased and the government sent an aerial reconnaissance and MAPPING expedition to DRONNING MAUD LAND with the intention of making a TERRITORIAL CLAIM, prompting NORWAY to formally claim the territory.

After World War II both East Germany (GDR) and West Germany (FRG) became involved with Antarctica. Scientists from GDR were involved with USSR expeditions from 1959. GDR acceded to the ANTARCTIC TREATY in 1974 and became a CONSULTATIVE PARTY in 1987. FRG acceded to the treaty in 1979, established a permanent base—the NEUMAYER STATION on the EKSTRÖM ICE SHELF—in 1981, and became a consultative party the same year. Both the GDR and FRG were invited to the final meeting of the negotiations for the CONVENTION ON THE CONSERVATION OF ANTARCTIC MARINE LIVING RESOURCES (CCAMLR) due to their interest in FISHING, especially for KRILL.

When the two German states merged in 1990, their place in ANTARCTIC TREATY SYSTEM meetings was taken by the united Germany.

Getz Ice Shelf Over 480 km (298 miles) wide, the Getz runs along the coast of MARIE BYRD LAND and several large islands—including SIPLE ISLAND—are embedded within it.

Giant isopod (*Glyptonotus antarcticus*) Growing up to 20 cm (8 in) long, and weighing up to 70 g (2 oz), these Antarctic isopods are giants of their kind. Distantly related to the common woodlouse, isopods are CRUSTACEANS, a mainly aquatic class of arthropods. *Glyptonotus antarcticus* is found from intertidal waters to depths of 790 m (2600 ft) throughout the waters of Antarctica, the SOUTH SHETLAND, SOUTH ORKNEY and SOUTH SANDWICH ISLANDS and SOUTH GEORGIA.

The giant isopod is both a large predator and

a scavenger of the BENTHOS, occupying the niche of crabs in other ocean environments. It, in turn, is preyed on when young by notothenioid FISH and OCTOPUS. When fully grown, however, it is protected by its armoured, spiny body. An omnivore, it will eat whatever food it finds, including smaller members of its own species. Possessing large, powerful mouth parts, it can dine on hard animals such as BRITTLESTARS, as well as gastropod MOLLUSCS, SEA URCHINS, KRILL, polychaete worms and carrion.

Glyptonotus antarcticus lives five to eight years and moults every 100 to 730 days. As an adaptation to the harsh environment of Antarctic waters, it incubates and raises its young in a brood pouch; the developing young eat nonviable eggs and maternal secretions. Breeding is non-seasonal and young are released throughout the year, after which the female usually—but not always—dies. In addition to the giant isopod, there are about 40 other species of isopod in Antarctic waters, most of which are small.

Giant petrel See SOUTHERN GIANT PETREL and NORTHERN GIANT PETREL.

Glaciers Great flowing 'rivers' consisting mostly of ICE with air bubbles, sometimes water and often some rock debris. Most Antarctic glaciers are located on the edges of the ICE SHEETS and in the DRY VALLEYS. The LAMBERT GLACIER, which flows into the AMERY ICE SHELF in EAST ANTARCTICA, is the largest glacier in the world.

Glaciers force their way over the Antarctic landscape, eroding and polishing rock surfaces, plucking at mountain sides and pushing up or depositing piles of debris, called MORAINES. The most dynamic landforms on the continent, they link the past to the present by shaping the landscape and leaving traces of ancient Antarctica in their wake. Within layers of accumulated ice,

Above: The Wright Glacier, in the Dry Vallleys region, is fed from the Polar Plateau via the Airdevronsix Ice Falls.

glaciers conceal evidence of past CLIMATES, and today the balance between their advances and retreats is our most effective indicator of climate change.

A glacier forms when enough SNOW survives summer melt to form a névé, or accumulation zone. More snowfall than is lost by melting or evaporation is accumulated in this zone, feeding the glacier. The snow at the base of the zone is compacted into ice by the weight of new snowfall and eventually becomes heavy enough to flow downhill under gravity.

The point when ABLATION (or net loss of snow) is greater than the supply of ice is called the 'terminal' or 'snout' of the glacier. The terminal will retreat in warmer climatic conditions, when less ice is supplied to feed the glacier, and advance in colder or snowier conditions. At the coast, the glacier's terminal may be a 'tongue' of ice that protrudes into the ocean. Periodically parts of this ice tongue will shear off to form ICEBERGS. In early 2000, the Ninnis Glacier, which flows from the EAST ANTARCTIC ICE SHEET, lost its entire tongue in just a few months, thus changing the shape of the eastern Antarctic coastline.

'Polar', or 'dry-based', glaciers occur when the total thickness of ice is below melting point through all seasons. In Antarctica, these are generally the small glaciers of the DRY VALLEY region. At the base of these glaciers, the ice is frozen or welded to the underlying BEDROCK or sediment. Any sliding at the base is almost impossible and the glacier can only move by 'deformation', which occurs when ice grains shear past each other, usually in a salt-rich layer of 'amber ice' at the bed of the glacier. The ice creeps forward slowly, only

about 10 to 50 cm (4 to 20 in) per year. It is estimated that it takes more than 5000 years for snow to turn to ice and travel the full length of a small polar glacier.

The large outlet glaciers that drain the Antarctic ICE SHEET and the glaciers on the ANTARCTIC PENINSULA are 'subpolar', or 'polythermal'. The ice is thick enough relative to the surface air temperature, and the weight of the glacier enough to generate sufficient shear stress or friction for melting point to be reached at its base, creating a thin layer of water; this allows the glacier to slide forward or move on its rocky basal bed. These partly wet-based glaciers move more quickly than dry-based ones. When a wet basal zone refreezes, the ice plucks sediment from the bed to form a thick 'dirty' layer of basal ice.

Glaciological research Early glaciological studies were simply recorded observations. Robert SCOTT's second expedition was the first to carry out a glaciological programme: geologists Charles Wright and Raymond PRIESTLEY made detailed observations of the glaciers in VICTORIA LAND. On the joint NORWEGIAN-BRITISH-SWEDISH EXPEDITION in 1949–52, glaciologists began studying glaciological processes rather than simply documenting what they saw. By digging SNOW pits they were able to document layers of seasonal snow accumulation and the gradual compaction of snow to ICE.

Until the INTERNATIONAL GEOPHYSICAL YEAR (IGY) any studies of snow and ice in Antarctica had been at very specific locations. One of the goals of the IGY was to calculate the total volume, or 'mass balance', of ice in Antarctica. Data

were collected from across the continent to map snow accumulation rates and the direction of flow of ice. By simply measuring snow accumulation against stakes on the ice surface, it was possible to make long-term measurements of snowfall, and estimate the depth of ice over Antarctica.

Glaciologists now track the behaviour of the Antarctic ICE SHEET using SATELLITE surveying techniques, and by measuring the movement over time of markers placed in the ice surface. Ice thickness is measured using ECHO SOUNDERS, which reflect a signal off the BEDROCK and back through the ice, and ice CORE SAMPLES, which provide an archive of environmental conditions over millions of years, are extracted by DRILLING. Computer modelling, using data gathered from the 'field', is being developed to determine whether Antarctica is gaining or losing ice and to predict the impact of future CLIMATE change on Antarctica.

One of the most challenging problems facing Antarctic glaciologists is to determine what causes the rapid movement of ICE STREAMS. These are inextricably linked to ice loss from the continent and sea-level rise. Scientists are probing the beds of ice streams using RADAR and by drilling through the ice to extract samples from the base of these streams.

Global Atmospheric Research Programme (GARP) (1967–82) A multinational global science project to study the interaction between the OCEAN and the ATMOSPHERE. It led to better weather forecasting and in-depth understanding of the Earth's CLIMATE, particularly of large weather systems. The First GARP Global Experiment

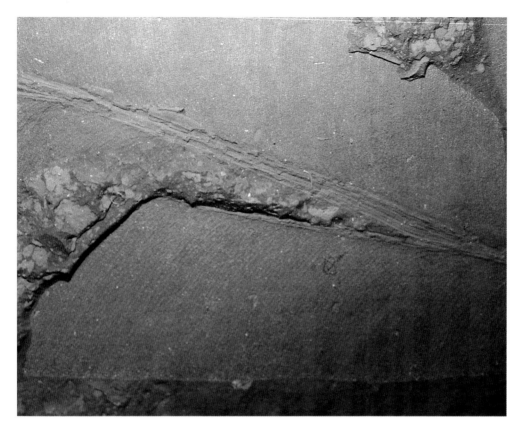

Above: Glossopteris *leaf fossils are a remnant from the flora of Gondwana.*

(FGGE) was carried out near Antarctica, when over 300 buoys were released into the SOUTHERN OCEAN equipped with instruments to measure surface temperature and pressure. Their positions were tracked by SATELLITE—in the winter of 1979 there was no more than 500 km (310 miles) between buoys over most of the ocean—and they provided data from areas that had not then been reached by ships, allowing scientists to build a clearer picture of the dynamics of the Southern Ocean. GARP operated alongside POLEX, a Soviet oceanographic experiment.

Global communications system In Antarctica a COMMUNICATIONS system provided through a network of low-earth orbit SATELLITES allows voice, data, fax and paging services between the continent and the rest of the world. This has proved useful for emergency operations, such as the winter evacuation of Dr Ronald Shemenski from the SOUTH POLE in 2001.

Glossopteris A deciduous conifer with thick, fleshy, fibrous, tongue-shaped leaves. The remains of FORESTS of *Glossopteris* have been fossilized in COAL seams of Permian age (about 245 to 286 million years ago) in the TRANSANTARCTIC MOUNTAINS. In 1912, Edward WILSON, a member of Robert SCOTT's polar party, collected FOSSILS of *Glossopteris* on the ill-fated return journey from the SOUTH POLE.

Glossopteris fossils were first found in INDIA and AUSTRALIA and have also been discovered in SOUTH AFRICA and South America. Leaf fragments have subsequently been found in marine rocks in eastern Southland, NEW ZEALAND. The discovery of these fossils on different continents is considered to be evidence of CONTINENTAL DRIFT and of the original composition of GONDWANA.

Glycoproteins Molecules made of repeating units of sugar and amino acids. In the supercooled waters of the Antarctic, FISH such as ANTARCTIC COD generate ANTIFREEZE glycoprotein (AFGP), which circulates in their blood and prevents ICE crystals from forming. Most proteins work by interacting with other proteins. AFGP is unusual in that it interacts directly with ice.

SOUTHERN OCEAN water temperatures are very low—for instance, the sea water in MCMURDO SOUND has a nearly constant mean annual temperature of –1.9°C (28.7°F)—and cold-blooded fish are constantly at risk of ice crystals forming in their blood. Once a single crystal forms, others can grow around it, setting off a potentially deadly chain reaction. Unlike the way antifreeze works in cars, fish antifreeze does not lower the actual freezing point of the fish's blood; instead, glycoprotein inhibits the growth of the ice crystals. When ice begins to form, AFGP surrounds the crystals, binding to their sides and stopping further crystal growth. AFGPs are not found within cells but are synthesized in the liver of Antarctic fish, then secreted into the circulatory system.

Goggles Used to protect eyes against SNOW BLINDNESS. Early explorers tried various types: Robert SCOTT opted for conventional small, round glass goggles, which tended to mist up easily and did not provide adequate protection. Roald AMUNDSEN preferred goggles based on an Eskimo model, which were developed by Frederick COOK for use in the Arctic. They had a wide visor with ventilation slits at the top, and lens made from photographic filters with a wide angle of vision. The modern version is glacier glasses, which feature UV-filtering lens and leather flaps to block out light from the sides or ski goggles, which cover and protect the entire eye area.

Gold Traces of gold, but no large deposits, have been found in Antarctica. The largest quantity of gold in the southern region probably lies beneath the ocean off the coast of the AUCKLAND ISLANDS, in the SHIPWRECK of the *General Grant*. This ship was returning to London from the Australian goldfields in 1866 when it was wrecked: officially, it was carrying 70 kg (152 lb) of gold, although rumours suggest that it may have been as much as 8 tonnes (8 tons). Over the decades, several attempts have been made to find and salvage the wreck; in 2001, a NEW ZEALAND entrepreneur was mounting a modern-day treasure hunt to try to recover the cargo.

Gondwana With Antarctica as its keystone, Gondwana linked the world's southern hemisphere continents into one giant land mass. This ancient supercontinent began to break apart about 183 million years ago with violent eruptions, strewing lava across the landscape, and the continents began their slow drift to different corners of the globe. Edward Suess first proposed the concept of a southern supercontinent in 1885. He named it Gondwana after an area in central INDIA, inhabited by a tribe called the Gonds; FOSSILS found in this area matched those from other continents, suggesting the land masses must have been linked in the past. However, Suess could not explain the mechanism by which the continents could have split apart and drifted such large distances. It wasn't until the theory of CONTINENTAL DRIFT, which had first been proposed in the early 1900s, was developed by German scientist Alfred Wegener in 1912 that there was a viable explanation of the break-up of Gondwana. Wegener

Below: Goggles that filter out the harsh glare from the snow and ice are essential for preventing snow blindness.

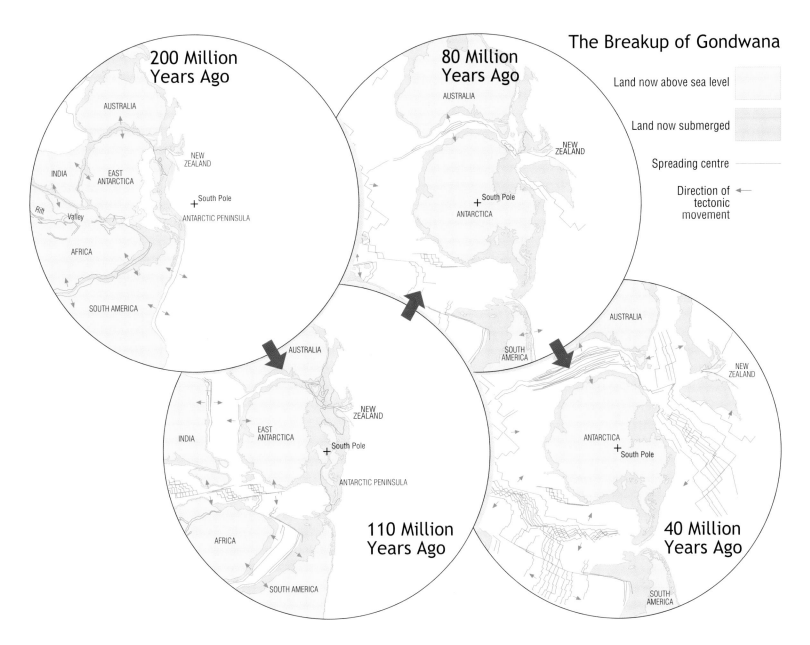

The Breakup of Gondwana

200 Million Years Ago

AUSTRALIA

NEW ZEALAND

INDIA

EAST ANTARCTICA

+ South Pole

ANTARCTIC PENINSULA

Rift Valley

AFRICA

SOUTH AMERICA

80 Million Years Ago

AUSTRALIA

NEW ZEALAND

+ South Pole

ANTARCTICA

SOUTH AMERICA

Land now above sea level

Land now submerged

Spreading centre

Direction of tectonic movement

110 Million Years Ago

AUSTRALIA

NEW ZEALAND

INDIA

EAST ANTARCTICA

+ South Pole

ANTARCTIC PENINSULA

AFRICA

SOUTH AMERICA

40 Million Years Ago

AUSTRALIA

NEW ZEALAND

ANTARCTICA

+ South Pole

SOUTH AMERICA

described a supercontinent, which he named 'Pangaea' ('All-lands'), that had been dismembered as continents drifted apart over the Earth's molten core. This concept was supported by geological evidence from Alexander du Toit in 1937, who stated that, 'The role of the Antarctic is a vital one ... the shield of EAST ANTARCTICA constitutes the "key piece" around which, with wonderful correspondences in outline, the remaining "puzzle pieces" of Gondwanaland can with remarkable precision be fitted.' These ideas were received with widespread scorn by the scientific community. It was not until the 1960s that GEO-LOGICAL RESEARCH in Antarctica provided conclusive evidence to show that a supercontinent had indeed existed. About 280 million years ago, thick layers of sediment were laid down over Gondwana during an ICE AGE. Geologists have mapped cross-sections of the sediment, finding perfect matches in layering between Antarctica and other continents. Also, fossils of species that would have been incapable of crossing oceans have been found in Antarctica, and matched with similar fossils from other continents—further proof that Antarctica's origin is Gondwana.

The break-up of Gondwana began during the Jurassic period (208 to 144 million years ago), when DINOSAURS still roamed the world. About 180 million years ago the Antarctic landscape was one of plains, rivers, lakes and swamps, with luxuriant stands of vegetation. Then, gradually, Antarctica sheared away from surrounding land masses.

A deep seaway formed between India and Antarctica about a 100 million years ago. About 40 million years ago AUSTRALIA and NEW ZEALAND broke away, and the final severing of South America from the ANTARCTIC PENINSULA occurred about 23 million years ago.

This last event opened up the DRAKE PASSAGE, and a strong circulation pattern was established around Antarctica. Deep water began to flow as the CIRCUMPOLAR CURRENT, which cut Antarctica off from warmer water and WEATHER to the north. Antarctica's CLIMATE grew colder, and light and dark cycles became more seasonal. Plants and animals that inhabited the continent became highly specialized to survive the harsh conditions or they became extinct. The once fertile continent was swept into an ice age.

Gough Island Some 340 km (211 miles) southeast of TRISTAN DA CUNHA in the Atlantic Ocean, Gough Island is home to vast numbers of PEN-GUINS and is the most northerly breeding ground for ELEPHANT SEALS and WANDERING ALBATROSSES. It is barricaded by steep cliffs, especially on the western side, where they reach up to 460 m (1500 ft). The vegetation is lush and green, due in part to the high rainfall (3m/10 ft) per year, and relatively high average temperatures of around 12°C (54°F). The island's resources were heavily exploited throughout the 19th century: in 1881, for example, one group reported taking 8 tonnes (8 tons) of guano, 4000 penguin eggs and 151 skins of FUR SEALS. The island is now a strategic base for meteorological observations. In 1995, Gough Island was designated the first subantarctic World Heritage Site.

Graham Land The northern section of the ANTARCTIC PENINSULA, bounded on the east by the LARSEN ICE SHELF and the WEDDELL SEA and on the west by a string of islands, including ANVERS and BISCOE. First charted by Jules-Sébastien DUMONT D'URVILLE in the summer of 1838.

Above China's Great Wall Base was established in 1985 on King George Island in the South Shetland Islands.

Grasses Only one grass species grows below 60°S in Antarctica itself: the ANTARCTIC HAIRGRASS (*Deschampsia antarctica*), which is one of only two FLOWERING PLANTS to survive in Antarctica. It grows in small clumps near the western coastline of the ANTARCTIC PENINSULA, and is also found, growing more vigorously, in the SOUTH SHETLAND and SOUTH ORKNEY ISLANDS.

The warmer SUBANTARCTIC ISLANDS typically have a coastal fringe of tall TUSSOCK grass (*Poa* spp.), which is frequented by BIRDS and SEALS. Thousands of burrowing PETRELS provide fertiliser to aid tussock growth. On SOUTH GEORGIA these tussocks can reach 2 m (6.5 ft) in height, and short grassland often covers dry, shallow soils.

Graves The grave of Nicolai Hanson, zoologist on Carsten BORCHGREVINK's 1898–1900 SOUTHERN CROSS EXPEDITION and the first person to be buried on the ANTARCTIC CONTINENT, is located on the top of Cape ADARE in a tomb Hanson's colleagues excavated by dynamite. The crash site of the 1979 Air New Zealand DC10 on Mount EREBUS has been declared a tomb by the ANTARCTIC TREATY members. Other burial sites are inaccessible, have never been relocated or were impermanent: for instance, Douglas MAWSON buried Xavier Mertz in his sleeping bag on a sledging trip in GEORGE V LAND in January 1913.

Crosses have been erected in memory of many who have died on the continent. Crosses can be seen on OBSERVATION HILL on ROSS ISLAND for Robert SCOTT and the four other members of his polar party who perished on the return from the SOUTH POLE; and on Wind Vane Hill, Cape EVANS for three members of Ernest SHACKLETON's ROSS SEA party who died in 1916. Shackleton himself is buried in SOUTH GEORGIA, in the cemetery at Grytviken, he died of a heart attack, aged 47, in his cabin aboard the *Quest* on 5 January 1922.

Gravity wind The most common source of WIND in Antarctica, its airflow is driven by gravity, rather than the pattern of atmospheric WEATHER. At the POLAR PLATEAU an inversion layer traps cold air at the surface. Heavier than the air above, this layer flows downhill as a light breeze known as an 'inversion wind'. This is the beginnings of the KATABATIC WIND, the ferocious gravity wind that surges from the Plateau downwards to the Antarctic coastline.

Grease ice Forms in open water from ICE crystals, giving the surface a greasy sheen with low reflectivity. It also occurs in POLYNYAS when FRAZIL ICE is herded into elongated plumes by WIND and WAVE action. When plumes of grease ice have PANCAKE ICE at the head, they are termed 'tadpoles.'

Great Ice Barrier James Clark ROSS's name for the ROSS ICE SHELF.

Great Wall Base Built by CHINA in 1985 on KING GEORGE ISLAND in the SOUTH SHETLANDS. The base has accommodation for 14 people to OVERWINTER, and for up to 22 in summer.

Great-winged petrel (*Pterodroma macroptera*) Shy and solitary, great-winged petrels rarely follow ships, preferring to stay close to land and displaying their strong, impetuous flight pattern, often wheeling in broad arcs. Their wingspan is 97 cm (38 in) and they are about 41 cm (16 in) long. These GADFLY PETRELS are almost circumpolar in distribution: they are seen in the subantarctic region but rarely around Antarctica. They are blackish-brown with stout black bills, dark feet, and long wedge-shaped tails.

Breeding takes place in winter on a variety of islands, including TRISTAN DA CUNHA and GOUGH ISLAND. Autumn-winter breeders, the single egg is laid in a burrow in May to June and the chicks fledge around November or December, just as the summer-breeding petrels are starting to breed. They move to and from their burrows at night to reduce the risk of being preyed upon by birds such as the ANTARCTIC SKUA and are night feeders, feeding on SQUID, FISH and CRUSTACEANS.

Greater Antarctica Another name for EAST ANTARCTICA, which makes up over two-thirds of the ANTARCTIC CONTINENT.

Greater sheathbill (*Chionis alba*) See SNOWY SHEATHBILL.

Greenhouse effect Warming of the ATMOSPHERE, caused by greenhouse gases. In the same way as a glass roof traps heat in a greenhouse, the gases let short-wave SOLAR RADIATION pass through to the surface, but prevent emitted long-wave radiation (from the Earth's surface) escaping. This is an essential natural process for supporting life beneath the 'roof', but the high levels of greenhouse gases produced by human activity, such as CARBON DIOXIDE, CHLOROFLUOROCARBONS (CFCs), methane and nitrogen oxides, raise surface temperatures beyond normal levels. In an extreme scenario, this global warming may cause the Antarctic ICE SHEET to lose ICE, which could raise sea levels by perhaps 3 m (10 ft) over the next 1000 years.

The study of air bubbles in ice cores has shown that greenhouse gas concentrations have been increasing steadily over the past two centuries, and that they are now at higher levels than at any time in the last 160,000 years. Although global warming and the OZONE HOLE have been linked, in fact they are two distinct processes: they are connected in that CFCs are greenhouse gases, and greenhouse gases stop radiation reaching the stratosphere (so that stratospheric temperatures remain lower for longer and more ozone is destroyed).

The break-up of the LARSEN ICE SHELF in 1995 was widely reported in the media, with headlines such as 'Worries about sea level as Antarctic glacier collapse' and 'Global warming sharply reduces Antarctic ice shelves'. However, parts of the ICE SHELVES constantly break away as large ICEBERGS around Antarctica, as part of a natural cycle, and there is no indication that they will collapse completely. The Larsen Ice Shelf is in the ANTARCTIC PENINSULA, part of the continent that is much warmer than the ice shelves on the ROSS SEA.

Greenpeace A pressure group, founded in 1971 and represented in over 30 countries, Greenpeace aims to protect biodiversity, prevent POLLUTION and end nuclear threats. It has lobbied on issues such as commercial WHALING in Antarctic waters and the banning of CHLOROFLUOROCARBONS, and has sailed vessels into the SOUTHERN OCEAN to protest against illegal, unregulated and unreported FISHING.

Greenpeace established the year-round World Park Base at Cape EVANS in January 1987 to monitor scientific bases and investigate the effects of human activities on the Antarctic ECOSYSTEM. The base was dismantled and removed in 1992.

Greenpeace also spearheaded an international campaign in the 1980s to prevent the CONVENTION ON THE REGULATION OF ANTARCTIC MINERAL RESOURCES (CRAMRA) coming into force; in its stead, the organization promoted a ban on all MINING, which was enacted for 50 years with the signing of the MADRID PROTOCOL in 1991. It continues to push for Antarctica to be designated a WORLD PARK. Greenpeace works with the ANTARCTIC AND SOUTHERN OCEAN COALITION (ASOC).

Greens Open areas of TUSSOCK found on SUBANTARCTIC ISLANDS. They are favoured habitats of WANDERING ALBATROSSES in search of mates. The birds gather on tussock greens that are large enough for them to spread their magnificent wings in dramatic courtship displays. Usually a green will contain just one female and the males who are hoping to impress her, but up to 13 young birds have been seen displaying together.

Grey petrel (Procellaria cinerea) Winter-breeding SHEARWATERS, they lay their single eggs in burrows on a number of SUBANTARCTIC ISLANDS including CAMPBELL ISLAND and the ANTIPODES. These large, solidly built PETRELS have dark underwings, white breasts and wedge-shaped tails. Their uniformly grey upperparts can appear dark at sea, and they have pale green or horn-coloured bills, and wingspans of 98 cm (39 in).

The distribution of these usually solitary birds is circumpolar and they are a familiar sight in subantarctic waters. High-flying birds, their ALBATROSS-like gliding is often interspersed with rapid wing-beats. Although little is known about their diet, grey petrels are renowned for pursuit-diving from a height of 5–10 m (16–33 ft) to catch prey. They sometimes gather in small groups to follow WHALES and FISHING boats.

Grey-backed storm petrel (Garrodia nereis) These STORM PETRELS are summer breeders on islands such as the FALKLANDS, SOUTH GEORGIA, Îles KERGUÉLEN, Îles CROZET and on several SUBANTARCTIC ISLANDS near NEW ZEALAND. They nest among TUSSOCKS and ROCKS. Although they are circumpolar and dispersed widely throughout subantarctic and subtropical waters, little is known about their range. They are small—only 17 cm (7 in) long—with dark upperparts, white bellies, grey rumps and white wings with conspicuous black leading margins.

Grey-headed mollymawk (Thalassarche chrysostoma) These mollymawks are distinguishable by their wholly grey head and black bill that has narrow yellow stripes on both the upper and lower mandibles. The white underwings are lined with black, especially on the leading edges, and the tail is grey. The birds are similar in size to BLACK-BROWED MOLLYMAWKS with a wing-span of 81 cm (32 in).

They breed every two years and nest in colonies on many of the SUBANTARCTIC ISLANDS, including SOUTH GEORGIA, MACQUARIE and CAMPBELL ISLANDS, and islands in the south INDIAN OCEAN. Nests are built on cliff edges and colonies may be shared with black-browed mollymawks. Both parents are involved in raising the single chick. The birds are circumpolar: they tend to fly alone and remain in subantarctic seas for most of the year, where they feed on FISH, SQUID and CRUSTACEANS. They do not usually follow FISHING vessels, yet will escort WHALES.

Grove Mountains Also known as Grove Nunataks or South Eastern Nunataks. A mountainous region covering an area of approximately 65 by 30 km (40 by 18½ miles), the Grove Mountains lie east of the Mawson Escarpment, which juts into the AMERY ICE SHELF in PRINCESS ELIZABETH LAND. They were named after Squadron Leader I L Grove, an Australian Airforce pilot, who landed there in November 1958.

Gulls (Laridae) Members of this family of birds are found everywhere in the world, from the Arctic to the Antarctic, and include eight genera and 42 to 44 species. The gulls are strongly represented in the Northern Hemisphere, and only one species—the KELP GULL, also known as the DOMINICAN or BLACK-BACKED GULL—is found in the Antarctic.

*Below: Grey-headed mollymawks (*Thalassarche chrysostoma*) nest on the cliffs of subantarctic islands.*

h

Haakon VII Sea Off the coast of DRONNING MAUD LAND, named by a 1929–30 Norwegian party led by Hjalmar RIISER-LARSEN for the incumbent king. In the first half of the 20th century, Haakon VII Sea was frequented by the Norwegian WHALING fleet.

Hägglunds Vehicles built by the Swedish military for Arctic conditions, Hägglunds have linked fibreglass cabs and rubber tracks, which float if the vehicles fall through SEA ICE. They have a top speed of 50 km (31 miles) per hour, can carry a team of four or five, a week's provisions and fuel. The Hägglunds are used extensively by a number of national Antarctic programmes.

Hairgrass See ANTARCTIC HAIRGRASS.

Hallett Station This base was established on the edge of the ROSS SEA during the INTERNATIONAL GEOPHYSICAL YEAR, in a joint USA–NEW ZEALAND operation. The site was a PENGUIN rookery, and 6000 birds were relocated. Any birds attempting to return to their breeding site were barricaded out with fuel drums.

The station was occupied year-round until 1964, when the laboratory was destroyed by FIRE. It was abandoned until the 1980s, when most of the buildings and equipment were removed. Recolonization by penguins has been slow—the compacted ground makes it difficult for them to find stones with which to make nests. In a 1994–95 clean-up, 100,000 litres (22,000 gallons) of fuel were removed from tanks and drums, but widespread debris and WASTE remained.

Halley, Edmond (1656–1742) English astronomer and mathematician, regarded as the founder of geophysics. Halley was educated at Oxford University, and published three astronomical papers while still an undergraduate. On voyages to the ATLANTIC he made the first catalogue of stars of the Southern Hemisphere and described his theory of magnetic variation. He also made significant contributions to cometary astronomy, including observing and predicting the future path of the comet named after him. He paid for the publication of Isaac Newton's *Principia Mathematica* in 1687.

From 1698 to 1700 he commanded the *Paramore* on two expeditions undertaken for scientific purposes. His instructions from the British Admiralty were to observe variations in compass readings in the South Atlantic, to determine accurate latitudes and longitudes for his various ports of call, and to search for *TERRA AUSTRALIS INCOGNITA*. Having crossed the ANTARCTIC CONVERGENCE on 1 February 1700, at 52°24'S, he became the first known person to sight—and sketch—tabular ICEBERGS. Cold and stormy weather, and fears of colliding with an iceberg, caused the expedition to return north. In 1720 he was appointed England's Astronomer Royal.

Halley Base Halley Base was established by BRITAIN during the INTERNATIONAL GEOPHYSICAL YEAR and is still occupied. Named after Edmond HALLEY, the base is located on the Brunt Ice Shelf in COATS LAND near an EMPEROR PENGUIN colony. The current base is the fifth to occupy this location, the first four having been abandoned after being crushed by overlying ice. Research activities include atmospheric sciences, surveying, geology and glaciology. Halley Base operates throughout the year, with around 15 people OVERWINTERING, and a maximum population of 65.

Halos These OPTICAL PHENOMENA occur when light from the sun or moon shines through tiny ICE crystals high in the ATMOSPHERE. They are more common over Antarctica's POLAR PLATEAU than anywhere else in the world. An indication of bad WEATHER, halos are formed from water vapour that is transported into the upper atmosphere by a low-pressure system. When light is scattered through the vapour, it may produce FOG BOWS. The vapour cools and condenses into solid ice, which results in wispy cirrus clouds moved ahead of the cold weather by high WINDS. Light refracting and reflecting on the ice crystals produces rings and arcs.

The most common halo over Antarctica is the 22° HALO: it is formed by light refracted by plate-shaped crystals. Bright luminous halos or spots—called PARHELION, SUN DOGS or MOCK SUNS—often form on either side of the sun, occurring when the longest edges of ice crystals hang vertically in the sky. Optical displays are particularly dazzling when these effects occur together. On a sledging journey, Edward WILSON described 'no less than

Below: Originally built by the Swedish military for Arctic conditions, Hägglunds are designed to float if they sink through sea ice.

Above: Helicopters have become an essential mode of transportation in Antarctica, particularly for moving field equipment and supplies.

nine mock suns ... and arcs of 14 or more different circles, some of brilliant white light against a deep blue sky, others of brilliant rainbow.'

The colours of the halos depend on the way that light is scattered through the crystals. When light is simply reflected off the tiny flat surfaces, the halo will appear white because colour bands in the light do not separate. However, when the light passes through crystals, beams of light are bent, or refracted, by the ice and the colours are split into different wavelengths, which are refracted at different angles. When the sun's rays are transmitted in a relatively straight line, the centres of halos and parhelion are usually dark and the inner edges outlined in red; this fades into yellow, green and white, and finally changes to blue on the outer rim.

Hanssen, Helmer (1870–1956) Norwegian explorer, born in the Vesterålen Islands. He took part in Roald AMUNDSEN's 1903–06 expedition in the *Gjoa*, on which the north-west passage was navigated for the first time.

He accompanied Amundsen on the 1910–12 NORWEGIAN ANTARCTIC EXPEDITION, and acknowledged that 'our goal, and only goal, was to reach the south pole'. He was in charge of the supply depot during the first winter. Hanssen was a member of the first party that reached the SOUTH POLE on 14 December 1911, on which his expert dog-handling skills were invaluable.

In 1918 he joined Amundsen's expedition in the *Maud*, navigating the Arctic's north-east passage. He later worked as a polar advisor to a film company and for the Norwegian customs service, and in 1936 published the book *Voyage of a Modern Viking*.

Hassel, Helge (1876–1928) Norwegian explorer. He had already served on the *FRAM* under Fridtjof NANSEN, when Roald AMUNDSEN asked him to join the 1910–12 NORWEGIAN ANTARCTIC EXPEDITION on the same ship. Known as the 'Managing Director of Framheim's Coal, Oil and Coke Company Ltd', he was responsible for maintaining fuel supplies at FRAMHEIM. Known for his adaptability—he was the expedition's navigator,

sail-maker, carpenter and saddler—Hassel was a member of Amundsen's party that reached the SOUTH POLE on 14 December 1911.

After the expedition, along with Helmer HANSSEN, he worked for the Norwegian customs service. Hassel died on a visit to Amundsen in 1928, a few months before his old leader went missing in the Arctic.

Heard Island Named in 1853 by John Heard, captain of an American sealer, Heard Island lies near the ANTARCTIC CONVERGENCE and 80 percent of it is covered in ice. It is made up of a circular VOLCANO, Big Ben, which last erupted in 1992. Like the neighbouring MCDONALD ISLANDS, Heard is a MARITIME ISLAND controlled by SOUTHERN OCEAN weather. There are no trees or shrubs, and the isolated patches of vegetation consist mainly of TUSSOCK, waterlogged peat bogs and 'cabbage' plants. Long Beach, on Heard's southern coast, is home to one of the largest colonies of MACARONI PENGUINS. Elephant Spit, a sand-shingle beach extending 7 km (4 miles) east, is a breeding site for about 40,000 elephant seals. It was also an easy hunting ground during the SEALING era.

In 1947 SOVEREIGNTY for Heard Island passed from BRITAIN to AUSTRALIA, and a scientific base was established by the AUSTRALIAN NATIONAL ANTARCTIC RESEARCH EXPEDITIONS (ANARE), which was closed in 1955.

Helicopters Used for the first time in Antarctica during OPERATION WINDMILL in 1947–48, helicopters are now an indispensable part of Antarctic logistics. Both ship- and land-based helicopters are used to transport parties to remote field camps. They can carry large loads of EQUIPMENT and SUPPLIES, sometimes hung beneath the aircraft in slings. Their low-flying and hovering ability makes them invaluable for search and rescue operations, and for use on unstable SEA ICE.

Herbivores ANIMALS which feed on plant material. In Antarctica, marine herbivores include organisms of the ZOOPLANKTON, which feed on PHYTOPLANKTON. Terrestrial herbivores include PROTOZOA, ROTIFERS, NEMATODES, TARDIGRADES, SPRING-

TAILS and some MITES, all of which are found in the MOSS, ALGAE and LICHEN communities. Because most terrestrial primary production (photosynthesis) in Antarctica is confined to moss and lichen, there are no large herbivores such as BIRDS or mammals because there is insufficient plant material to support them.

Heritage Antarctica A coalition of national Antarctic heritage trusts that coordinates and promotes 'the restoration, preservation and protection of the structures, artefacts and records which reflect the history of human endeavour in Antarctica,' including HISTORIC SITES and HUTS.

Heroic Age (1895–1922) The period in which the unknown ANTARCTIC CONTINENT was extensively explored on land, its coastline and typography charted, and the SOUTH POLE reached. It begins with the ANTARCTIC EXPEDITION landing at Cape ADARE, and encompasses heroic adventures of survival, bravery and endurance, from the enforced wintering-over of the group led by Otto NORDENSKJÖLD in 1902–03, through the tragic death of Robert SCOTT's party after having been beaten to the South Pole by Roald AMUNDSEN, to Ernest SHACKLETON and the *ENDURANCE* expedition's extraordinary survival after shipwreck. Shackleton's death at SOUTH GEORGIA marks the end of the Heroic Age.

Hillary, Edmund (1919–) NEW ZEALAND mountaineer and explorer. Born in Auckland, New Zealand, Hillary began mountain climbing as a teenager. During World War II, he was a navigator in the Royal New Zealand Air Force. He worked as a beekeeper, and climbed mountains in the European Alps and the Himalayas. On 29 May 1953, Hillary and the Nepalese mountaineer

Below: Edmund Hillary, who led a team to the South Pole in tractors as part of the 1955-58 Commonwealth Trans-Antarctic Expedition.

Above: Ernest Shackleton's hut, below Mount Erebus, is one of around 75 historic sites listed under the Madrid Protocol.

Tensing Norgay became the first to reach the summit of Mount Everest.

Hillary led the ROSS SEA party of the 1955–58 COMMONWEALTH TRANSANTARCTIC EXPEDITION, headed by Vivian FUCHS, which aimed to make the first successful overland crossing of the ANTARCTIC CONTINENT, from the WEDDELL SEA to the Ross Sea. Hillary's team built SCOTT BASE, and successfully scouted out a route for Fuchs's Weddell Sea party, laying down supply depots. On 3 January 1958 Hillary's party became the first to drive with VEHICLES—farm TRACTORS—to the SOUTH POLE, arriving with just 76 litres (20 gallons) of fuel remaining. Fuchs's Weddell Sea party successfully crossed the continent in 99 days. Fuchs's and Hillary's book about the expedition, *The Crossing of Antarctica*, was published in 1958.

Hillary set up the Himalayan Trust in 1964, raising funds for the construction of hospitals, schools and airstrips in Nepal. In 1977 he led the first jet-boat expedition up the Ganges River and climbed the Himalayas to reach the river's source. From 1984 to 1989 Hillary served as New Zealand's High Commissioner to India. In 1999 he published his autobiography, *The View from the Summit*.

Historic sites Under the MADRID PROTOCOL, approximately 75 sites of historical significance are listed and earmarked for preservation. These sites include HUTS, memorial crosses, plaques and cairns and GRAVES.

Some of the oldest historic sites in the Antarctic region are those of the 18th- and 19th-century SEALING industry. The most important sites are the huts built by early expeditions as bases for exploration and shelter over the extreme Antarctic winters. Most historic huts are located in the ROSS SEA region or on the ANTARCTIC PENINSULA. Although cold temperatures and low precipitation in the Antarctic help to preserve historic relics, strong winds, snowdrifts and, in coastal regions, salt from sea spray can do much damage to BUILDINGS

that were, after all, only meant to be temporary.

Interest in conserving Antarctica's history was raised during the INTERNATIONAL GEOPHYSICAL YEAR. In the early 1960s expeditions sponsored by the NEW ZEALAND government restored three British huts on Ross Island. In the following years BRITAIN and the USA sent expeditions to restore and protect sites in the East Antarctic, MARIE BYRD LAND and Antarctic Peninsula regions. Today, organizations such as HERITAGE ANTARCTICA carry out regular preservation work.

Hodges, William (1744–97) English artist, the only son of a London blacksmith. Hodges was a pupil of William Shipley's School and was apprenticed to Richard Wilson, the English painter of landscapes in the style of French artist Claude Lorrain, before being appointed artist on James COOK's 1772–75 voyage of discovery. His paintings of the SOUTHERN OCEAN are the first illustrated records of the area; they include the *Resolution* and *Adventure* among ICEBERGS, 'Ice Islands', sailors taking blocks of ice aboard to be melted for drinking water, and numerous seascapes.

After his return, he was employed by the British Admiralty to oversee the engraving of his drawings in preparation for the publication of Cook's journal. His *Travels in India* recorded journeys from 1780 to 1783, and he undertook an extensive tour of Europe as far as St Petersburg. He exhibited at the Royal Academy until 1794, was named a Royal Academician in 1787 and became a banker in Devonshire, where, after the failure of his bank, he is said to have committed suicide by overdosing on opium.

'Holes in the Poles' theory In 1818 the American John Cleves Symmes announced his theory that the Earth is hollow, with concentric spheres within and holes at the poles (and he provided a certificate of his sanity to accompany it). His memorial in Hamilton, Ohio, states 'as a Philosopher, and the originator of Symmes Theory of Concentric Spheres and Polar Voids; He contended that the Earth is hollow and hospitable within.' He suggested that an expedition should investigate, an idea that was supported by some New England sealers and whalers, who petitioned Congress. Charles WILKES regarded this as the impetus for the UNITED STATES EXPLORING EXPEDITION of 1838–42. Symmes died in 1829, long before his theory was conclusively disproved.

Hooker, Joseph (1817–1911) British naturalist, the son of William Hooker, who was professor of botany at Glasgow University and first director of Kew Gardens. Hooker was 21 and had just graduated in medicine when he joined James Clark ROSS's 1839–43 expedition to Antarctica as botanist. Francis CROZIER was most impressed with him, writing that he was 'never idle, making perfect sketches of all he collects.' He collected samples of sea water and its microorganisms

Below Hooker's sea lion (Phocarctos hookerii) frequents subantarctic waters and is one of the world's rarest sea lions.

*Above: Humpback whales (*Megaptera novaeangeliae*), summer visitors to the Southern Ocean, are easily distinguished from other whales by their long pectoral fins.*

at different latitudes and longitudes, and was the first to recognize the significance of PHYTO-PLANKTON.

Hooker published the six-volume *Flora Antarctica*, which included an almost complete description of the botany of the SUBANTARCTIC ISLANDS, undertook botanical expeditions to North Africa, INDIA and North America, and became a friend and supporter of Charles Darwin. In 1855 he was appointed assistant to his father at Kew Gardens, and in 1865 succeeded him as director. In November 1893 he was appointed to the Antarctic Committee of the ROYAL GEOGRAPHICAL SOCIETY.

Hooker's sea lion (*Phocarctos hookerii*) Also known as the New Zealand sea lion, these otariid (eared) SEALS are found on the AUCKLAND ISLANDS, the SNARES and the CAMPBELL ISLANDS as well as on the NEW ZEALAND mainland. Hooker's sea lions move with agility on land, as well as in water. One of the world's rarest and most provincially contained sea lions, their population is estimated at around 12,000 animals. Males swim slightly further afield in search of food, but females remain close to their breeding locations throughout the year. Their diet includes FISH, CRUSTACEANS and CEPHALOPODS.

Males are incapable of breeding until they are around six years, and females become sexually mature at three to four years. Breeding usually takes place on sandy beaches and gestation is thought to be around 11 months. The life expectancy of Hooker's sea lions is at least 23 years.

Named for Joseph HOOKER, the sea lions were

killed for their pelts in the early 1800s. Although populations are recovering, some sea lions are caught in trawling nets: the New Zealand government closes the fishery when deaths exceed 80 per year.

Hope Bay On the northern tip of the ANTARCTIC PENINSULA, facing JOINVILLE ISLAND. Three members of the SWEDISH SOUTH POLAR EXPEDITION led by Otto NORDENSKJÖLD were stranded here over the winter of 1903, sheltering in a stone hut.

The site of a large ADÉLIE PENGUIN rookery, Hope Bay is where the first BIRTH in Antarctica took place, on 7 January 1978 at ESPERANZA STATION, which was established by ARGENTINA in 1951.

Horlick Mountains Consisting of the Wisconsin Range, Long Hills and Ohio Range, they lie west of QUEEN MAUD MOUNTAINS in the TRANSANTARCTIC MOUNTAINS. First sighted by BYRD'S SECOND EXPEDITION of 1933–35, the entire Horlick mountain group was surveyed by the USA between 1959 and 1964.

Richard BYRD named the mountains after William Horlick of Horlick's Malted Milk Company, who was a supporter of the expedition.

Hourglass dolphin (*Lagenorhynchus cruciger*) Hourglass dolphins are TOOTHED WHALES and are named for the distinctive white 'hour-glass' side markings. They are small—less than 2 m (6½ ft) in length—and are usually seen in open water, well away from land. Little is known about their biology.

Humpback whale (*Megaptera novaeangliae*) Humpbacks are plumper than other members of the family Balaeopteridae. Like other BALEENS, they follow the migration routine of southern summers in cold, productive waters at high latitudes, and winters in temperate or subtropical zones.

Humpbacks are relatively slow swimmers and are more likely to approach boats than other species, characteristics that made them easy prey for whalers, and favourites with whale-watching tourists. Because they stay close to coastlines, more is known about humpbacks than other whales.

Humpbacks have extremely large flippers that can measure as much as one-third of their total body length. The males are about 15 m (49 ft) long, and females slightly larger at 16 m (52 ft); both sexes can weigh up to 60 tonnes (60 tons). Humpbacks feed on small schooling FISH and KRILL.

Their breeding waters are warm (about 25°C/76°F) and are located off the coasts of AUSTRALIA, South America and SOUTH AFRICA. They migrate south in spring at about 15 degrees latitude per month. Although the groups mix in the Antarctic feeding grounds, there is a distinct segregation during migration: pregnant and non-lactating females arrive first, followed by immature males, mature males and lactating females with their calves. The pregnant females leave the feeding grounds last, having spent over six months there. At birth, calves are around 4–5 m (13–16½ft) long and weigh about 1350 kg (2977 lb).

The most acrobatic of the large whales, humpbacks are often seen making spectacular leaps and splashes: breaching, as this behaviour is known, sometimes involves catapulting out of, and clearing, the water's surface. Although it is not known exactly why humpbacks breach, one theory suggests that the resulting immense splashes are means of communicating with other whales. Humpbacks also communicate vocally, possessing a complex repertoire of rumbles, chirps and whistles. Scientists believe vocalising is mainly used by adult males to attract females and that different populations of humpbacks have different 'dialects'. Their slow cruising speed and stocky body structure made these whales prime targets for commercial WHALERS: in the years between 1949 and 1962 over 20,000 slaughtered humpbacks were processed in Australian shore stations. Populations are increasing but the population is still only a small percentage of what it used to be. Humpback whales are regarded as vulnerable.

Hurley, Frank (1885–1962) Australian photographer and explorer, born in Sydney, AUSTRALIA. He took up photography at 17 and became a partner in a postcard business.

Hurley was 24 when selected as PHOTOGRAPHER on the 1911–14 AUSTRALASIAN ANTARCTIC EXPEDITION. Together with Bob Bage and Eric Webb, he travelled inland by SLEDGE towards the SOUTH MAGNETIC POLE, taking photographs as they went. 'Although we did not reach the Magnetic Pole,' Hurley wrote, 'our records are unique and will form some of the most valuable scientific data of the expedition.'

He then joined the 1914–17 IMPERIAL TRANS-ANTARCTIC EXPEDITION. Shortly before the expedition ship, the ENDURANCE, sank in PACK ICE, Hurley dived underwater to rescue photographic plates. Together with Ernest SHACKLETON, he chose which to save and smashed over 400 glass negatives to ensure he could not change his mind: 'I had a painful hour,' he later wrote. He saved a camera and some film, and preserved the remaining photographic plates during five months on the ICE, the boat journey in wild seas to ELEPHANT ISLAND, and a winter on the island itself. On Elephant Island, Hurley put his metal-working skills to use to construct a blubber-burning stove.

Hurley's dramatic images of the *Endurance* being crushed by ice, and of the stranded men being rescued from Elephant Island on 30 August 1916, are some of the most famous of the 20th century. Greenstreet, one of the expedition members, said that Hurley was 'a warrior with his camera and would go anywhere or do anything to get a picture.'

In 1917 Hurley visited SOUTH GEORGIA to finish shooting a film and photographs for a book about the expedition. He attempted to repeat Shackleton's epic crossing of the island but, despite being well equipped, was unable to do so.

An official Australian war photographer in World Wars I and II, Hurley returned to Antarctica on the 1929–31 BRITISH-AUSTRALIAN-NEW ZEALAND ANTARCTIC RESEARCH EXPEDITION.

Huskies See DOGS and SLEDGING.

Huts Early explorers' huts were well constructed, and many still remain (in various states of disrepair) dotted around the Antarctic coastline. Constructed from wood, they were sealed against the elements, usually with tar. Some parties spent long WINTER months in these SHELTERS, and DOGS and PONIES were usually kept in 'lean-tos' on the lee of the hut. Conditions were cramped; there was no communication with the outside world; no running water and poor light and cooking facilities.

The oldest huts on the continent are the two built in 1899 at Cape ADARE by the BRITISH ANTARCTIC EXPEDITION. ROSS island is the location of three huts—at Hut Point, Cape EVANS and Cape ROYDS—the bases for Robert SCOTT's two expeditions and Ernest SHACKLETON's expedition of 1907–09 respectively. On the ANTARCTIC PENINSULA, the oldest huts date back to the SWEDISH SOUTH POLAR EXPEDITION which left behind shelters at HOPE BAY, PAULET ISLAND and SNOW HILL ISLAND.

Modern huts, on the other hand, can be flown into a field camp by HELICOPTER. They have more space and provide better shelter than a TENT, and can be constructed in a couple of hours.

Hypothermia If a person loses heat more quickly than his or her body can replace it, the core body temperature may drop. Cold temperatures combined with WINDCHILL are constant dangers in Antarctica and contribute to this dangerous condition, along with fatigue and hunger. Initially it may cause muscular weakness and other symptoms, such as a lack of coordination, disorientation, lethargy and irrational behaviour. It may prevent the sufferer from realizing anything is wrong, and if the body is not warmed up, hypothermia can quickly lead to unconsciousness and possibly death.

Below: Scott's hut at Cape Evans, like others of Antarctica's heroic explorers, has been preserved much as it was the day the surviving expedition party left. The historic huts are under the care of the New Zealand Antarctic Heritage Trust.

ice Almost 98 percent of Antarctica is covered with ice, which has accumulated over millions of years. According to astronauts, from space Antarctica 'radiates light like a great white lantern across the bottom of the world.' This ice cover is moving continuously. Great ICE STREAMS drain the interior of the continent, forming ICE SHELVES that extend over large areas of the huge embayments covering much of the coastal seas. If all Antarctic ice were returned to the oceans, it would raise global sea levels by about 60 m (197 ft). If it were spread over AUSTRALIA, it would form a layer 3 km (1.86 miles) thick.

Most ice in Antarctica is formed by a process called 'firnification'. When SNOW falls it has a lot of air between the crystals, and a low density. After accumulating, the snow undergoes metamorphism (its crystal structure breaks down), and gradually consolidates under its own weight to form 'firn'. Firn—snow that has survived one or more summers—is made up of individual granules and has less trapped air. With time and increasing depth and pressure, the crystals weld together and form GLACIER ice, which looks white at a distance because it contains many small air bubbles. When the ice is compressed further, and all the air is squeezed out, it turns into clear hard ice. This often appears blue because the ice is not able to filter out the blue section of the light spectrum.

The time it takes for snow to change to ice varies, as does the depth at which this process occurs. On the ice shelves near the coast, glacier ice forms relatively quickly: in about 200–300 years, at depths of around 50 m (164 ft). In inland areas the process occurs much more slowly because PRECIPITATION is considerably lower. In 1909 Douglas MAWSON described the horizontal layers (stratification) of ice and compressed snow on an ice shelf face, which often mark annual snowfall, and noted that even at a significant depth it was possible to distinguish between them.

In some conditions, ice can also be superimposed, layer upon layer. This often occurs in windy environments, when water vapour that is supercooled below 0°C (32°F) freezes to cold surfaces. The air temperature must be above freezing level, so this process is limited to areas such as OASES that are subject to summer melting. Spectacular hoar frosts form in calm conditions, when snow at the surface is colder than the air of the sub-surface snow: water vapour migrates to the cold surface forming delicate, flat crystals that can be as large as dinner plates and stand on their edges, stacked into intricate structures.

Right: Loose snow crystals that fall onto the Antarctic Ice Sheet are gradually compacted down and compressed into 'firn', the first stage in the transformation of snow into glacier ice.

Above: *Microorganisms have been found in the ice of lakes such as Vanda.*

Above: Eerie colours and forms lend an ice cave a cathedral-like quality.

Ice Age Glacially sculptured landforms now dominate the Antarctic landscape, as they must have some 300 million years ago, during the earlier ice ages of GONDWANA. The present ice age began when other Gondwana continents sheared away from Antarctica, forming deep seaways that isolated the continent from warmer water.

About 34 million years ago, an unstable ICE cover developed, as GLACIERS began to advance down valleys, spreading into fringing ICE SHELVES along the coastline. The lush vegetation gave way to cool temperate woodland. As glaciation intensified, the ice shelves became grounded on the ocean floor and local ICE CAPS developed. They covered the islands that make up the geological basement of WEST ANTARCTICA, as well as the MOUNTAIN ranges of EAST ANTARCTICA. Results from the DRILLING project at Cape ROBERTS record a change in coastal vegetation about 25 million years ago, from FORESTS to small herbs and MOSSES, closer to the present species in these areas. Then, about 15 million years ago, the ice caps welded into the great ICE SHEET that binds East and West Antarctica into a single continent: CORE SAMPLES drilled in sea-floor sediment and volcanic ash layers from South VICTORIA LAND show changes that indicate the appearance of the permanent ice cap at this time.

The Ice Sheet has grown in size and may have reached its present state only recently, in geological terms. Many researchers believe that swings in CLIMATE are caused by slight changes in the orbit of Earth around the sun. This may cause Antarctica to receive slightly more SOLAR RADIATION, which explains warm interglacial periods and the dynamic behaviour of the Ice Sheet. These interglacials punctuated the Ice Age throughout the Pleistocene era, between 5 and 1.5 million years ago; recent findings suggest that some regions of Antarctica may have been mainly free from ice as recently as about 3 million years ago. Fossilized wood, identified as *Nothofagus*, or southern beech, was found in the glacial till of the BEARDMORE GLACIER area, for example. This species may have survived in isolated enclaves along the ROSS SEA coast, and recolonized inland areas during an interglacial period. Shells drilled from the seabed indicate that even up to about 1.5 million years ago, the Ross Sea was 6 to 7°C (about 13°F) warmer than it is now.

Ice Cap See ICE SHEET.

Ice cave These occur naturally at the base of GLACIERS and ICEBERGS. When moving ice pushes against an obstacle, a space may be created in the 'lee' of that obstacle. Beneath glaciers, caves are produced by water tunnelling through the ICE, creating sculptured, smooth cavities. In icebergs, wave action working on CREVASSES can form caves at the waterline. The ice often 'transmits' blue light, giving the caves and their walls a translucent blue appearance. Reginald Ford, a member of the 1901–04 NATIONAL ANTARCTIC EXPEDITION, was awed by '... arches that glistened in the sunlight with the most beautiful and delicate iridescent hues, whilst the interiors of the caves displayed rich shades of azure and violet.'

Ice core See CORE SAMPLES.

Ice floe Flat sheets of SEA ICE that together comprise PACK ICE. Some floes may be tens of kilometres (miles) across. When floes are rafted together, or forced edge-to-edge, they form pressure ridges, which are very hazardous to SHIPS.

Ice Sheet The accumulation of millions of years of snowfall, ICE in the Antarctic Ice Sheet flows slowly from the POLAR PLATEAU to the coast. The most important element of the world's glacial system, it interacts with oceanic and atmospheric circulation, and changes in this huge ice mass will affect the CLIMATE on a global scale. The Ice Sheet contains an estimated 32.4 million cubic km (12.5 million cubic miles) of ice and covers 97.6 percent of the continent. By far the largest body of ice on Earth, its surface area is 1.7 times that of AUSTRALIA and 1.4 times the size of the USA. Seventy percent of the world's freshwater is locked up in the Ice Sheet.

At the SOUTH POLE, the Ice Sheet is nearly 3 km (1¼–2 miles) thick and is constantly moving—about 9 m (30 ft) per year.

EAST ANTARCTICA contains 88 percent of the Ice Sheet. Stable and slow moving, the EAST ANTARC-TIC ICE SHEET is land-based or 'terrestrial', meaning that its base is located primarily above sea level. The thickest ice lies in a subglacial trench, which is 400 km (248 miles) from the TERRE ADÉLIE coast and is 4776 m (15,665 ft) deep—about half the height of Mount Everest. The average height of the Ice Sheet above sea level is greater than that of any other continent, making Antarctica the highest continent.

It is also the lowest continent. The WEST ANTARCTIC ICE SHEET, which has expanded and contracted many times since its formation 20 million years ago, is the world's last marine-based ice sheet: the weight of the ice has pushed its base below sea level. The lowest point is the BENTLEY SUBGLACIAL TRENCH, which is 2539 m (8328 ft) below sea level. The West Antarctic Ice Sheet is more dynamic and responds more quickly to climate change than its eastern counterpart: if the ice melted, the land beneath the West Antarctic Ice Sheet would become buoyant and rise up again over thousands of years.

Ice shelves Formed at the edge of the Antarctic ICE SHEET, ice shelves are attached to land but float out over the ocean surface, and usually end in impregnable ice cliffs. Although ice shelves are likely to have been significant in shaping ancient landscapes, such as the fiords in SWEDEN, they are a phenomenon now found only in Antarctica (apart from some small Arctic remnants composed mostly of multi-year SEA ICE). They make up half the coastline of Antarctica, and cover 11 percent of the Antarctic terrestrial ice area.

Delicately balanced between ice gain and ice loss, ice shelves gain mass from the flow of inland ice to the coast, by snow accumulation on their upper surfaces and, in some areas, by seawater ('marine ice') freezing on to their lower surfaces. Small shelves are composed almost entirely of locally formed ice and they respond quickly to changes in snowfall. Larger shelves are fed by inland ice: as this ice travels a vast distance to the coast, it may take many centuries for changes in PRECIPITATION at the POLAR PLATEAU to affect these shelves. Ice is lost from ice shelves primarily by ICEBERG calving at the seaward ice cliff, and also by melting at the lower surface. A 'grounding' or 'hinge line' is the region where the Ice Sheet loses contact with the ground and begins to float as an ice shelf. This line is actually a zone about 100 km (62 miles) wide where the ice is attached to the sheet in some places and detached in others. The seaward margins of an ice shelf and the grounding line are the two most important boundaries in determining whether an ice shelf is advancing or retreating.

Ice shelves flow over the ocean surface at a rate that increases with the distance from the coastline. As the ice flows away from the coast, the shelf spreads faster. This process slows where the ice shelf is confined along its edges by friction or pinned on islands or sea-floor ridges.

Ice streams Dynamic, rapidly flowing rivers of ICE, ice streams 'pull' ice from the interior of the Antarctic ICE SHEET. Although accounting for only 10 percent of the Ice Sheet's volume, ice streams

Above: At the edge of an ice shelf, a section of ice leans towards the Ross Sea. Eventually this slab will calve off and with the thaw will drift in the Southern Ocean.

are sizeable features: up to 50 km (31 miles) wide, 2000 m (6560 ft) thick and hundreds of kilometres (miles) long. At the margins of the ice streams, shearing causes extensive bands of CRE-VASSES, about 5 km (3 miles) wide. Nearly 90 percent of the ice flowing across WEST ANTARCTICA

Below: The National Science Foundation's icebreaker Nathaniel B Palmer.

converges into three main drainage systems: the ice streams of the Siple Coast that flow into the ROSS ICE SHELF; Rutford, Carlson and Foundation Ice Streams that flow into the RONNE ICE SHELF, and the large ice streams that flow directly into the AMUNDSEN SEA. They drain ice through the

marine-based WEST ANTARCTIC ICE SHEET to the coast at the rate of around 1 km (½ mile) a year. This is very rapid compared with the flow of sur-rounding ice. The main reason ice streams move so rapidly is the presence of water at the base of the Ice Sheet. This creates a lubricating layer of slurry-like sediment that is saturated in water, on which the streams slide and move.

Ice tongue At the coast, the terminal of a GLACIER may form a 'tongue' of ice that protrudes into the ocean. Periodically, parts of the ice tongue shear off to form ICEBERGS.

Icebergs *See following pages.*

Icebreakers With strengthened steel hulls and powerful engines, icebreakers carve paths through thick PACK ICE to access the coastline of Antarctica. The first icebreakers to work in the SOUTHERN OCEAN were *Northwind* and *Burton Island*, both operated by the USA during the 1946–47 OPERATION HIGHJUMP. Both RUSSIA and JAPAN use nuclear-powered icebreakers for research, and to resupply bases. Most icebreakers have advanced scientific facilities, including spe-cialized laboratories, echo-sounding equipment to map the sea floor, and a range of devices that can be used to take sedimentary and biological sam-ples from the ocean.

Continued page 106

Icebergs

Formed when ICE breaks or 'calves' off from the Antarctic ICE SHEET, ICE SHELVES and GLACIERS along the coastline. About 2 million tonnes (tons) of ice calve off the Ice Sheet every year to form icebergs. Icebergs contain a great deal of records—about snowfalls, air temperatures, sediments—and provide information about ice loss.

The first recorded sighting of icebergs, or 'ice hills', in Antarctic waters was in January 1774 at a latitude of 71°10'S by James COOK: 'The Clowds near the horizon were of a perfect Snow Whiteness and were difficult to be distinguished from the Ice hills whose lofty summits reached the Clowds. The outer or Northern edge of this immense Ice field was compose[d] of loose or broken ice so close packed together that nothing could enter it; about a mile in began the field ice, in one compact solid boddy and seemed to increase in height as you traced it to the South; In this field we counted Ninety Seven Ice Hills or Mountains, many of them vastly large ...' More poetic is the 1840 description by James Savage, armourer on the *Erebus*: 'The Fragments as I call the floating Islands though Large Enough to build London on their Summit must through a Long Succession of years have parted from the Barrier they never could accumulate to Such an Enormous hight otherwise.'

Icebergs are calved, or 'berged', by processes that include ocean swells, tsunamis, TIDES and vibrations from colliding icebergs. Factors that may instigate calving include weaknesses in ice shelves and glacier tongues at hinge lines, CREVASSES, and the topography of the sea floor.

Calving can be spectacular. Alan Villiers watched the process from a whaling ship in 1923: 'the face of the ice cliffs suddenly crumbled away for lengths of miles at a time, in a beautiful waterfall of ice ...' Richard BYRD noted that this phenomenon was 'always accompanied by clouds of white "ice smoke", which rose in the air and hung about the Barrier [the ROSS ICE SHELF] surface for sometimes half a day.'

Once calved, many icebergs remain grounded on the ocean floor and stay close to the continent, lasting many years. Drifting icebergs have a much shorter life span. They tend to move north and westward around the coast before heading north again, then east, breaking and melting as they go. Driven by OCEAN CURRENTS, tides and WIND, they drift at speeds up to 5 km (3 miles) an hour, covering up to 20 km (12½ miles) per day. Occasionally they cross the ANTARCTIC CONVERGENCE and melt in warmer waters. In 1894, an iceberg was sighted in the Atlantic at 26°S; however, most break up quickly into smaller pieces, called 'bergy bits' or 'growlers', which have sharp jagged edges submerged beneath the surface, and can rip open a ship's hull with little warning.

Regarded with fear and caution by early sailors, icebergs also inspired awe. Charles WILKES of the 1838–42 UNITED STATES EXPLORING EXPEDITION wrote of '... lofty arches of many coloured tints, leading into deep caverns open to the swell of the sea, which rushing in, produced loud and distant thunderings. The flights of birds passing in and out of these caverns recalled the recollection of ruined abbeys, castles and caves, while here and there a bold projecting bluff, crowned with pinnacles and turrets, resembled some gothic keep.'

Icebergs are classified according to shape: as tabular, irregular or rounded. Tabular ('table-top') icebergs are formed from calving ice shelves and glacier tongues. A large tabular berg can be 200 to 300 m (656–984 ft) thick, weigh 400 million tonnes (tons), tower '10 storeys' above the ocean surface, and contain enough fresh water to supply 3 million people for a year.

Iceberg shapes include the 'tabular' or 'table top' (below) and rounded and irregular (above). Hollows are often sculpted by wave action (above left).

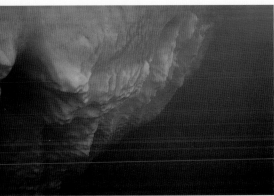

Above: Physical and chemical processes acting on the underside of an iceberg create an eerie, undulating surface.

On 28 December 1933 BYRD'S SECOND EXPEDITION sighted 8000 icebergs in 24 hours, the largest a tabular berg that it took two hours to pass travelling at 10 knots. In the 1960s a monster iceberg 20 times bigger than Manhattan Island sheared off from a glacier jutting into the BELLINGSHAUSEN SEA.

After being aground for decades, it broke in two and, in 1999, drifted into shipping lanes off ARGENTINA—a 66 km (41 mile) slab, which rose 55 m (180 ft) out of the sea and reached a depth of 275 m (900 ft). The largest iceberg recorded broke off the Ross Ice Shelf in March 2000. Code-named B-15, it measured 298 by 37 km (185 by 23 miles).

Irregular icebergs calve off small, crevassed floating glaciers and are sculpted into angular shapes by melting below the ocean surface—these features are revealed when the bergs become top heavy and flip over. In early 2000 the Ninnis Glacier, which flows from the EAST ANTARCTIC ICE SHEET, lost its entire tongue in just a few months and, in the process, changed the shape of the East Antarctic coastline.

Some icebergs are bottle green or jade and contain organic material, which reflects green light (unlike pure glacier ice, which reflects blue light) and comes from seawater frozen to the underside of ice shelves as 'marine ice'. The material is exposed when an irregular iceberg tilts over. Because marine ice forms at pressure beneath the surface where air is very soluble, the ice contains few bubbles and is incredibly clear. In the same way 'stratifications', or layers of compacted SNOW, white bubbly ice and clear blue ice from ice shelves and glacier tongues are exposed to form striped icebergs.

The ice lost from the continent to icebergs every year is equivalent to about half the world's freshwater. Schemes to 'harvest' icebergs for drinking water, involving towing them to warmer areas to melt, have been suggested from time to time. An Antarctic committee on mineral exploitation once commented that 'theoretically icebergs could be floated to any point accessible by a water route with depths of at least 200 m.' Such schemes are, of course, hindered by enormous logistical difficulties.

Above: Icefish (Channichthyidae) are the only vertebrates without haemoglobin, the red oxygen-carrying pigment normally found in blood.

Icebreakers can generally cut smoothly through young pack ice. BRITAIN's *James Clark Ross* (99 m/325 ft), for example, can be driven at a steady 2 knots through level sea ice a metre (3 ft) thick. As they cut, compressed air systems are used to roll the ships back and forth, preventing ice from squeezing their hulls. Where the ice is particularly thick or is concentrated into ridges, the ships repeatedly ram the ice to break through. When the ice becomes too thick to break, ships moor at its edge and cargo is unloaded for 'over-snow' transport.

Often icebreakers must wait for the right WEATHER and SEA ICE conditions to get to their destinations. Thick ice under pressure from WIND and OCEAN CURRENTS can stop, or even crush, the strongest and most powerful vessels.

Icefish (Channichthyidae) The 16 species of icefish belong to the suborder NOTOTHENIOIDEI, which live exclusively in the SOUTHERN OCEAN. Like most notothenioidei, they are sluggish, bottom-dwelling FISH that feed on KRILL, other CRUSTACEANS, worms, MOLLUSCS and smaller fish.

They are the only vertebrates without haemoglobin, the red oxygen-carrying pigment normally found in blood. Instead, oxygen is carried around the icefish's body in plasma. As a consequence, their blood is translucent and their gills and internal organs are colourless.

The pale, ghostly appearance of icefish, along with large mouths and many long teeth, led Antarctic whalers to call them 'white crocodile fish'.

Because icefish lack haemoglobin, their blood has only 10 percent of the normal oxygen-carrying capacity. Without haemoglobin, blood viscosity is lower and less energy is involved in circulating the blood. A large heart and high blood volume increases blood circulation and helps to make the transfer of oxygen from blood to tissues more efficient.

Icefish have a natural ANTIFREEZE that prevents their body fluids from freezing in Antarctic seas. These antifreeze molecules circulate in the blood and impede ice-crystal growth. Exactly how this works is not yet fully understood.

Igloos A dome-shaped SHELTER made from blocks of ice, traditionally used by Eskimos in the Arctic region. Like ICE CAVES, they provide good protection from Antarctica's elements. Field parties are trained in how to construct and live in one of these shelters.

Imperial shag (*Phalacrocorax atriceps*) See BLUE-EYED CORMORANT.

Imperial Transantarctic Expedition (1914–17) British expedition led by Ernest SHACKLETON that turned into one of the greatest rescues in Antarctic history. Shackleton, who marketed it as the 'greatest polar journey ever attempted', aimed to cross the ANTARCTIC CONTINENT from the WEDDELL SEA via the SOUTH POLE to MCMURDO SOUND in the ROSS SEA.

The *ENDURANCE*, captained by Frank WORSLEY, left Plymouth on 8 August 1914 with 28 men and reached the Weddell Sea in December. The 10-man team in the *AURORA*, commanded by Aeneas Mackintosh, headed for Cape EVANS in the Ross Sea. Their task was to lay supply depots for the transantarctic team along 170°E as far as the foot of BEARDMORE GLACIER.

Heavy PACK ICE slowed the progress of the *Endurance*, which by 19 January 1915 was frozen in. The party had to spend winter on board, but *Endurance* was being crushed by the ice and Shackleton gave the order to abandon ship on 27 October. A camp was set up on the ICE nearby and supplies and lifeboats recovered before the ship sank, including photographic plates rescued by Frank HURLEY. *Endurance* eventually sank on 21 November, after drifting 2410 km (1500 miles) on a zigzag course.

The camp drifted northwards with the pack ice. Shackleton wrote: 'I pray God I can manage to get the whole party safe to civilization.' According to Hurley, 'the precarious residence on the floe' was 'preferable to the continuance of the anxiety we ... endured ... on board ship'. In total the men lived on the ice floes for five months. On 9 April 1916, as the ice broke around them, they boarded the three lifeboats. Under the expert navigation of Worsley, they sailed through freezing, heavy seas for five days to ELEPHANT ISLAND. They had no water to drink and chewed raw SEAL meat for the blood.

From there, Shackleton decided to take one of the boats to SOUTH GEORGIA, 1483 km (920 miles) away, across one of the stormiest seas in the world. He set off with Worsley and four others in the boat *James Caird* on 24 April, leaving Frank WILD in charge at Elephant Island.

The *James Caird*, only 7 m (22.5 ft) long with a 1.8 m (6 ft) beam, took 16 days to reach South Georgia. After a few days' rest Shackleton, Worsley and Tom CREAN made a transalpine trek across the island to the WHALING settlement of Stromness. It took them 36 hours to cross 27 km

Below: Few Antarctic expeditions from the Heroic Age have captured modern imaginations as much as Ernest Shackleton's 1914–17 Imperial Transantarctic Expedition.

(17 miles) of icy, mountainous island. Shackleton made four attempts to reach Elephant Island, which was engulfed in pack ice, finally rescuing all the men on 30 August in the *Yelcho*, which was lent by the Chilean government.

At South Georgia, Shackleton learnt that the ROSS SEA PARTY had been stranded in Antarctica since 6 May 1915 when the *Aurora* had been dragged from its moorings at Cape Evans during a blizzard (then had drifted northwards to NEW ZEALAND). When he and Worsley arrived in the *Aurora* to pick them up on 10 January 1917, they discovered that, despite the stranding, the group had successfully laid depots for the planned Antarctic crossing. Three men had died, including Mackintosh, on the depot-laying journey, which had taken 160 days and covered 2872 km (1783 miles).

In the Footsteps of Scott Expedition (1985–87) Private British-Canadian expedition to retrace Robert SCOTT's 1911–12 trek to the SOUTH POLE. Robert Swan, Roger Mear and Gareth Wood left Cape EVANS in October 1985 and 70 days later reached the Pole, from where they were flown to MCMURDO STATION. The team's support vessel was crushed by PACK ICE and sank in the ROSS SEA and crew members were rescued by the USA.

India India first expressed interest in Antarctica in 1956, when it attempted to discuss internationalization of the region at the UNITED NATIONS. It withdrew the request when the ANTARCTIC TREATY was adopted in 1959. India's first scientific expedition to Antarctica was in 1982–83. It acceded to the Antarctic Treaty in 1983 and several weeks later became a CONSULTATIVE PARTY. India's inclusion in the ANTARCTIC TREATY SYSTEM weakened the campaign initiated by countries such as Malaysia in support of the COMMON HERITAGE PRINCIPLE.

India is a member of the Commission for the CONVENTION ON THE CONSERVATION OF ANTARCTIC MARINE LIVING RESOURCES (CCAMLR). Despite being a signatory to the MADRID PROTOCOL, some sections of the Indian government still express interest in the long-term possibilities of exploiting Antarctic OIL and other MINERAL reserves.

India operates two bases in Antarctica, DAKSHIN GANGOTRI BASE and MAITRI BASE.

Indian Ocean The body of water bordered by Africa to the west, the Asian continent to the north and AUSTRALIA and Indonesia to the east. To the south, the Indian Ocean merges with the SOUTHERN OCEAN near the coast of EAST ANTARCTICA. The southern Indian Ocean also surrounds a number of SUBANTARCTIC ISLANDS, including CROZET, Îles KERGUÉLEN, HEARD, MARION and PRINCE EDWARD ISLANDS.

Inexpressible Island Originally known as Evans Cove, Inexpressible Island in TERRA NOVA BAY was renamed by the Northern Party of the 1910–13 BRITISH ANTARCTIC EXPEDITION after they had been stranded there for eight and a half months, sheltering in an ICE CAVE and surviving on the meat and blubber of SEALS and PENGUINS.

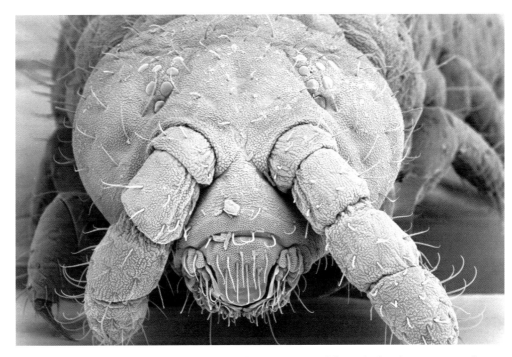

Above: Most Antarctic insects are parasites and so are protected from the harsh environment by a warm-blooded host. Of the few free-living insects, the largest—the springtail—copes with climatic extremes by living under rocks in moist regions.

Insects The majority of the 67 insect species recorded from the ANTARCTIC CONTINENT and the MARITIME ISLANDS are PARASITES on warm-blooded ANIMALS. They include the world's largest flea.

Of the free-living insects, only SPRINGTAILS (Collembola) are found on the continent and there are two species of MIDGE in the maritime zone. Biting LICE are the most numerous of the parasitic insects, living at the base of feathers on Antarctic BIRDS, and a few species of sucking lice are found on SEALS. *Glaciopsyllus antarcticus*, the world's largest flea, is found on FULMARS and PETRELS. Their warm-blooded hosts protect the parasitic insects from the harsh CLIMATE and conditions.

The insect FAUNA of the SUBANTARCTIC ISLANDS is much richer and includes flies, beetles, aphids, thrips and parasitoid wasps.

Instituto Antártico Argentino Created in 1951, the Institute coordinates ARGENTINA's Antarctic scientific and technical RESEARCH.

International Association of Antarctic Tour Operators The International Association of Antarctic Tour Operators (IAATO) was founded in 1991 and includes members and associate companies from 10 countries. IAATO's objectives are to advocate and promote safe, environmentally sound, private-sector travel to Antarctica. As well as being involved in TOURISM, IAATO members often provide logistic and scientific support to national Antarctic programmes and organizations. In return, scientists provide on-board lectures. IAATO has attended ANTARCTIC TREATY CONSULTATIVE MEETINGS as an observer and presented information papers on tourism.

International Biomedical Expedition (IBE) Organized by the SCIENTIFIC COMMITTEE ON ANTARCTIC RESEARCH. Working Group on Human Biology and Medicine, the IBE was a series of

multinational experiments carried out in 1980–81 to test the response of humans to the Antarctic environment.

International Centre for Antarctic Information and Research (ICAIR) Formed in 1992 and based at the International Antarctic Centre in Christchurch, New Zealand, ICAIR is an independent, non-profit international organization involving NEW ZEALAND, the USA and ITALY. ICAIR's areas of expertise include Antarctic environmental management, digital technologies and information management and geographical information systems. It aims to increase the accessibility of Antarctic environmental data to scientists and to make that information easily understood and more widely known to decision-makers, educators and the general public.

International Council of Scientific Unions The organization endorsed the 1957–58 INTERNATIONAL GEOPHYSICAL YEAR and set up a special committee to plan the scientific programme and invite participants. Its support of permanent international scientific cooperation was pivotal in the formation and success of the SCIENTIFIC COMMITTEE ON ANTARCTIC RESEARCH.

International Geophysical Year (1957–58) Initiated by the International Council of Scientific Unions (ICSU), the International Geophysical Year (IGY) was a cooperative global effort by scientists to understand the Earth's environment. Much of the field activity took place in Antarctica. Laurence Gould, the USA's chief scientist, wrote that 'No field of geophysics can be understood or complete without specific data available only from this vast continent and its surrounding oceans.'

Twelve nations participated in the Antarctic research, establishing about 60 RESEARCH STATIONS, OBSERVATORIES and laboratories. The USSR built VOSTOK STATION at the POLE OF GREATER INAC-CESSIBILITY. During OPERATION DEEPFREEZE, the USA established five bases, including MCMURDO STATION and the AMUNDSEN-SCOTT SOUTH POLE STATION. BRITAIN maintained 14 stations, and ARGENTINA, CHILE, FRANCE, AUSTRALIA, BELGIUM, JAPAN, NORWAY, SOUTH AFRICA and NEW ZEALAND also participated.

Data gathered by geologists, glaciologists, seismologists, oceanographers and atmospheric scientists were complemented by simultaneous observations around the planet. The first overland crossing of Antarctica was made during the IGY, by the COMMONWEALTH TRANSANTARCTIC EXPEDITION. Long-distance flights by AIRCRAFT surveyed millions of kilometres (miles). Along with support flights, and overland journeys to build bases and conduct geophysical surveys, little of Antarctica was left unseen.

The IGY was a benchmark for international scientific cooperation, and on its completion in 1958, US President Dwight Eisenhower invited the 11 other Antarctic IGY nations to the WASHINGTON CONFERENCE that led to the drafting of the ANTARCTIC TREATY.

International Polar Commission Initiated by Arctic scientist and explorer Karl Weyprecht, who believed that 'The key to many secrets of Nature ... is certainly to be sought for near the Poles. But as long as Polar Expeditions are looked on merely as a sort of international steeple-chase ... these mysteries will remain unsolved ... Decisive results can only be obtained through a series of synchronous expeditions, whose task it would be to distribute themselves over the Arctic regions and to obtain one year's series of observations made according to the same method.'

The Commission was established in 1879, with members from European and Scandinavian nations and RUSSIA and the intention to establish a programme of scientific investigations in both polar regions. During the FIRST INTERNATIONAL POLAR YEAR, only two expeditions travelled south; of these, only the GERMAN INTERNATIONAL POLAR YEAR EXPEDITION sailed into the Antarctic region.

The final meeting of the Commission was held in 1891, but it was revived between 1905 and 1913.

International Solar Terrestrial Physics Programme Collaborative research programme involving NASA, the European Space Agency (ESA), and JAPAN's Institute of Space and Astronautical Science (ISAS) to study the transfer of energy from the sun, through the upper ATMOSPHERE, to the Earth's surface. The clear skies above the continent make Antarctica a strategic location for this work, which uses around 20 research spacecraft, a wide range of ground-based facilities and computer models.

International Transantarctic Expedition (1989–90) International expedition led by Jean-Louis Etienne and Will Steger. The first crossing of Antarctica using DOG teams and the fourth traverse of the continent; they chose the longest route possible—a 6400 km (3968 mile) trek along

the ANTARCTIC PENINSULA to the SOUTH POLE, via VOSTOK to MIRNYY STATION.

International Union for the Conservation of Nature and Natural Resources (IUCN) Also known as the World Conservation Union, the organization was founded in 1948 to encourage the preservation of wildlife, natural environments and living resources. The Conservation Assessment and Management Plan (CAMP) is a tool developed by the organization to evaluate the status of various FAUNA in Antarctica, and to determine CONSERVATION priorities. In the case of PENGUINS, for example, the IUCN has undertaken survey and census work, intensive research, and captive breeding programmes for nine species and subspecies.

International Whaling Commission (IWC) When the International Convention for the Regulation of Whaling was revised in 1946, the International Whaling Commission (IWC) was established to govern the conduct of WHALING. The IWC has a secretariat in Cambridge, England, and holds an annual meeting at which the Commission takes advice from its scientific committee and votes on proposed regulations. To be adopted, decisions require a three-quarters majority, but members can opt out of being bound by a regulation by registering a formal objection.

Membership is open to any country that ratifies the Whaling Convention. Initially, members were nations involved in commercial whaling with little interest in CONSERVATION. But, as some original members have ceased whaling and new states interested in conservation have joined, the Commission's philosophy has changed, from being primarily concerned with the exploitation

of WHALES to having a majority of members more interested in conserving species. A new management policy, adopted by the IWC in 1975, was designed to bring all whale stocks to the levels providing the greatest long-term harvests, and catch limits for individual stocks were set below their sustainable yields. But in 1982 the IWC voted to stop all commercial whaling starting from 1985–86. This does not affect 'aboriginal subsistence' whaling, which is confined to certain Northern Hemisphere nationals and permitted in limited numbers. Despite the total ban, commercial whaling has not ceased, although it has been significantly curtailed, and IWC members are subject to ongoing, intensive lobbying from several countries intent on continuing to hunt whales.

To underline the ban and emphasize the importance of various whale habitats, several sanctuaries have been established. In 1994 a sanctuary was declared in the SOUTHERN OCEAN (the second sanctuary—the first was established in the INDIAN OCEAN in 1979). The sanctuary's northern boundary follows the 40°S parallel, except in the Indian Ocean sector where it joins the southern boundary of the first sanctuary at 55°S, and around South America and into the South Pacific where the boundary is at 60°S. Commercial whaling is prohibited in the sanctuaries, but whales may be taken for scientific purposes. JAPAN objected to the prohibition on the MINKE WHALE, and has conducted a significant amount of 'scientific' whaling of minkes. This has led to PROTESTS from governments and environmental organizations, which believe that Japan is in fact harvesting the whales commercially rather than taking them for scientific research. Several attempts, led by NEW ZEALAND, to extend the sanctuary limits have failed.

Below: Introduced species, such as these rabbits on Enderby Island, can severely damage delicate subantarctic ecosystems.

Above: Antarctica's marine environment is rich in invertebrates such as these sea stars (Odontaster validus) *and sea urchins* (Sterechinus neumayeri).

Introduced species HUSKIES were introduced to Antarctica as SLEDGE DOGS, but the ANTARCTIC TREATY nations decided that all dogs should be removed from Antarctica by 1994 for environmental reasons. Introduced MICROORGANISMS, including disease-causing species, are potentially major environmental problems and now certain items are prohibited—for example, poultry products are banned in the field because of fears that poultry viruses may infect PENGUINS or other birds nesting near research bases.

In 1953, an introduced weed (*Poa annua*) was discovered growing on DECEPTION ISLAND, just north of the ANTARCTIC PENINSULA, where it survived for several years before succumbing to the hostility of its surroundings. Although the harshness of the CLIMATE has been too much for species introduced to the ANTARCTIC CONTINENT, accidentally and deliberately introduced organisms have had a profound effect on the ECOSYSTEM of SUBANTARCTIC ISLANDS.

Only HEARD and MCDONALD ISLANDS are free of introduced mammals. Eight mammal species— Norway rat, black rat, house mouse, rabbit, sheep, reindeer, mouflon, cat—have established populations on the other subantarctic islands. Rats and mice were accidentally introduced from ships and among foodstuffs from the 18th century on; other mammals were deliberately introduced for food or sport. Alien GRASSES have also been spread, mainly for grazing, and become dominant in some areas.

Feral cats and rats prey on ground-nesting BIRDS on a number of islands, and rabbits and other grazing animals destroy vegetation, causing soil erosion and loss of cover for burrowing birds. Reindeer were introduced to SOUTH GEORGIA and Îles KERGUÉLEN for sport and remain today, but an attempt to introduce mink to Îles Kerguélen failed.

INVERTEBRATES, such as slugs, snails, earthworms and aphids, have also been introduced to the subantarctic, where they too can affect the ecological balance. On MARION ISLAND, for example, introduced cabbage moths have attacked the Kerguelen cabbage (*Pringlea antiscorbutica*).

Invertebrates Animals which lack a backbone. Antarctica's marine invertebrate FAUNA is rich and diverse, and includes BRYOZOANS, CORALS, SPONGES, SEA ANEMONES, CRUSTACEANS, MOLLUSCS, SEA SPIDERS and ECHINODERMS.

The terrestrial life of Antarctica is dominated by invertebrates—most VERTEBRATES such as BIRDS and SEALS depend on the ocean and essentially belong to the marine ECOSYSTEM. Terrestrial invertebrates are small: the largest is the WINGLESS MIDGE, which grows about 12 mm (½ in) long. Even smaller terrestrial invertebrates include PARASITES such as MITES and LICE; other groups are PROTOZOA, SPRINGTAILS and NEMATODES, which inhabit MOSS, ALGAE and LICHEN communities and ROTIFERS and TARDIGRADES, which may also be found in aquatic environments.

Islands Antarctica is ringed with COASTAL ISLANDS that are predominantly ICE covered and share the climatic conditions of the continent. The major influence on the MARITIME ISLANDS around the northern ANTARCTIC PENINSULA and on the Atlantic side of the continent is the SOUTHERN OCEAN. Further north, the cold, wind-swept SUBANTARCTIC ISLANDS support a wider variety of plants and animals. A number of non-existent, or PHANTOM ISLANDS, have been charted in the Southern Ocean.

Italy Italian scientific expeditions visited Antarctica in 1968–69, often working with other states already active there. Italy acceded to the ANTARCTIC TREATY in 1981, became a CONSULTATIVE PARTY in 1987 and is a signatory to the MADRID PROTOCOL. It is also a member of the Commission for the CONVENTION ON THE CONSERVATION OF ANTARCTIC MARINE LIVING RESOURCES (CCAMLR). The first official Italian expedition was in 1985–86. A scientific station was built at TERRA NOVA BAY in the 1986–87 season, and it operates during the summer.

Italian National Antarctic Research Program ITALY signed the ANTARCTIC TREATY in 1981. In 1985 it established the Italian National Antarctic Research Program, which constructed a permanent research station at TERRA NOVA BAY in 1986 and became a member of the SCIENTIFIC COMMITTEE ON ANTARCTIC RESEARCH in September 1988.

j

James Ross Island Off the northeastern tip of the ANTARCTIC PENINSULA, between SNOW HILL and SEYMOUR ISLANDS, James Ross Island is mountainous and rises to 1630 m (5350 ft) at its highest point. It was charted in 1903 by Otto NORDENSKJÖLD and named after James Clark ROSS.

Japan The 1910–12 JAPANESE ANTARCTIC EXPEDITION was the first Japanese interest in Antarctica. Although no formal TERRITORIAL CLAIM was made, Japan announced unspecified rights in Antarctica before World War II. When Japan signed the 1951 Treaty of Peace, it was forced to renounce all rights to and interests in any Antarctic claims.

Japan participated in the INTERNATIONAL GEOPHYSICAL YEAR and was one of the original CONSULTATIVE PARTIES to the ANTARCTIC TREATY. It has a permanent research base, SYOWA STATION, on East Ongul Island, off the coast of DRONNING MAUD LAND.

Japan has significant interests in the exploitation of Antarctic RESOURCES. Although it has ratified the MADRID PROTOCOL, the state-run Japanese National Oil Corporation has searched for OIL and gas in the Antarctic region. As a member of the Commission for the CONVENTION ON THE CONSERVATION OF ANTARCTIC MARINE LIVING RESOURCES (CCAMLR), Japan usually adopts a pro-fishing

Below: Japan's Mizuho station, established in 1970.

position and its vessels are involved in FISHING for KRILL in the SOUTHERN OCEAN. It is also one of the last nations to conduct commercial WHALING and its 'scientific' whaling of MINKE WHALES is a source of political controversy.

Japanese Antarctic Expedition (1910–12) Japanese government-sponsored exploratory expedition. Led by Nobu SHIRASE, the expedition sailed from Tokyo in the *Kainan Maru* under Captain Nomura on 1 December 1910.

The expedition reached Cape ADARE then sailed further south but, unable to make landfall, returned to Sydney for the winter, where they experienced hostility and racism from the Australian and NEW ZEALAND press until Edgeworth DAVID offered his support. Desperately short of money, Captain Nomura and some of the crew returned to JAPAN in an attempt to raise funds.

In the spring, the expedition returned south, this time heading for the BAY OF WHALES where they found the *FRAM* awaiting the return of Roald AMUNDSEN and his party from the SOUTH POLE; however, the language barrier precluded much communication.

Sailing further west along the edge of the ROSS ICE SHELF to a point they named KAINAN BAY, the expedition then cut a zigzag path up the shelf face and set up a base camp. A party of three—the 'Dash Patrol'—set out from here with DOGS and travelled about 185 km (115 miles) inland to 80°5'S. Meanwhile, the *Kainan Maru* sailed further west to Okuma Bay, from where a shore party explored the southern coast of 'King Edward VII Land'—now known as the Shirase Coast. The expedition returned to Japan to great acclaim.

Above: *The crew of the Japanese Antarctic Expedition crew on board the* Kainan Maru.

Japanese Antarctic Research Expedition (JARE) A cooperative Antarctic research project involving scientists from a number of Japanese government agencies. JARE approves all Japanese Antarctic research projects, which are carried out through the NATIONAL INSTITUTE OF POLAR RESEARCH.

Joinville Island About 64 km (40 miles) long, Joinville is the largest island in the group lying northeast of the ANTARCTIC PENINSULA. It was discovered in 1838 by Jules-Sébastien DUMONT D'URVILLE.

Joyce, Ernest (1875–1940) British seaman and explorer, member of 1901–04 NATIONAL ANTARCTIC EXPEDITION and took part in a number of SLEDGE journeys. In 1907 Joyce joined the BRITISH ANTARCTIC EXPEDITION after Ernest SHACKLETON saw Joyce pass by the expedition office window. Joyce was in charge of depot-laying for Shackleton's attempt to reach the SOUTH POLE.

When working for the Sydney Harbour Trust, he agreed to join the 1914–17 IMPERIAL TRANS-ANTARCTIC EXPEDITION as part of the depot-laying ROSS SEA PARTY. Despite being stranded at Cape EVANS in May 1915 when the *AURORA* was dragged from its moorings, the group spent seven months—from September 1915 to April 1916—laying depots for Shackleton's party, which of course never used them. On the journey back to the Cape Evans base, the Ross Sea Party suffered terribly from scurvy and lack of food and were caught in blizzards: 'We speak often of my late chief, of Captain Scott and his party ... If we had prolonged our stay in the tent another day I feel sure we would have remained powerless to get under way again. We would have shared a similar fate. ...' The Ross Sea Party was rescued in January 1917, and Joyce was awarded the Albert Medal for bravery.

k

Kainan Bay On the ROSS ICE SHELF near the BAY OF WHALES, named by the JAPANESE ANTARCTIC EXPEDITION, led by Nobu SHIRASE, in 1912.

Katabatic winds Persistent and powerful winds that reach speeds of around 300 km (186 miles) an hour. The word 'katabatic' is from the Greek *katabatikos*, meaning 'go down' or 'downward'. Cold, dense air near the surface of the ICE CAP rolls down from the POLAR PLATEAU under the influence of gravity (see GRAVITY WINDS). More persistent in winter when the air is colder, the wind may appear 'from nowhere' and its speed and direction vary considerably, dictated by the shape of the ice cap rather than by the overall pattern of atmospheric pressure.

The wind's source is the Polar Plateau and its onset is sudden. At its birthplace, the katabatic is merely a light breeze, but the cold air is dense and it flows down the sloping polar ice dome in all directions in a series of surges that gather speed in a fall towards the coast 1600 km (992 miles) away. A katabatic BLIZZARD rages close to the ground in a layer around 300 m (984 ft) deep, and

picks up loose SNOW and granules of ice, sweeping them across the landscape. It gathers momentum as it follows the ice down through the polar ICE STREAMS. In its wake, the wind leaves the landscape carved into ornately sculptured ridges. The lower and faster it gets, the warmer it becomes.

By the time the katabatic reaches the GLACIERS that drain to the coast, it has gathered full momentum. The landscape is worn down by constant blasting from the grit-laden air. In the final 160 km (100 mile) rush to the coast, the wind drops 3000 m (9840 ft), building to raging gusts of about 300 km (186 miles) an hour. The coastal PACK ICE is broken up and, as the katabatic interacts with warmer, moister marine air, it gathers heavy low-lying cloud.

Kayaking Wade Fairley and Angus Finney made some exploration of Antarctica's icy waters by kayak in 1994. After attempting to circumnavigate SOUTH GEORGIA, the pair hitched a ride down the ANTARCTIC PENINSULA, and paddled north again through the Lemaire Channel.

In 2001, a New Zealand team of Graham Charles, Marcus Waters and Mark Jones undertook the first major Antarctic kayaking expedition. They began paddling at HOPE BAY on 15 January, and over the following 34 days covered 850 km (530 miles), and paddled into the ANTARCTIC CIRCLE before being picked up and transported back to South America: 'Our route was quite straightforward—we just had to paddle

south,' Charles said. On the longest day they paddled for 19 hours, covering a distance of 92 km (57 miles). The custom-built kayaks had to carry all the FOOD and EQUIPMENT needed, and their CLOTHING was specially designed for the conditions.

Among the ever-present dangers were KATABATIC WINDS that could easily overturn a kayak, rolling ICEBERGS and ICE falls.

Kelp A long brown SEAWEED found in relatively shallow waters near the nutrient-rich shores of SUBANTARCTIC ISLANDS. Giant kelp floats as tall vertical plants attached to the sea bottom and provides habitats and shelter for FISH and INVERTEBRATES. The strap-like fronds of some species can grow as much as 30 cm (1 ft) a day during summer.

Kelp gull (*Larus dominicanus*) Otherwise known as dominican or black-backed gulls, these are the only gulls that regularly appear in the Antarctic. Striking in appearance and quite large—58 cm (23 in) long with a wingspan of 1.3 m (4½ ft)—they are all white except for the black saddle pattern across the back and upper wings. Despite their name, they do not feed on KELP, although their ground nests consist of bowls that they adorn with a variety of debris, including kelp and other SEAWEED.

Nesting sites occur on coastlines from the southern subtropical area to 65°S on the ANTARC-

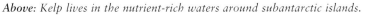

Above: Kelp lives in the nutrient-rich waters around subantarctic islands.

Above: Kerguelen fur seals (Arctocephalus tropicalis) *have a circumpolar distribution and are closely related to the Antarctic fur seal.*

TIC PENINSULA. Known for their opportunistic feeding strategies, these birds readily adapt to human habitation, and their range is expanding; they first appeared in AUSTRALIA in the 1940s and now seem to have settled there. They are also widespread in New Zealand and the SUBANTARCTIC ISLANDS.

Two, sometimes three, eggs are laid in summer and are incubated for three or four weeks. The chicks grow rapidly, fledging just a few weeks after hatching; although they may wander from their nests at this time, they remain dependent upon their parents for food. Kelp gulls scavenge whatever they can from almost any source, including carrion and scraps from SKUA kills. They also eat fresh food, such as earthworms, MOL-LUSCS, CRUSTACEANS and, sometimes, their own chicks and eggs when food is scarce.

Kemp, Peter (d. 1834) English WHALING captain. Nothing is known of Kemp prior to 1813. After that he was master of at least seven different vessels, undertook a dozen voyages to the SOUTHERN OCEAN, and was aboard the *Dove* when the SOUTH ORKNEY ISLAND group was discovered.

In July 1833, employed by a London SEALING company, possibly the ENDERBY BROTHERS, he sailed from London in the *Magnet* with a crew of 18 in search of new sealing grounds in the southern INDIAN OCEAN. He made directly for the Îles KERGUÉLEN and proceeded south, hoping to reach the area discovered by John BISCOE. In November, at 60°S, he encountered ICEBERGS and FUR SEALS, and was trapped in PACK ICE from which he broke free on 14 December. Further pack ice was encountered on Christmas Day when just north of the ANTARCTIC CIRCLE, and on the following day he saw land to the south. Because even denser pack ice prevented him from approaching any

closer, it was decided 'the filling of casks with [seal] oil was of greater importance than weeks wasted in unrewarded effort.' This part of the Antarctic coast, now called KEMP LAND, was not reached again until the BRITISH-AUSTRALIAN-NEW ZEALAND ANTARCTIC RESEARCH EXPEDITION of 1929–31. Back at KERGUÉLEN, a course was set for Simonstown, SOUTH AFRICA. But on 21 April 1834 Kemp fell overboard and drowned.

Kerguélen, Iles Also known as Kerguélen Island, an archipelago consisting of one main island and numerous islets which lies north of the Antarctic mainland in the INDIAN OCEAN and is named for Yves-Joseph KERGUÉLEN-TRÉMAREC, who first sighted it and claimed it for FRANCE in February 1772.

On his return to France, Kerguélen-Trémarec exaggerated the extent of his discovery, describing the island as temperate, suitable for agriculture and probably inhabited. In reality, they are subantarctic, lie in the path of ferocious winds and are partly covered by an ICE CAP. GLACIER action has carved deep valleys and high-altitude lakes into the mountainous interior of the island, and its northeastern and southeastern coasts are notched with fiords.

Kerguélen was a SEALING base from 1791. Over a century later, France formally annexed the island and granted an exclusive 50-year lease to Frères Bossière, who established a sealing and WHALING station, which operated from 1909 to 1914, then again in the 1920s. In 1874 expeditions from three countries landed there to observe the transit of Venus.

Kerguelen cormorant (*Phalacrocorax verrucosus*) The smallest of the BLUE-EYED CORMORANTS, they have black upperparts with blue-black heads and

necks and white underparts. They are found on Îles KERGUÉLEN, in the subantarctic zone, where there is a seemingly stable breeding population of about 6000–7000 pairs. Heavy, graceless fliers, they have wingspans of only 1.1 m (3½ ft) and stay close to the coast when feeding. Nesting tends to occur on rocky cliffs in sheltered bays, where food is readily available.

Kerguelen fur seal (*Arctocephalus tropicalis*) A close relative of ANTARCTIC FUR SEALS, Kerguelen fur seals (also called subantarctic fur seals) are eared or otariid seals. They have a circumpolar distribution and are found on the SUBANTARCTIC ISLANDS, notably KERGUÉLEN, MACQUARIE and MARION ISLANDS. Their diet varies, but consists mainly of KRILL, FISH, SQUID and, occasionally, PENGUINS.

Kerguelen tern (*Sterna virgata*) These are one of the world's rarest TERNS and breed on only three island groups: Îles CROZETS, Îles KERGUÉLEN and PRINCE EDWARD ISLANDS. The total breeding population is estimated to be around 2400 pairs and their main threat is from feral CATS. They are smaller than the ANTARCTIC and ARCTIC TERNS, growing to about 33 cm (13 in), and have brownish heads and grey backs. Like all terns, they are incredibly graceful in flight—their wing-span (75 cm/30 in) is about two-and-a-half times their body length.

Kerguélen-Trémarec, Yves-Joseph de (1734–97) French naval officer and explorer. Born in Brittany, he entered the French Navy at the age of 16 and between 1752 and 1767 sailed in Canada, the West Indies and Iceland.

On 1 May 1771 he set out in command of *Fortune*, accompanied by *Gros Ventre*, to survey an alternative trade route to eastern Asia. On 3 February 1772 the ships sailed southeastwards from Port Louis, Mauritius, and 13 days later sighted 'a continuation of land stretching uninterruptedly from the north-east to the south'—Îles KERGUÉLEN. No landing was made, in the fog the two ships were separated, and Kerguélen-Trémarec wrongly recorded the longitude by several degrees. Despite having written that 'I can confirm that never was felt a cold so bitter' in his diary, his imagination began to run wild on his return and he presented a glowing report of 'La France Australe', a warm land of lush vegetation, to the government. Kerguélen-Trémarec led a second, unsuccessful southern voyage and was court-martialled on his return to France for not relocating the islands. After the French Revolution (1789), he was appointed to a government position in the Mariru Ministry, but was then imprisoned during The Terror (1793–94).

Keyser, Dirck Gerritsz (c. 1600–55) In 1622 this Dutch pilot reported having been driven off-course to 64°S, where he discovered a land with snow-covered mountains, similar in appearance to those in Norway. It is thought he may have sighted the SOUTH SHETLAND ISLANDS, the first to do so.

Killer whale (*Orcinus orca*) One of the TOOTHED WHALES, the killer, or orca, is the largest member of the DOLPHIN family and is easily recognized by its bold black-and-white markings. Males are larger than females, and reach up to 8–9 m (26–30 ft) in length, with a weight of about 8 tonnes (8 tons). Both sexes have a prominent dorsal fin, which reaches almost 1.8 m (6 ft) in males. Their distribution is probably more widespread than any other cetacean; the Antarctic population is thought to be around 70,000. They have strong teeth that they use to tear apart their prey. Formidable predators, they hunt in cooperative groups, taking SEALS, PENGUINS and other cetaceans, including large BLUE WHALES; they are known to kill great white sharks. Orcas have been timed at speeds of 25 knots.

Unlike SPERM WHALES, they have limited diving ability and take their prey at or near the surface, from ice floes and even by smashing through thin ice. In a few populations—for example, Patagonia, FALKLAND ISLANDS and Îles CROZET—killer whales have been observed taking seals directly off the beach. Orcas swim in large pods of up to 100 or more, and have been recorded following ICEBREAKERS. Although little is known about their reproductive cycle, adults and small calves have been sighted within the PACK ICE in winter, and it is thought they may breed in Antarctic waters. Research on non-Antarctic populations has shown Orcas have complex social systems and rich behavioural repertoires.

King penguin (*Aptenodytes patagonicus*) Second largest of the penguins, at 90 cm (35 in), it is 10 cm shorter than its close relative the EMPEROR PENGUIN, but at 16 kg (35½ lb) is only half the weight. Although similar in appearance to emperors, king penguins have more brightly hued upper breasts. They are found on SUBANTARCTIC ISLANDS, including the FALKLANDS, SOUTH GEORGIA, MARION, HEARD and MACQUARIE. The breeding population is estimated at between 1 and 1.5 million pairs.

Breeding takes place in large colonies on ROCK or TUSSOCK flats near the shore, close to abundant supplies of SQUID and FISH. The eggs are laid around November. King penguins do not have nests. Instead, the incubation method is similar to

*Below: Killer whales (*Orcinus orca*), also known as orca, cruise the ice edge, hunting for food.*

that of emperors, with a single egg being carried on the feet of the brooding parent; however, unlike emperors, both male and female kings take turns incubating the egg. It takes 55 days for the egg to hatch, then the chick is brooded by the parents for 30 days and remains dependent for almost one year; this means parents can only raise two chicks every three years.

Kites Used in early meteorological experiments, both Robert SCOTT and Erich von DRYGALSKI flew kites, along with making BALLOON ascents in Antarctica. They attached instruments to the kites to record TEMPERATURE, pressure and humidity, but had little success in collecting data. Kites were also used by Wilhelm FILCHNER's expedition in 1912, and later at LITTLE AMERICA. Modified kites,

or 'parasails', are now commonly used on SKIING expeditions to speed progress.

Krill 'Krill', a Norwegian word meaning 'small fry', is a general term used to describe about 85 species of open-ocean CRUSTACEANS belonging to Euphausiacea. Antarctic krill are the major food source for WHALES, PENGUINS, SEALS and many BIRDS, and one of the most abundant and successful animal species on Earth. Female Antarctic krill are believed to lay up to 10,000 eggs at a time, several times in a season; there is estimated to be about 500 million tonnes (500 million tons) of Antarctic krill in the SOUTHERN OCEAN. Without krill, the ocean's ECOSYSTEM would collapse.

Concerns about commercial FISHING of Antarctic krill, which began in the 1970s, led to

Above: Krill is a general term to describe about 85 species of crustacean. One of the world's most abundant animals, they are a major food source for whales, fish and birds.

Below: King penguins (Aptenodytes patagonicus) *are the second largest living penguin species.*

the signing of the CONVENTION ON THE CONSERVATION OF ANTARCTIC MARINE LIVING RESOURCES (CCAMLR) in 1981, which restricts—but does not ban—krill fishing; its stated aim is conservation of stocks.

Antarctic krill (*Euphausia superba*) are the dominant species north of the PACK ICE; a smaller species *Euphausia crystallorophias* is the main krill type found under the ice and was first described from specimens collected through holes cut in the ice by the 1901–04 NATIONAL ANTARCTIC EXPEDITION. Three other krill species are also found in Antarctic waters.

Euphausia superba grows approximately 6 cm (2½ in) long and weighs over 1 g (⅓ oz). It forms dense swarms that can spread over several square kilometres (miles) and extend to a depth of up to 200 m (650 ft). With densities as high as 30,000 individuals per cubic metre (22,000 per cub yd), these swarms turn the water red or orange. However, most of the time krill remain unseen in daylight hours, rising to the surface to feed at night. They are mainly HERBIVORES, feeding on PHYTOPLANKTON, but also feed on other ZOOPLANKTON.

Krill are able to swim both forwards and backwards, using their five pairs of rear, paddle-shaped legs to swim forwards and their tail for backwards thrust. Another six pairs of forward legs are used for feeding, gathering minute DIATOMS from the water. Unlike most zooplankton, krill are heavier than water and must paddle constantly to maintain their position.

Antarctic krill spawn mainly in summer. Their eggs are thought to sink and hatch at depths of 2000 m (6500 ft); from there the larva gradually rise towards the surface to feed, a journey that takes up to 10 days. Larvae are believed to take two to three years to mature.

Krill do not build up fat reserves for winter, but have been known to survive over 200 days' starvation under laboratory conditions. They retain the ability to moult their exoskeleton in adulthood and, thus, are able to 'downsize' in times of low food availability, using their own body proteins as a source of fuel. The age of krill cannot, therefore, be told from size alone, making age-determination difficult. Nevertheless, scientists estimate that Antarctic krill may live between five and 10 years.

l

Lakes These occur mainly in the DRY VALLEYS or OASES. Some are located in the TRANSANTARCTIC MOUNTAINS and a number in the VESTFOLD HILLS and the BUNGER HILLS. Like the GLACIERS in these areas, many of the lakes were more extensive in the past. Until 8000 years ago, the glacial Lake WASHBURN filled the Lower Taylor Valley to about 308 m (1010 ft) above present sea level.

Current depths vary from a few centimetres (inches) to over 60 m (197 ft). The lakes are mostly capped by permanent ICE cover between 2–5 m (7–16 ft) thick; in the summer, the ice melts at the shoreline, forming a moat. The lakes are wet-based or frozen through to their beds. Lakes that freeze more slowly than the sediment around them have internal pools of water that eventually freeze and form ice domes on the lake surface. Uplifted lake ice is also formed when lakes freeze completely to the lake bed.

As well as the freshwater lakes, such as lakes Chad and Popplewell, which are fed by streams of glacial meltwater during summer, Antarctica has a number of SALINE LAKES in closed basins. Both types can support life. Microbial mats at the base of lakes in the Dry Valleys have been intensively sampled and studied. They resemble MICROORGANISMS that lived at the bottom of oceans 3000 million years ago, and are thought to be one of the oldest life forms on Earth and, possibly, one of the first sources of oxygen on the planet. Tiny pockets of melted water within the ice cover of some lakes may also support microbial life.

A number of SUBGLACIAL LAKES lie hidden below thick layers of ice in the Antarctic ICE SHEET. Their extent is being investigated using SATELLITE surveying and radar probing of the ice.

Lambert Glacier The largest valley GLACIER in the world, the Lambert is 400 km (248 miles) long and covers more than 1 million sq km (386,100 sq miles). Much like a major RIVER system, the Lambert Glacier is fed by a complex series of tributaries. It flows through the Prince Charles Mountains into the AMERY ICE SHELF, draining about a quarter of EAST ANTARCTICA. It moves about 230 m (755 ft) through the mountains each year and by the time it reaches the ice shelf it has sped up to about 1 km (½ mile) per year.

Larsen, Carl Anton (1860–1924) Norwegian whaling captain and entrepreneur. In 1892 he was sent by a Hamburg firm in search of RIGHT WHALES and, in the ship *Jason*, set a new farthest south record—68°S—sailing along the east coast of the ANTARCTIC PENINSULA, and discovered the LARSEN ICE SHELF.

On the SWEDISH SOUTH POLAR EXPEDITION of 1901–04, Larsen captained the *Antarctic* and, when the ship became ice-bound and eventually crushed in February 1903, he and 19 crew wintered over in a makeshift stone hut on PAULET ISLAND. They survived on penguin meat and blubber, made their way back to base camp by boat and on foot, and were rescued by the *Uruguay*.

In the following years, Larsen established WHALING stations in SOUTH GEORGIA and the SOUTH SHETLAND ISLANDS. In 1923–24 he commanded *Sir James Ross*, a whaling factory ship that operated around MACQUARIE ISLAND and was the largest ship to sail in the SOUTHERN OCEAN up to that time, and established the first whaling operation in the ROSS SEA.

Larsen Ice Shelf Lies in the northwest WEDDELL SEA and extends along the east coast of the ANTARCTIC PENINSULA. It is named after Carl LARSEN. The Larsen Ice Shelf has been retreating gradually since the 1940s, and there was a dramatic loss of ice in early 1995, when a huge ICEBERG, 70 km (43 miles) long and 25 km (16 miles) wide, calved off. Although periodic loss of ice is a normal part of the lifecycle of an ICE SHELF, in this instance the whole northern section—an area of 2000 sq km (772 sq miles)—disintegrated into small 100–200 hectare (250–500 acres) icebergs. This may have been due to surface melting over a number of warm summers, leaving water-filled CREVASSES that created a network of wounds.

The break-up of the ice shelf occurred simultaneously with an impressive calving event on the FILCHNER ICE SHELF. The two combined represented the biggest calving of ice in recorded history, and were equivalent to three times the average annual ice loss from the entire Antarctic ICE SHEET.

Over a five-week period between January and March 2002, Larsen B Ice Shelf—with an area extending 3250 sq km (1255 sq miles), containing an estimated 720 billion tonnes (tons) of ice—collapsed, releasing thousands of icebergs into the Weddell Sea.

Latitude Antarctica is the most southern land mass, lying at the opposite end of the globe to the Arctic. The continent's latitudinal boundary is generally considered to be 60°S, which is the official boundary of the SOUTHERN OCEAN and the approximate position of two natural boundaries around the continent: the ANTARCTIC CONVERGENCE and the ANTARCTIC CIRCLE.

Law Dome A large ice dome that rises 1395 m (4575 ft) above sea level. It lies immediately inland from the Budd Coast in WILKES LAND and was named after Philip LAW, scientist and head of AUSTRALIA's Antarctic operations in the 1950s.

Law of the Sea The international law used to regulate activity on the high seas of the world, including the seabed. The law has developed from customary practice, and more recently through the UNITED NATIONS Convention on the Law of the Sea (UNCLOS) and other international treaties dealing with the high seas.

The UNCLOS was signed on 10 December 1982 and entered into force on 16 November 1994. It led to the creation of an international seabed authority to administer the exploitation of minerals on the bottom of the oceans. However, UNCLOS made little reference to Antarctica because establishing an international authority over Antarctica was seen as too difficult. Under the Law of the Sea, the CONVENTION ON THE CONSERVATION OF ANTARCTIC MARINE LIVING RESOURCES (CCAMLR) is treated as a regional fisheries agreement. In the 1990s several new agreements were negotiated to deal with environmental

Below: Most of the lakes on the continent are located in the Dry Valleys and near mountain ranges. They include a number of saline lakes, as well as mysterious subglacial lakes that lie hidden under the Ice Sheet.

problems on the high seas. These included agreements on straddling stocks (one stock straddling two fishing areas)and migratory species, the use of flags of convenience, and approaches to dealing with illegal, unregulated and unreported FISHING.

See also ECONOMIC EXCLUSION ZONES and LAWS, INTERNATIONAL, REGARDING ANTARCTICA.

Law, Philip (1912–) Australian scientist and explorer. Born in Victoria, AUSTRALIA, and educated at Melbourne University, Law was appointed advisor to the AUSTRALIAN ANTARCTIC EXPEDITIONS in 1947, and from 1949 to 1966 headed the Australian Antarctic Division. He undertook the mapping of 6000 km (3720 miles) of Antarctic coastline, and some 1.3 million sq km (502,000 sq miles) of new territory.

Laws, international, regarding Antarctica The body of rules and principles that are binding on states in their relations with each other. International law is created by consensus among states from a variety of sources, including judicial decisions, international conventions and writings about international law.

Because of the DISPUTED CLAIMS over Antarctica, the question of which states have jurisdiction in particular areas is unresolved. The compromise reached in the ANTARCTIC TREATY is that official observers, scientists and their support staffs remain under the jurisdiction of the state of which they are nationals. Problems arise when people not associated or involved in a national research programme, such as tourists and adventurers, get into legal trouble in the Antarctic. For these people, it is not clear which court, if any, would have jurisdiction. With the increasing commercial interest in Antarctic TOURISM this issue could become a problem.

The development of international law regarding ENVIRONMENTAL PROTECTION is relatively recent, largely dating from the decades after the Antarctic Treaty was signed in 1959. The 1972 Stockholm Declaration of the UNITED NATIONS Conference on the Human Environment declared, as one of its principles, that a state's activities should not damage the environment outside its own jurisdiction. The 1992 United Nations Conference on Environment and Development (UNCED) developed 'Agenda 21', an agreement that incorporated the idea of sustainable development, reiterating central principles contained in the various CONVENTIONS regarding Antarctic marine resources.

Despite calls from environmentalists for Antarctica to be made into a WORLD PARK, the Antarctic Treaty members have not supported greater internationalization of Antarctica; instead, they have expanded the ANTARCTIC TREATY SYSTEM with new agreements such as the MADRID PROTOCOL.

See also the LAW OF THE SEA.

Lazarev, Mikhail Petrovich (1788–1851) Russian naval officer and explorer. Born near Moscow, Lazarev entered the Imperial Naval Cadet Corps aged 12 and three years later was 'loaned' to the British Navy for four years. After serving in the Baltic Fleet, from 1813 to 1816 he commanded the *Suvorov*, the supply ship for the Rossiysko-Amerikanskaya Kompaniya serving the Aleutians and Alaska.

From 1819 to 1821 Lazarev captained the *Mirnyi* on the Russian exploratory expedition to Antarctica led by Thaddeus BELLINGSHAUSEN. From 1822 to 1825 he circumnavigated the globe with a Russian scientific expedition, and later served with the Russian Navy in the Mediterranean and the Baltic before his appointment in 1833 as commander of the Black Sea Fleet.

Leads Elongated ice-free areas in the PACK ICE. Unlike POLYNYAS, leads open and close over a few hours with changes in WIND direction and speed. They form important arteries for SHIPS and for marine MAMMAL and BIRD migration. Like polynyas, these wet OASES help species to survive in the pack ice through the depths of winter.

Le Maire, Jacob (1585–1616). Dutch explorer. Along with Willem and Jan SCHOUTEN, Le Maire sailed through DRAKE PASSAGE, around the southern tip of TIERRA DEL FUEGO, which they named CAPE HORN, thus proving that Tierra del Fuego was not part of *TERRA AUSTRALIS INCOGNITO*. The steep-sided Lemaire Channel, between Booth Island and the ANTARCTIC PENINSULA, is not named after Le Maire, but after a Charles Lemaire, who explored the Congo.

*Below: A leopard seal (*Hydrurga leptonyx*), shortly after giving birth on an iceberg.*

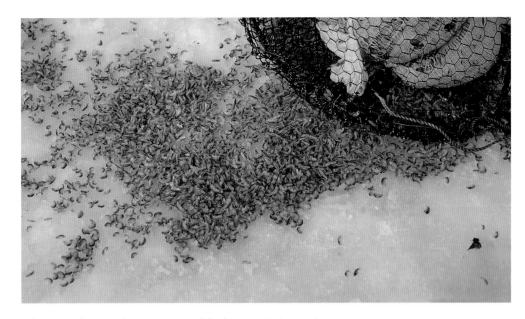

Above: Sea lice are the most successful of Antarctica's parasites.

Legends According to Polynesian legend, around AD650 the navigator Ui-te-Rangiora set out from Fiji—then a major centre of Polynesian culture—in *Te Iwi-o-Atea* accompanied by seven other canoes to explore the Pacific Ocean. Several years of active exploration followed, during which 86 islands were discovered or visited, and described, ranging from New Hebrides to NEW ZEALAND, Hawaii and Rapanui (Easter Island). During one of these voyages, as he was heading for the island of Rapa (which is about 1770 km/1100 miles southeast of Raratonga), Ui sailed into Antarctic waters and, according to oral tradition, reported 'monstrous seas, white "rocks" that grow out of the sea, kelp waving in the water, mushy frozen sea ice resembling pia [arrowroot], strange animals that dive to great depths and swim underwater, some with tusks protruding from their mouths, the foggy, misty and dark atmosphere, and a barren land not shone on by the sun, with rocks and peaks piercing the skies but completely bare without anything growing on them and no sign of habitations.' Another great Polynesian explorer, Te Aru-tanga-nuku-Ariki, led an expedition south to verify Ui's claims about 250 years later.

It seems likely that other Polynesian voyagers may have sailed into the SOUTHERN OCEAN. Tongan tradition reports the existence of ice-covered oceans to the south—the phrase *tai-fatu* means congealed or frozen sea.

Leopard seal (*Hydrurga leptonyx*) These phocid SEALS have spotted coats, which are dark grey along their backs and pale on their bellies, with neck and sides covered in 'leopard' spots. They have long slender bodies designed for underwater speed, with long tapered flippers, large heads, massive lower jaws and wide ominous gapes. Males can grow to 3 m (10 ft) and weigh up to 300 kg (660 lb); the females are slightly larger at 3.6 m (12 ft) and 450 kg (992 lb).

They are circumpolar in distribution and inhabit the Antarctic PACK ICE. However, juvenile individuals have been known to travel further north during the winter, to the SUBANTARCTIC ISLANDS, and have been seen occasionally off the coasts of Patagonia, SOUTH AFRICA and southern parts of AUSTRALIA and NEW ZEALAND.

Leopard seals are well known for their predatory habits. Their diet includes PENGUINS, pups of CRABEATER, WEDDELL and FUR SEALS, FISH and SQUID, but by far their most important food is KRILL. They themselves are preyed on by KILLER WHALES, but little else.

Leopard seals are sexually mature by the age of about six, and have a one-year reproductive cycle, giving birth to a single pup around November. It is thought these seals may have a life span of about 25 years. The Antarctic population is estimated at around 222,000.

Lesser Antarctica Another name for WEST ANTARCTICA, which is about one-third the size of EAST ANTARCTICA.

Lewis, David (1919–) NEW ZEALAND navigator. Educated at Otago Medical School, Lewis climbed mountains and competed in the 1960 trans-Atlantic single-handed YACHT race. In 1964 he sailed across the Pacific Ocean using traditional Polynesian navigational methods (and without instruments). In 1972–73 he sailed in his 10 m (33 ft) yacht *Ice Bird*, from Stewart Island, New Zealand, to PALMER STATION on the ANTARCTIC PENINSULA, the first solo voyage to the continent; after spending the winter in the north, he returned and sailed the *Ice Bird* to Cape Town, SOUTH AFRICA, in January 1974.

Lewis has made further voyages to Antarctica, with crews, and he helped to establish the independent, Sydney-based OCEANIC RESEARCH FOUNDATION, which has undertaken a series of research expeditions.

Lice Biting LICE are the most numerous of the parasitic INSECTS found in Antarctica. They avoid the harsh conditions by living in the warm microclimate at the base of feathers on BIRDS. A few species of sucking lice are also found, and these are restricted to SEALS.

Lichens PLANTS formed from the symbiotic association of FUNGI and ALGAE or CYANBACTERIA. Lichens occur in many different habitats and are able to occupy extreme environments. Some 300 to 400 species are known in Antarctica, including some that survive in the cracks and pores inside ROCKS under conditions of extreme cold and drought in the DRY VALLEYS. Many Antarctic species are also found in the Arctic, but not in regions in between.

Lichen distribution extends as far south as 86° 30'. They are able to photosynthesize while frozen at temperatures as low as –20°C (–4°F) and can absorb water when covered by SNOW. During long droughts, they survive in a dry and inactive state.

There are three main types of lichens found in Antarctica: crustose lichens, which form a thin crust on the surface they grow on; foliose lichens

Below: Lichens are not individual plants but a symbiotic union of an algae or cyanobacteria and a fungus.

*Above: Graceful fliers, light-mantled sooty albatrosses (*Phoebetria palpebrata*) breed on Antarctic and subantarctic islands.*

with leaf-like lobes; and fruticose lichens, which have a shrubby growth form.

Lichen growth rates are extremely slow. In the northern part of the ANTARCTIC PENINSULA and nearby islands they grow up to 1 cm (½ in) per century. In the harsher Dry Valleys region, they may take 1000 years to grow this much. Some lichens can survive for at least 2000 years.

Lichens are sensitive to atmospheric POLLUTION: Antarctic lichens have been found to contain traces of DDT and organochlorines. They are, however, undamaged by increased levels of ULTRAVIOLET RADIATION.

Light-mantled sooty albatross (*Phoebetria palpebrata*) Graceful, sooty-brown birds, they have a conspicuous pale mantle and back, long slender wings, and pointed tail. At close range, a semicircle of white feathers is readily visible behind the eyes. The feet are pale grey.

Extremely graceful fliers with a circumpolar range and wingspans of 1.8–2 m (6–6½ ft), the birds intersperse strong wing-beats with smooth soaring and gliding, and are known for following ships for long periods. They breed on various Antarctic and SUBANTARCTIC ISLANDS every second or, on some islands, third year and on average only produce single chicks every five years—making it one of the lowest breeding of all albatrosses. They fall prey to longline FISHING, which kills up to 4000 birds annually.

Limpets The Antarctic limpet *Nacella concinna* is the only animal visible in the intertidal zone of the ANTARCTIC PENINSULA, and it dominates the shore zone. Few organisms can exist in the harsh environment of ice-scoured, exposed coastal areas (there are almost no barnacles or mussels in Antarctica), and the Antarctic limpet survives by migrating. It spends summer feeding in shallow water, grazing on mats of benthic DIATOMS that form in tide pools and shallow waters, then it moves out into deeper water for winter. The limpets, in turn, are preyed on by nearby colonies of KELP GULLS, which swallow these MOLLUSCS whole, digesting the edible portions, then regurgi-

tating the shells. Some Antarctic limpet species are extremely long-lived and are known to survive for over 100 years.

Little America Five bases with the name Little America were established by Richard BYRD on the ROSS ICE SHELF. Because snowfall rendered an old base unusable, the Little America base had to be reconstructed for every new expedition.

Little America I was built for BYRD'S FIRST EXPEDITION adjacent to the BAY OF WHALES, and it was from here that Byrd made his first Antarctic flights in 1929. Little America II was established to support BYRD'S SECOND EXPEDITION of 1933–35. Little America III (also known as West Base) was built for the 1939–41 UNITED STATES ANTARCTIC SERVICE EXPEDITION, along with EAST BASE. Little America IV was constructed for OPERATION HIGHJUMP of 1946–47. The last station, Little America V, was built for the INTERNATIONAL GEOPHYSICAL YEAR and operated from 1955 to 1959.

All the Little America bases were lost at sea in the 1960s on an ICEBERG that calved off the Ross Ice Shelf.

Little shearwater (*Puffinus assimilis*) Small black-and-white SHEARWATERS, they have a wing-span of only 62 cm (24 in) and are found in the subtropical and SUBANTARCTIC zones. From above, they look black apart from white about the face; in contrast, the underparts are all-white and the feet lilac-blue with flesh-coloured webbing.

Their diet consists of FISH and SQUID, which they catch by pursuit-plunging and surface-diving. Their flight pattern consists of short glides, often interspersed with rapid wing-beats. Small groups of just a few BIRDS can often be seen close to their winter breeding areas, which are located on many islands in subtropical and subantarctic areas.

Long-finned pilot whale (*Globicephala melas*) 'Globicephala' means 'globe-headed' and 'melas' translates as 'black', a fair description of these medium-sized whales that are widely distributed in the SUBANTARCTIC latitudes of the SOUTHERN OCEAN. They do venture into shallow waters, and are one of the whales commonly found stranded. Their low, long-based dorsal fin is set about one-third along their body, and they have long pointed flippers. Extremely social, they move in pods of 20 to 100 animals, although groups of over 1000 have been sighted. They feed mainly on SQUID and on shoaling FISH. Females, which grow to a maximum size of 5.5 m (18 ft) and weigh around 1 tonne (1 ton), become sexually mature at seven to 10 years; males may grow to over 6.3 m (21 ft) and 1.75 tonnes (1.75 tons).

Longitude Located concentrically around the SOUTH POLE, Antarctica spans all longitudes—standing directly on the south pole, it is impossible not to face north. On James COOK's 1772–75 expedition to the SOUTHERN OCEAN, the astronomer Edmond HALLEY was charged by the ROYAL SOCIETY with testing instruments invented to determine longitude.

*Below: Little shearwaters (*Puffinus assimilis*) nest in crevices rather than digging their own burrows, and will occupy and defend their nesting site several months before laying eggs.*

m

Macaroni penguin (*Eudyptes chrysolophus*) Named after an 18th-century men's hairstyle favoured by members of the London Macaroni Club. The penguin's crest of orange plumes starts between the eyes and sweeps over the back of the head. Macaronis average about 71 cm (28 in) tall and weigh about 4 kg (9 lb).

With an estimated population of around 12 million pairs, macaronis are also the most abundant of the subantarctic penguins. They are found throughout the SUBANTARCTIC ISLANDS and there are populations numbering millions on SOUTH GEORGIA, ÎLES CROZET, HEARD and MACDONALD ISLANDS.

Like ROCKHOPPERS, macaronis feed mainly on CRUSTACEANS and lantern fish. They lay two eggs, the first of which is about half the size of the second and is kicked out of the nest once the larger egg is laid. ROYAL PENGUINS, which are similar in appearance and have identical breeding and feeding patterns, are usually classified as a subspecies of macaronis.

McDonald Islands First sighted in 1853, the three small McDonald Islands lie near the ANTARCTIC CONVERGENCE and, like the neighbouring HEARD ISLAND, are MARITIME ISLANDS controlled by SOUTHERN OCEAN weather. The islands are Australian territory.

McMurdo Sound In the ROSS SEA, bounded by the eastern coast of ROSS ISLAND, the ROSS ICE SHELF and South VICTORIA LAND and named by James Clark ROSS in 1841 after an officer on HMS

*Above: The most abundant of the subantarctic penguin species, macaronis (*Eudyptes chrysolophus*) derive their name from the crested feathers on their heads, which reminded early subantarctic visitors of a hairstyle in vogue with members of the London Macaroni Club.*

Terror. McMurdo Sound is the location of the USA's MCMURDO STATION and NEW ZEALAND's SCOTT BASE. Several GLACIERS terminate in the Sound, which is edged by Capes ROYDS and EVANS.

McMurdo Station Located on the southern end of ROSS ISLAND, McMurdo was built in 1955–56 by the USA as part of OPERATION DEEPFREEZE. Originally called Naval Air Facility McMurdo, it was renamed McMurdo Station in 1961. It is also known as 'Mactown'.

An important staging facility for operations throughout Antarctica, McMurdo has a harbour, outlying AIRSTRIPS and HELICOPTER pad. Air transportation to and from NEW ZEALAND is common between October and February. McMurdo is visited every year by ICEBREAKERS and resupply SHIPS.

It is a research base in biomedicine, geology and geophysics, glaciology and glacial geology, marine and terrestrial biology, meteorology and upper atmosphere physics.

McMurdo has an OVERWINTERING population of around 250, but in SUMMER may house over 1000 people—the largest community in Antarctica. Sara Wheeler described it in *Terra Incognita* as resembling 'a small Alaskan mining town with roads, three-story buildings [and] the ill-matched architecture of a utilitarian institution ...' It includes laboratories, bars, gymnasium, water distillation plant and a fire station. It had the only operational NUCLEAR POWER PLANT in Antarctica, from 1962–72. The reactor, nicknamed 'Nukey Poo', tended to break down and was eventually shut down after a coolant leak. The reactor and thousands of tonnes of NUCLEAR WASTE were shipped back to the USA.

Macquarie Island Midway between Tasmania and the Antarctic continent, Macquarie is a narrow ridge of land, about 34 km (21 miles) long and 2.5–5 km (1½–3 miles) wide, belonging to AUSTRALIA. The island was charted and named in 1810 by Captain Frederick Hasselborough, in the *Perseverance*, who reported seeing a SHIPWRECK off the coast. About 4 million penguins live on the island, which is the only breeding ground of ROYAL PENGUINS, and about 100,000 ELEPHANT SEALS.

The island is geologically unique: the only place where hot magma from deep within the Earth's mantle is pushed to the surface. This occurs at a plate boundary (see PLATE TECTONICS) called the Macquarie Ridge. This crustal movement causes regular earthquakes, which trigger mud slides and rock falls. Strong westerly winds continually buffet the steep cliffs around Macquarie's coastline, and sweep across the

Left: Spring at McMurdo Station: awaiting the first sunrise in four months.

grassland plateaux. The TEMPERATURE remains stable all year, but PRECIPITATION is high; it rains, snows or hails on more than 300 days annually. There are no trees or shrubs; instead, Macquarie is blanketed by thick TUSSOCK, and in waterlogged areas there are unusual 'featherbeds', or 'quaking bogs'—floating patches of vegetation that may be more than 6 m (20 ft) deep and can support the weight of a person. The Macquarie Island cabbage (*Stilbocarpa polaris*) has bright yellow flowers and was eaten by stranded sailors to prevent scurvy.

Bird and mammal populations have only recently recovered from over a century of exploitation, from the island's discovery in 1810 to the final ban on commercial licences in 1919. In 1812 one ship reported taking 14,000 FUR SEAL skins, and by 1830 the resource was obliterated. After 1870, ships returned to harvest ROYAL and KING PENGUINS for oil. Douglas MAWSON was appalled at the devastation when he visited during his 1911–14 expedition, and lobbied the Australian government for the island's preservation. It was made a wildlife sanctuary in 1933. While on the island, Mawson set up the first two-way RADIO communication with the Antarctic continent.

Macquarie Station The first Macquarie Station was established on MACQUARIE ISLAND as part of the 1911–14 AUSTRALASIAN ANTARCTIC EXPEDITION and was abandoned in 1915. The current station was built by AUSTRALIA in 1948 and has operated continuously since. Biology, botany, auroral physics, meteorology and medical research is conducted at the station. In SUMMER there are over 40 people at the station, and around 20 through WINTER.

Mac.Robertson Land In EAST ANTARCTICA, it was discovered and named by Douglas MAWSON during an aerial-survey flight in January 1930. It extends to the LAMBERT GLACIER and AMERY ICE SHELF.

Madrid Protocol The Protocol on Environmental Protection to the Antarctic Treaty (known as the Madrid Protocol) was signed on 4 October 1991 in Madrid. It came into force on 14 January 1998 after it was ratified by all of the CONSULTATIVE PARTIES to the ANTARCTIC TREATY. The Madrid Protocol was negotiated following the collapse of the 1988 CONVENTION ON THE REGULATION OF ANTARCTIC MINERAL RESOURCE ACTIVITIES (CRAMRA) and represented a major change in Antarctic POLITICS in favour of environmental protection and making the ANTARCTIC TREATY SYSTEM more open and accountable.

The Madrid Protocol is a comprehensive framework for the environmental protection of Antarctica and dependent and associated ecosystems. It effectively prohibits the exploitation of MINERALS and OIL in Antarctica for a 50-year period. It includes a schedule on arbitration and five annexes dealing with: environmental impact evaluations, CONSERVATION of FAUNA and FLORA, WASTE DISPOSAL, prevention of marine POLLUTION and area protection and management. A conference can be called to review the Madrid Protocol in 2041.

As required by the Protocol, a COMMITTEE FOR ENVIRONMENTAL PROTECTION (CEP) has been established to provide advice and recommendations to the consultative parties. CEP activities include: maintaining lists of ENVIRONMENTAL IMPACT ASSESSMENTS, audits and reviews; and acting as a forum for discussion on such issues as introducing non-native diseases into Antarctica and protection of specific species.

Some problems with the implementation of the Protocol have arisen. Despite lengthy negotiations, the consultative parties have been unable to reach a consensus on a liability annex, and, at times, they have been slow to implement their obligations. In addition, not all the parties to the Antarctic Treaty have ratified the agreement or all of its annexes. Although environmental impact assessments are required for all activities in Antarctica, compliance with the requirements has been uneven.

Magellan, Ferdinand (*c.* 1480–1521) Portuguese navigator and leader of the first expedition to circumnavigate the world. Sailing for Spain in search of a westerly route to the Indies, he set out from Seville on 10 August 1519 with five ships and 270 men. The expedition sailed south along the coast of South America and through the strait that bears his name, into the Pacific Ocean, which he named. Magellan reported land to his south, later identified as TIERRA DEL FUEGO. He sailed across the Pacific, but was killed in the Philippine Islands. One of his five ships completed the circumnavigation, returning with 20 of the original crew.

Magellan Strait Strait between Chile and TIERRA DEL FUEGO, connecting the Atlantic and Pacific Oceans. Discovered in 1519 by Portuguese navigator Ferdinand MAGELLAN.

Magnetic Crusade In the early to mid-19th century European scientists and scientific associations, particularly Britain's ROYAL SOCIETY and ROYAL GEOGRAPHICAL SOCIETY, and the GERMAN MAGNETIC ASSOCIATION, were preoccupied with the problem of magnetism. The lack of magnetic observations in the Southern Hemisphere was identified as a major obstacle to verifying Karl Frederick Gauss's theory of terrestrial magnetism. The pursuit of knowledge about magnetism led to three expeditions: Jean-Sébastien DUMONT D'URVILLE'S 1837–40 French expedition in search of the SOUTH MAGNETIC POLE; the 1838–42 UNITED STATES EXPLORING EXPEDITION led by Charles WILKES, and, most successfully, the British government-sponsored 1839–43 voyage to the SOUTHERN OCEAN of HMS *Erebus* and HMS *Terror* under the command of James Clark ROSS. The latter expedition, using methods developed by a group of international experts, carried out regular observations synchronized with others taken throughout the globe, and returned with a great amount of magnetic data.

Magnetic South Pole See SOUTH MAGNETIC POLE.

Maitri Base In 1989 INDIA established Maitri, the only year-round base in central DRONNING MAUD LAND. Research conducted here involves geology, glaciology, human biology and global change.

Maps See CARTOGRAPHY.

Marie Byrd Land Lies between ELLSWORTH LAND and EDWARD VII LAND in WEST ANTARCTICA. It is named after Richard BYRD's wife.

Marion Island The larger of the two PRINCE EDWARD ISLANDS, located in the south Atlantic Ocean, southeast of SOUTH AFRICA.

Maritime islands Although they are located within the ANTARCTIC CONVERGENCE, the CLIMATE of these maritime-zone islands is influenced more by the SOUTHERN OCEAN than the Antarctic continent itself. Mainly covered by permanent ice caps and GLACIERS that flow to sea level, many maritime islands experience frequent volcanic activity. In summer TEMPERATURES may rise above freezing level, but maritime islands are surrounded by PACK ICE in winter.

The largest maritime archipelago in Antarctica, the SCOTIA ARC, reaches from the end of the ANTARCTIC PENINSULA 540 km (335 miles) into the Southern Ocean. Consisting of the SOUTH SHETLAND, SOUTH ORKNEY, SOUTH SANDWICH and SOUTH GEORGIA ISLAND groups, the islands are the outcropping spine of a submarine MOUNTAIN range that links Antarctica to the Andes Mountains on the west coast of South America. Other maritime island groups include the isolated BOUVETØYA in the Atlantic; and the BALLENY, SCOTT and PETER I ØY in the Pacific sector of the Southern Ocean.

Markham, Clements Robert (1830–1916) British geographer, born in Yorkshire, England. He entered the British Navy at the age of 14, and in 1850–51 took part in a search for the lost Arctic explorer John Franklin. After leaving the navy, he travelled in Peru and elsewhere, worked in the civil service, and published a number of books, many of them on travel.

From 1893 to 1905 Markham was president of the ROYAL GEOGRAPHICAL SOCIETY and used the opportunity of his presidency to pursue Antarctic research and exploration. After a determined, and at times bitter, campaign for funding, in 1899 he announced the NATIONAL ANTARCTIC EXPEDITION, for which he championed Robert SCOTT as leader and insisted the British Navy had control. He worked tirelessly on preparations for the expedition, and organized the two relief expeditions sent to rescue the ice-bound DISCOVERY.

When Ernest SHACKLETON returned to Britain in 1909, having led members from his independently funded BRITISH ANTARCTIC EXPEDITION to a point just 180 km (100 miles) from the SOUTH POLE—the furthest south anyone had reached—Markham suggested Shackleton had falsified his results.

Mars fragments Only a handful of Martian meteorites—which originate when large asteroids collide with Mars, catapulting fragments into the solar system—have been found on Earth. The most important find was made in the ALLAN HILLS in 1984: a Martian meteorite, now known as ALH84001, that plunged to Earth about 15 million years ago. In 1996 a team of NASA scientists announced the meteorite appeared to contain microscopic FOSSILS of bacteria, the 'fingerprints' of ancient microbial life, which could be evidence that life had flourished on Mars. ALH84001 was subsequently divided into pieces and sent to different research teams. After analysing samples, some scientists concluded that the worm-like fossils were, in fact, mineral grains the same shape as bacteria.

Marsupials The 1981 discovery of the FOSSIL jawbone of a primitive marsupial on SEYMOUR ISLAND near the ANTARCTIC PENINSULA supports the theory that marsupials migrated from origins in the Americas and Antarctica to Australia via GONDWANA. Named *Antarctodolops dailyi*, the small, berry-eating marsupial was probably about 30 cm (1 ft) long. It lived 40 to 45 million years ago (during the late Eocene period), when Antarctica was covered in dense FORESTS and had a moist, temperate CLIMATE. *Antarctodolops* was the first fossilized mammal ever recovered from Antarctica and belongs to the family Polydolopidae, previously known only from South America.

Mawson, Douglas (1882–1958) Australian geologist and explorer, born in Yorkshire, England. His family moved to AUSTRALIA when he was an infant. Mawson studied geology at Sydney University under Professor Edgeworth DAVID, who became his lifelong friend and mentor. In 1905

Above: *The influential Australian explorer and scientist Douglas Mawson.*

Mawson was appointed lecturer in mineralogy and petrology at Adelaide University.

Mawson's first trip to the Antarctic, with David, was on the BRITISH ANTARCTIC EXPEDITION of 1907–09. Together with David and Dr Alistair Mackay, he made the first ascent of Mount EREBUS in March 1908. The same group was the first to reach the SOUTH MAGNETIC POLE, on 16 January 1909—a journey covering 2028 km (1260 miles) and believed to be the longest unsupported sledging journey undertaken in the Antarctic. David praised him highly: 'Mawson was the soul of the march to the Magnetic Pole. ... In him we had ... a man of infinite resource, splendid spirit, marvellous physique and an indifference to frost and cold that was astonishing.'

Declining an invitation to join the 1910–13 BRITISH ANTARCTIC EXPEDITION, Mawson organized the AUSTRALASIAN ANTARCTIC EXPEDITION of 1911–14. In November 1912 Mawson set off to explore the GEORGE V LAND coast, accompanied by Lieutenant Belgrave Ninnis and Dr Xavier Mertz. On 14 December Ninnis fell down a crevasse to his death, together with a sledge carrying most of the supplies.

About 506 km (315 miles) from their Cape DENISON base, the two remaining men started the journey back, killing their dogs for food. Mertz died on 7 January, and it wasn't until years later that it was discovered he had been poisoned by Vitamin A from the dogs' livers.

Mawson trekked the final 160 km (100 miles) alone with very little food, surviving a fall down a crevasse and a blizzard. 'My physical condition was such that I might collapse at any moment,' he wrote. '... it was easy to sleep on in the bag, and the weather outside was cruel.' He arrived back at Cape Denison on 8 February to see the expedition ship AURORA leaving. Six men had stayed behind to search for Mawson's party, and they spent another winter in Antarctica before the ship picked them up the following spring. Edmund HILLARY described Mawson's solo journey as 'probably the greatest story of lone survival in Polar exploration.'

Mawson was professor of geology at Adelaide University from 1921 to 1931 and made two more voyages to the Antarctic as leader of the 1929–31 BRITISH-AUSTRALIAN-NEW ZEALAND ANTARCTIC RESEARCH EXPEDITION, which carried out extensive scientific investigations and mapped 2500 km (1555 miles) of coastline.

Mawson advocated international controls on WHALING, SEALING and the slaughter of PENGUINS. As a result of his efforts, MACQUARIE ISLAND was declared a wildlife sanctuary in 1933. When he died in 1958, he was given a state funeral.

Mawson Base Mawson Base, named after Douglas MAWSON, was established by AUSTRALIA in 1954, in Horseshoe Harbour on the coast of MAC.ROBERTSON LAND.

Mawson is the oldest continuously operating RESEARCH STATION south of the ANTARCTIC CIRCLE. Meteorological, geophysical, atmospheric and space physics research, as well as EMPEROR and ADÉLIE PENGUIN monitoring, is conducted at Mawson.

Mechanical Age (1923–present) The first successful flights over Antarctica—made in 1928–29 by Carl Eielson and Hubert WILKINS—marked a change from the HEROIC AGE of exploration to an era where improved TECHNOLOGY has allowed humans to live year-round in the Antarctic.

Medical research Doctors on early expeditions to Antarctica undertook basic physical monitoring, and early research was mostly concerned with exposure conditions, such as HYPOTHERMIA and FROSTBITE.

In 1980–81 the INTERNATIONAL BIOMEDICAL EXPEDITION carried out a series of studies of how the human body reacts to polar conditions, research of interest to space researchers as it is believed that astronauts face similar extreme conditions. Other research areas include immune systems and hormonal adaptation to polar conditions. The hormone melatonin, for example, regulates body patterns, and is associated with light-dark cycles. Over an austral winter, secretion of melatonin is disturbed, and researchers have been experimenting with replacement therapy.

Antarctica is also a unique place to study the human immune system. Living away from sources of infection appears to lower people's natural immunity to disease. Consequently, when the first staff arrive at the end of winter, colds and influenza spread around bases. Researchers are isolating every strain of microbe in every individual at the beginning of the season, and then tracking their movement from person to person to show which strains become dominant and which die out.

Medicine Although all staff, scientists and tourists must be in good health before they travel to Antarctica, doctors must be prepared to cope with any medical problem. They must also be ready to improvise. Ingenious medical solutions have included dentures repaired with parts of a seal tooth and glasses ground out of Perspex to match a prescription.

Antarctic ailments have changed greatly since the early days of exploration. Many medical problems suffered on those expeditions were a result of poor DIETS and meagre PROVISIONS. In 1912, during a long overland trek on the AUSTRALASIAN ANTARCTIC EXPEDITION, Douglas MAWSON and Xavier Mertz began losing large amounts of skin. Mertz eventually died. This was caused by a condition called hypervitaminosis, or Vitamin A poisoning, contracted by eating their dogs' livers. The lack of Vitamin C also caused many cases of scurvy. Frederick COOK, medical officer on the BELGIAN ANTARCTIC EXPEDITION, described the symptoms: 'We became pale, with a kind of greenish hue; our secretions were more or less suppressed. The stomach and all the organs were sluggish and refused to work. Most dangerous of all were the cardiac and cerebral symptoms. The heart ached as if it had lost its regulating influence ... The men were incapable of concentration, and unable to continue prolonged thought.' Dental problems—including cracking teeth, fillings falling out and severe toothache—were also frequent because of exposure to the cold.

Doctors now have the ability to cope with

Above: Meteorites buried within the ice sheet long ago are sometimes released in glacial thaw or pushed to the surface as the ice moves up against mountains and moraines. They offer valuable insight into the evolution of our solar system and of the Earth.

most medical problems, and station facilities generally contain consulting rooms, an operating theatre, dental equipment, X-ray facilities and a medical laboratory. Some of the large bases have fully equipped hospitals. The most common medical problem is injuries (cuts and wounds) caused by accidents. Over winter, doctors may have to perform emergency SURGERY since they generally cannot evacuate a patient.

Melbourne, Mount A young VOLCANO in northern VICTORIA LAND, near TERRA NOVA BAY. It erupted less than 200 years ago and still has steam seeping from its summit, where there is an ice-free area on which ALGAE, a very unusual liverwort, a distinctive leafy MOSS and a 'testate' amoeba (which lives within a shell) occur.

Melchior Islands A group of 16 islands in the PALMER ARCHIPELAGO on the west coast of the ANTARCTIC PENINSULA, each island is named after a letter in the Greek alphabet. ARGENTINA erected a lighthouse on one of the islands, Lambda, in 1942, which is now protected as an HISTORIC SITE.

Messner-Fuchs Crossing In 1989–90 the Austro-Italian Reinhold Messner and German Anved Fuchs made the third crossing of the Antarctic continent. Using SKIS and towing SLEDGES, the pair set out from the WEDDELL SEA coast, reached the SOUTH POLE on New Year's Day 1990 and ROSS ISLAND on 12 February. The crossing took 92 days.

Meteor **Expedition (1925–26)** German scientific expedition led by Captain F Spiess following the death of expedition organizer Dr Alfred Merz. The *Meteor* sailed for the SOUTHERN OCEAN in 1925, equipped with ECHO SOUNDERS. The voyage is recognized as an important milestone for OCEANOGRAPHIC RESEARCH. Investigating the Atlantic sector of the ocean, the expedition collected data on water temperature, salinity and oxygen levels, and related them to different ocean layers, the circulation of OCEAN CURRENTS and the topography of the sea floor.

Meteorites Antarctica is the world's richest source of extraterrestrial matter. Thousands of meteorites, mostly from asteroids and comets, that

have plunged through the Earth's ATMOSPHERE have been found buried in Antarctica—meteorites are more difficult to find in other parts of the world, which have been disturbed by human habitation. They offer important clues to the origin of the solar system and the age of our planet. Some meteorites are fragments left over from the birth of the solar system about 15 billion years ago. Others are the primitive building blocks of small planetary bodies that have been shattered in space by violent collisions. Rare fragments of rocks from MARS and the moon have also been found. Like all specimens, meteorites may be collected for scientific purposes only, and are protected by the ANTARCTIC TREATY.

The first meteorite found in Antarctica was discovered by a party from the AUSTRALASIAN ANTARCTIC EXPEDITION in 1912 in TERRE ADÉLIE; according to Douglas MAWSON, 'It measured approximately five inches by three inches by three and a half inches and was covered with a black scale which in places had blistered; three or four pieces of this scale were lying within three inches of the main piece. Most of the surface was rounded, except one face which looked as if it had been fractured. It was lying on the snow, in a slight depression, about two and a half inches below the mean surface, and there was nothing to indicate that there had been any violent impact.'

The first concentration of meteorites was discovered in 1969 by Japanese scientists, near the Queen Fabiola Mountains in EAST ANTARCTICA. Since then thousands of meteorites have been plucked off the ICE SHEET. An American programme that has been operating since 1976 has recovered nearly 10,000 specimens, mainly from the ALLAN HILLS and Yamato Mountains. Most seem to have reached Earth between 10,000 and 70,000 years ago, and are concentrated in places where the flowing ice—acting as a conveyor belt—runs into an obstruction. Mountain ranges tend to funnel ice, or block ice flow, which squeezes and churns meteorites towards the surface. Strong winds then scour the surface layer, exposing the meteorites.

Compared with those collected in warm regions, Antarctic meteorites are well preserved and many still have a 'fusion crust', a glossy black coating left over from their plunge to Earth. Because meteorites absorb cosmic radiation on their flight through space, the time of their impact with Earth can be determined from laboratory studies. The time that they have rested on the ice, the meteorites' 'terrestrial residence age', also provides information that may be useful in studies of the Antarctic Ice Sheet.

Meteorological research The study of the processes that control the CLIMATE of Antarctica involves understanding WEATHER on a short time scale and linking this to patterns in the global climate system.

The first detailed climatological records of Antarctica were kept by William COLBECK and Louis BERNACCHI on the 1898–1900 SOUTHERN CROSS EXPEDITION. More complete records and observations from the 1950s on, about conditions such as TEMPERATURE, humidity, air pressure, and

WIND speed and direction, have been used to show long-term changes to weather systems in the ATMOSPHERE and their relationship to the climate.

In the past, meteorological data were sparse, gathered in convenient locations, rather than places most representative of Antarctica's weather. In 1950 there were only 20 meteorological stations in Antarctica and until the 1960s most weather information came from WHALING fleets. In the INTERNATIONAL GEOPHYSICAL YEAR, the first Automatic Weather Stations (AWS) were placed in remote regions to 'fill in the gaps' and provide fuller information about weather across the whole continent, allowing comparative analysis of data about the extreme weather of inland Antarctica and variations of climatic region. Now, every Antarctic base records weather conditions; this information, along with AWS data, is sent to 'supercomputers' at meteorology offices all over the world for global weather predictions.

During World War II techniques were developed to make meteorological measurements high in the atmosphere using hydrogen-filled BALLOONS. In 1947, OPERATION HIGHJUMP provided the first network of atmospheric observations to produce twice-daily weather maps of Antarctica. These maps, intended as air navigational aids, shed new light on the movement of coastal weather systems. Balloons are still sent into the Antarctic atmosphere regularly and are tracked across the continent using RADAR. Since the early 1970s SATELLITES have been used in Antarctic meteorology, most commonly to show CLOUD FORMATIONS associated with weather systems. Instruments on board satellites can measure the amount of sunlight reflected from the cloud cover, the infrared RADIATION emitted from the clouds and PRECIPITATION.

Microbial activity In Antarctica, microbial life and activity continues even within frozen SEA ICE,

inside translucent ROCKS in the seemingly sterile DRY VALLEYS and possibly within the subglacial Lake VOSTOK, which may have been sealed under the ICE SHEET for over a million years.

As sea ice forms, DIATOMS, other PHYTOPLANKTON and BACTERIA are trapped, initially in the FRAZIL ICE and finally in the PACK ICE, where measurements of increasing chlorophyll concentrations show that these MICROORGANISMS continue to actively photosynthesize and grow. At times their numbers become so dense that the underside of the ice is stained brown. The sea-ice community of diatoms is usually much denser than that in the water below, and the 'captured' phytoplankton are protected from grazing HERBIVORES until the pack ice melts.

In the Dry Valleys region, some microorganisms—called endoliths—live inside the rock and survive on warmth, moisture and the small amount of light that seeps between rock crystals. These LICHEN, ALGAE and FUNGI can only be seen when the rocks are split open.

The discovery of Lake Vostok, the huge SUBGLACIAL LAKE of liquid water the size of Lake Ontario trapped beneath the Ice Sheet, was confirmed as recently as 1996. It is believed that microbial life may have developed in this incredibly cold, dark environment in isolation from the rest of the biosphere. In fact, scientists believe that conditions in Lake Vostok may be comparable to those on Europa, one of Jupiter's frozen moons. If so, these organisms will have evolved in a very specialized way.

Microbial research In the late 19th century, scientists began to realize that microbes play an important role in recycling nutrients through ecosystems. However, few believed that there was much MICROBIAL ACTIVITY in Antarctica. A breakthrough occurred on the 1901–03 SWEDISH SOUTH POLAR EXPEDITION, when bacteria was discovered

in surface soil. This had not then been achieved in the Arctic. On the AUSTRALASIAN ANTARCTIC EXPEDITION, bacteria was found in falling SNOW, glacier ice and meltwater. The first coordinated microbial research programme was a survey of soil microbiology on Signy Island in the SOUTH SHETLANDS. The 1970s discovery that MICROORGANISMS, such as BACTERIA, ALGAE and FUNGI, survive inside translucent rocks in the DRY VALLEYS aroused international interest, and these species have been the focus of a long-term research programme.

Microorganisms Any organism not large enough to be seen with the naked eye but visible under a microscope, such as BACTERIA, DIATOMS, single-celled ALGAE, microfungi, PROTOZOA, PHYTOPLANKTON and smaller animals within the ZOOPLANKTON. Microorganisms are found in both the marine and terrestrial habitats of Antarctica and are important as primary producers (photosynthesizers), particularly in the marine FOOD WEB. Microorganisms are also decomposers and therefore contribute to NUTRIENT RECYCLING, and CYANOBACTERIA are important fixers of atmospheric nitrogen. MICROBIAL ACTIVITY has been found to occur even within the extreme environments of the PACK ICE and within the pores and fissures of rocks in the desolate and seemingly sterile DRY VALLEYS.

Midge The WINGLESS MIDGE (*Belgica antarctica*) is the largest INVERTEBRATE in Antarctica. Found on the ANTARCTIC PENINSULA, where it grows to approximately 12 mm (½ in) in length, it may be seen crawling over MOSS on warm days. A second, flying, species of midge is established at Signy Island after having been accidentally introduced from SOUTH GEORGIA.

Midwinter Following a tradition that began with the early explorers, on the midwinter solstice, on 21 or 22 June (depending on the year), those OVERWINTERING celebrate the mid-point of their working year with banquets and parties across Antarctica, and messages are exchanged between bases.

Military activities Before the ANTARCTIC TREATY came into force in 1961, there had been several military skirmishes and one WAR fought in the Antarctic area. Article I of the Antarctic Treaty prohibits in Antarctica (defined as below 60°S) 'any measure of a military nature, such as the establishment of MILITARY BASES and fortifications, the carrying out of military manoeuvres, as well as the testing of any type of weapon.' This does not prevent military assistance for SCIENTIFIC RESEARCH or other peaceful purposes, which is specifically permitted by the treaty.

Military bases Some early Antarctic bases were of a military nature—for example, those built by ARGENTINA on Gamma and DECEPTION ISLANDS in 1947–48. However, the ANTARCTIC TREATY prohibits MILITARY ACTIVITIES and bases, although a RESEARCH STATION may be managed by the particular state's military forces.

Below: Some microorganisms, such as rotifers, have the ability to undergo cryptobiosis—in which they enter a state of reduced metabolic activity—when environmental conditions are unfavourable. This has enabled them to survive in the harsh Antarctic environment.

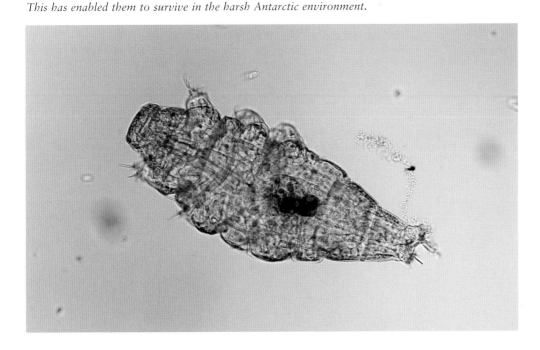

Mineral resources As most rocks in Antarctica are submerged beneath the ICE SHEET, only a small proportion of the continent has been investigated for mineral deposits. However, rich mineral reserves found in other GONDWANA continents that were once attached to Antarctica suggest that large deposits may be hidden, undiscovered, beneath the Ice Sheet.

The first mineral discovery was made by Frank WILD, who recorded COAL seams near the BEARD-MORE GLACIER during the 1907–09 BRITISH ANTARCTIC EXPEDITION. Although low grade, the TRANSANTARCTIC MOUNTAINS contain considerable quantities of coal, and substantial iron deposits are found in the slopes of the PRINCE CHARLES MOUNTAINS. Other mineral resources are only known to be present in small quantities. Traces of GOLD, titanium, tin, copper, cobalt and uranium on the ANTARCTIC PENINSULA are extensions of the rich mineral deposits in the Central Andes of South America. Geologists speculate that the PEN-SACOLA MOUNTAINS may contain some precious metals, and the Dufek Massif may have a relatively sizeable quantity of platinum, cobalt and nickel, along with copper and iron deposits. Mineral nodules on the floor of the SOUTHERN OCEAN contain small quantities of iron, manganese, copper, nickel and cobalt.

The most sought-after resources potentially lie offshore, in hydrocarbon (OIL and gas) deposits on the CONTINENTAL SHELF surrounding Antarctica. The geological evolution of the Antarctic continental margin has resulted in the develop-ment of large sedimentary basins that have been gently deformed to create hydrocarbon reservoirs. In 1973 gaseous hydrocarbons were found in the ROSS SEA, sparking considerable international interest, and it is believed that further gas and oil deposits may be located in the ROSS, AMUNDSEN, BELLINGSHAUSEN and WEDDELL SEAS.

The MINING of mineral resources was banned for 50 years by the MADRID PROTOCOL to the ANTARCTIC TREATY, which was signed in 1991, and came into force in 1998.

Mining Antarctica's isolation and environment make the extraction and transportation of MINER-AL RESOURCES extremely expensive and complex. However, as more efficient techniques develop (and as reserves are depleted in more accessible areas), mining in the Antarctic may become commercially viable. The associated technical problems are very similar to those encountered, and solved, in the Arctic. Nevertheless, major environmental hazards—such as KATABATIC WINDS, ICEBERGS and storms—along with a short working season and the isolation of Antarctica, will always be obstacles to mining operations. The environmental sensitivity and the potentially disastrous consequences of OIL spills in the SOUTHERN OCEAN puts mining high on the CONSERVATION agenda.

The ANTARCTIC TREATY makes no reference to minerals because, at the time it was drafted (1958), treaty nations believed that agreement was impossible. In 1982 negotiations began on the CONVENTION ON THE REGULATION OF ANTARC-TIC MINERAL RESOURCE ACTIVITIES (CRAMRA). A strong environmental lobby opposed CRAMRA, arguing that the convention would only provide protection once mining began, and that treaty nations should be discussing whether minerals should be mined at all. The *Exxon Valdez* oil spill in Alaska, and the 1989 sinking of the *Bahia Paraiso* near PALMER STATION, which leaked oil into the Southern Ocean, drove home the potential environmental impact of mineral exploitation in Antarctica. Instead of ratifying CRAMRA, the treaty nations opted to draft the PROTOCOL ON ENVIRONMENTAL PROTECTION to the Antarctic Treaty; this was signed in Madrid in 1991 and came into force in 1998. Also known as the MADRID PROTOCOL, it places a ban on all mining activity for at least 50 years. Article 7 of the Protocol reads: 'Any activity relating to mineral resources, other than scientific research, shall be prohibited.' The treaty nations may amend this at any time, but it seems unlikely that any mining ventures will be permitted to operate in Antarctica in the near future.

Minke whale (*Balaenoptera acutirostrata*) The smallest and commonest of the Antarctic BALEEN WHALES, minkes grow to around 11 m (36 ft) in length and weigh up to 10 tonnes (10 tons). Although populations have been identified in subtropical waters, there is a theory that the minkes may winter in the cold Antarctic waters rather than migrate north to breed as other species do. There is also uncertainty about whether they have

Below: *The smallest of the baleen whales, minkes* (Balaenoptera acutirostrata) *feed on krill and copepods.*

a one- or two-year reproductive cycle, although it is known that gestation is about 10 months. They feed on KRILL and COPEPODS.

As a small, fast-swimming species, minkes were not targeted by commercial whalers until the larger whales were depleted. Because minke whale populations are quite large in many areas, some scientists feel they can be sustainably harvested on a small scale. Minkes have also been targeted by countries that want to whale commercially; Japan, for example, continues to kill minke whales for 'scientific research' despite international protest.

Mirages Conditions over Antarctica are ideal for the formation of these OPTICAL PHENOMENA. The ICE chills a layer of air over the surface, above which the air TEMPERATURE often increases with height (called a temperature inversion). Differences in the way light is transmitted through the layers of air in an inversion produce confusing distortions: for instance, mountains may seem to loom above the horizon when they are, in fact, well beyond it. Other features may be flipped upside down to form 'inverted mirages': as Ernest SHACKLETON wrote, 'Icebergs hang upside down in the sky, land appears as cloud. Cloud-banks look like land.'

W Burn Murdoch, artist on the 1892–93 SCOTTISH NATIONAL ANTARCTIC EXPEDITION, described the surrealness of mirages: 'Later in the evening a mirage strangely affected the appearance of the bergs. Round the horizon they divided into a circle of pale druidical pillars with yellow light shining between them, and seemed to support the canopy of faintly grey sky. Then a soft white fog fell and they disappeared and the ships began calling to one another like partridges in the evening when the mist lies low on the winter field.'

When the inversion layer is at high altitude, objects may seem much closer. This effect is exaggerated by the crystal clarity of polar air, and many navigation errors have resulted.

Charles WILKES 'discovered' land in 1840; he claimed to be within 15 km (9 miles) of the Antarctic coast. Later explorers, who plotted his claimed position as 700 km (434 miles) from the ANTARCTIC PENINSULA, discredited his findings; although Wilkes was accused of falsely mapping land and court-martialled, he may have been misled by a mirage. Similarly, in 1841 James Clark ROSS 'discovered' mountains, now believed to be a mirage of a range about 400 km (248 miles) further away.

Mirnyy Base Established with an OBSERVATORY in February 1956 in preparation for the INTERNATIONAL GEOPHYSICAL YEAR, the base was named after one of Thaddeus BELLINGSHAUSEN's ships. It was the main USSR Antarctic base until 1963. Mirnyy is on the WILHELM II LAND coast, by the DAVIS SEA, and lies between DAVIS and CASEY STATIONS. RESEARCH conducted here includes meteorology, geophysics, seismology, glaciology, medicine and climatology. After the break-up of the USSR, Mirnyy was transferred to RUSSIA.

Mites (Acari) Mites and SPRINGTAILS are the dom-

Above: *Mirages such as this one of rough sea ice are caused by differences in the way light is transmitted through the layers of air.*

inant land INVERTEBRATES in Antarctica. Mites have eight legs and look like colourful, miniature versions of their relatives, spiders. Some may reach only 0.5 mm (0.02 in) as full-grown adults.

The 528 mite species found in Antarctica and the SOUTHERN OCEAN include parasitic, freshwater and marine species. Some 150 are terrestrial free-living species, which inhabit MOSS; some are carnivorous, preying on other mites and springtails. Weighing in at only 0.1 g (1.5 grains), the predatory mite *Gamasellus racovitzai* is the largest terrestrial predator in Antarctica: springtails are its main source of prey.

Alaskozetes antarcticus, a species common on the ANTARCTIC PENINSULA, can sometimes be found in clusters of thousands. Some of the highest population densities are found among ALGAE growing on the feathers of dead PENGUINS and in moss and algal communities in areas enriched by penguin colonies. One mite species, *Nanorchestes antarcticus*, can be found as far south as 85°S. Many avoid freezing by supercooling, keeping their body fluids liquid at temperatures as low as –35°C (–31°F) with the aid of glycerol ANTIFREEZE synthesized within their body. Because food in their gut increases the risk of ice forming, they must find a balance between starving and freezing in order to survive.

Many Antarctic mites are PARASITES on other ANIMALS, particularly VERTEBRATES, with some living in the nasal passages of SEALS. Others—parasitic feather mites—flourish within the warm microclimate of the plumage of BIRDS.

Mock sun A name for PARHELION, types of HALOS.

Molecular biological research Molecular biologists study the structure and function of biological molecules, especially nucleic acids and proteins. This highly specialized research area has only been applied to Antarctic studies since the late 20th century, but has already yielded some interesting results. Molecular biologists are able, for example, to glean vital DNA information about WHALES from small skin samples. Molecular studies have also been carried out on mummified SEAL carcasses from the DRY VALLEYS, and similar techniques have been applied to Antarctic MOSSES, ALGAE, terrestrial INVERTEBRATES and FISH that produce ANTIFREEZE compounds to protect their tissues.

Molluscs The Antarctic marine FAUNA includes almost 900 species of molluscs, a family of soft-bodied INVERTEBRATES, the bodies of which are usually protected by a shell. Gastropods, bivalves and CEPHALOPODS are all molluscs. NUDIBRANCHS are gastropod molluscs that have dispensed with their protective shell.

The LIMPET, *Nacella concinna*, is a gastropod mollusc and the most abundant animal in the inter-tidal fauna of Antarctic shores. Chitons and small winkle-like snails may also be present, but mussels and barnacles, commonly found among temperate inter-tidal fauna, are absent.

Antarctic bivalves and gastropods are usually small. Among the exceptions is the Antarctic yoldia, *Yoldia eightsi*, the shell of which reaches a length of 3.5 cm (1½ in) at an estimated age of 65 years—based on growth-ring calculations, yoldia with a 4.5 cm (1¾ in) shell length are about 150

Above: *An ancient moraine (at right) formed by a granite escarpment.*

years old. The Antarctic scallop, *Adamussium*, grows to 10 cm (4 in) and another bivalve, *Lateruula elliptica*, grows to about 7 cm (2¾ in). In the SOUTH ORKNEY ISLANDS, 75 percent of the gastropod species found at depths from the surface to 100 m (330 ft) are less than 10 mm (½ in) long.

The cephalopod class of molluscs includes OCTOPUS and SQUID, but little is known about the Antarctic species.

Mollymawk The common name for some of the smaller ALBATROSSES. Nesting albatrosses and mollymawks are very tame and allow humans to walk right up and even stroke them. Because of this, Dutch sailors used to call them 'mollymawks', or 'foolish gulls'.

Montreal Protocol An international agreement, signed in 1987 and amended in both 1990 and 1992, on reducing the production and consumption of compounds that deplete OZONE in the stratosphere.

Moraine Accumulations of glacial debris, or till,

that have been 'bulldozed' by GLACIERS into mounds and hummocks. Rows of moraines undulate across previously glaciated areas in Antarctica, such as the DRY VALLEYS. They are most commonly formed by glacier retreat, when rock material is carried far from its origin at the glacier base and deposited; these 'end moraines' are often dated to give clues about past glacial advances and retreats. 'Lateral moraines' form parallel to the direction of ICE flow, and are built up by debris that tumbles down the side of the glacier to the ice edge. When two glaciers meet, their lateral moraines are also pushed together, and become 'medial moraines'.

Morrell, Benjamin (1795–1839) American sealer and explorer. The son of a Rye, New York, shipbuilder, Morrell went to sea at the age of 16. In 1821 he was first mate on the *Wasp* on a SEALING expedition to the SOUTH SHETLAND ISLANDS. In 1822, as commander of the *Wasp*, he made the first known landing on BOUVETØYA, visited Îles KERGUÉLEN, and from there made his way to the SOUTH SANDWICH ISLANDS. Morrell always maintained that he reached 70°S in the WEDDELL SEA,

but discrepancies in his accounts and serious miscalculations of his positions have placed doubts on many of his claims. Morrell died of fever in Mozambique.

Mosses To date, about 100 species of mosses have been recorded on the Antarctic continent. The southernmost moss species that has been collected is *Ceratodon purpureus*, found at 84°30'S at Mount Kyffin in southern VICTORIA LAND.

In places where LICHENS grow, a little sandy SOIL will have been established, and this provides a starting point for mosses. Many Antarctic mosses have tightly packed stems and shoots, which prevent water loss. Mosses are not as hardy as lichens and thus are less widespread.

The most productive regions of the ANTARCTIC PENINSULA, where most Antarctic moss species are found, are the western areas north of 70°S. Here, on clear days in the heart of summer, the temperatures of the exposed rock faces on which the mosses grow can rise to 40°C, even though the air temperature may be close to zero. Production here can be high, although growth phases are brief.

For most of the year mosses and other plants

survive the cold, drought and nutrient deprivation in a resting state of suspended animation. Moss banks may build to several metres (yards) in depth, with all but the top 25 cm (10 in) permanently frozen and turned into peat. Such peat banks may be thousands of years old—moss growth is measured in hundreds of years and a footprint in a moss bank will remain there for centuries.

Motor cars An innovation on Ernest SHACKLETON's 1907–09 IMPERIAL ANTARCTIC EXPEDITION was motorized transport—the first on the Antarctic continent—in the form of a 12–15 horsepower, four-cylinder, air-cooled Arrol-Johnson car with special oil, ordinary petrol, a reinforced frame, and recycled exhaust to help heat the engine.

On its first trial, the car travelled about 90 m (95 ft), but after tuning managed 48 km (30 miles) at between 5 and 24 km (3–15 miles) per hour. The car proved of great worth in laying depots until it disappeared into a crevasse; although it was rescued, it was not used again.

Today, many Antarctic programmes use commercially available cars and trucks around their bases, modified with specialized fuels and lubricants to cope with the environmental conditions.

Mottled petrel (*Pterodroma inexpectata*) Of the GADFLY PETREL family, mottled petrels are solitary birds, medium sized with wing-spans of around 85 cm (33 in). Their flight pattern, which consists of fast, high arcing and sustained gliding, is typical of most gadfly petrels.

Easily recognized at sea by the conspicuous patterned underparts, consisting of black leading margins on white underwings and dark grey patches on the lower breast and belly, the petrels' backs are grey except for dark 'M'-shaped markings that stretch across the wings. They have white faces with mottled grey foreheads and short black bills, and their feet are flesh-coloured with dark webs and edges.

A summer breeder endemic to the islands in the NEW ZEALAND region and the SNARES, mottled petrels have an extraordinary distribution: they migrate to the northern Pacific for the winter, but during the summer they are often encountered in the Pacific region of the SOUTHERN OCEAN as far south as the ice edge. They usually stay clear of ships. In the Antarctic they feed on SQUID, FISH and CRUSTACEANS.

Mountaineering The 3795 m (12,450 ft) high Mount EREBUS, the southernmost active VOLCANO in the world, was first climbed in 1908 by a group of six men, including Douglas MAWSON, from the 1907–09 expedition led by Ernest SHACKLETON. However, there was no serious mountaineering on the continent until a 1966 American Alpine Club expedition led by Nicholas Clinch, in the course of which VINSON MASSIF, Antarctica's highest mountain, along with a number of other high ranges, was climbed. The following year a NEW ZEALAND team climbed in northern VICTORIA LAND.

A further series of mountaineering expeditions took place from the mid-1980s. A team led by Colin Monteath ascended peaks in the DRY VALLEYS region. A 1994 Norwegian expedition led by Ivar Tollefsen ascended 36 peaks in DRONNING MAUD LAND, including Antarctica's first 'big-wall' rock climb on a towering red spire of granite called Ulvetanna (the 'Wolf's Fang'), which involved 11 days of climbing. When they reached the summit, Tollefsen flew off using a parapente. In the summer of 1997–98 ADVENTURE NETWORK INTERNATIONAL began flying climbing groups into Dronning Maud Land.

Mountains Although much of Antarctica has been overwhelmed by ICE, mountainous outcrops of ROCK protrude from the white surface cover in many parts of the continent. In some places, only isolated spires of rock called NUNATAKS appear above the ice; in others, long spines of rock form the exposed backbone of partly hidden mountain ranges. The longest mountain chain in Antarctica is the TRANSANTARCTIC MOUNTAINS. Extending for more than 3200 km (2000 miles), it separates EAST ANTARCTIC from WEST ANTARCTICA. The SENTINEL RANGE of the ELLSWORTH MOUNTAINS is Antarctica's tallest, and includes the VINSON MASSIF, which reaches 4897 m (16,062 ft) above sea level.

Roald AMUNDSEN was enraptured by the mountainous landscape: 'Peaks of the most varied forms rose high into the air, partly covered with driving clouds ...' he wrote. He became almost poetic in describing Mount Helmer Hanssen: '... its top was as round as the bottom of a bowl, and covered by an extra-ordinary ice-sheet, which was so broken up and disturbed that the blocks of ice bristled in every direction like the quills of a porcupine.'

Mountain building in Antarctica is a complex process of uplift and erosion. Between the Devonian period (approximately 408 to 360 million years ago) and the middle part of the Jurassic period (about 180 million years ago), sediments were laid down in lakes and basins, where former mountain chains had been carved away by erosion. Known as the Beacon Sandstone, this formation of platform sediments contains a rich FOSSIL record of extinct Antarctic life forms. Over long periods, the sedimentary material was uplifted and deformed into mountains, which have been carved by glacial erosion and covered in ice. Some ranges, such as the Gamburtsev Mountains, remain completely buried and have only been mapped by seismic reflections (see GEOLOGICAL RESEARCH).

Murray, John (1841–1914) Scottish naturalist, born in Canada. Murray studied at Edinburgh University, spent some months on an Arctic whaler, and joined the *CHALLENGER* EXPEDITION in 1872. During the four-year voyage he was in charge of bird specimens, and worked on the classification of deep-sea deposits. After dredging material at 66°33'S, he noted that, 'fragments of mica schists, quartzites, sandstones, compact limestones, and earthy shales ... leave little doubt that within the Antarctic Circle there is a mass of continental land quite similar in structure to other continents.'

Following the death of Wyville THOMSON, Murray took over the editorship of the 50-volume *Report on the Scientific Results of the Voyage of the H.M.S. Challenger*, which was published between 1881 and 1895.

Murray was an influential lobbyist for Antarctic research and developed a plan for

Below: Although close to 100 moss species have been identified on the Antarctic continent, they are not widespread.

Above: Stretching 3200 km (2000 miles), the Transantarctic Mountains form a north-south barrier separating East and West Antarctica.

extensive and systematic scientific research, which was followed by several later expeditions. The Sir John Murray Glacier at the head of Robinson Bay, Cape ADARE, was discovered and named after him by Carsten BORCHGREVINK.

Museums Collections of Antarctic material in museums usually include artefacts from early expeditions, polar equipment, rock specimens, fossils, animal and underwater dioramas and models, diagrammatic displays and geographical models, as well as journals, documents and portraits.

Museum exhibitions at the traditional GATE-WAY ports vary in quality. The Canterbury Museum in Christchurch, NEW ZEALAND, holds an extensive collection of HEROIC AGE artefacts, including items used at the SOUTH POLE by Roald AMUNDSEN's party, the motor-sledge used by Ernest SHACKLETON's Ross Sea Party in 1914–16, items from Robert SCOTT's 1910–13 expedition, a

SNO-CAT from the COMMONWEALTH TRANS-ANTARCTIC EXPEDITION and material from OPERA-TION DEEPFREEZE. Two museums in Hobart, Tasmania—Tasmanian Museum and Gallery and the Maritime Museum of Tasmania—have holdings of artefacts from early expeditions and from whaling and sealing voyages. Antarctic displays at the Museo Naval y Maritime in Punta Arenas, CHILE, and Museo de Ushuaia, ARGENTINA, include some artefacts, photographs, model ships and natural history displays.

Museums with collections in the Northern Hemisphere are located in countries with strong Antarctic connections. The largest collection in the USA is in the Navy Museum in Washington DC. The SCOTT POLAR RESEARCH INSTITUTE in Cambridge, England, holds an extensive collection of journals, manuscripts and documents, as well as artefacts from Scott's and Shackleton's expeditions. Whaling museums at Sandefiord and

Tønsberg, NORWAY, and museums in Oslo have material from Norwegian expeditions, including Amundsen's ship FRAM. The Shirase Antarctic Expedition Museum, near Konoura, JAPAN, features items from Nobu SHIRASE's 1910–12 expedition. RUSSIA's Arctic and Antarctic Museum is located in St Petersburg.

Music Ralph Vaughan Williams's SINFONIA ANTARTICA, probably the best known music composed about Antarctica, originated as the score for the 1948 Charles Frend FILM *Scott of the Antarctic*. Likewise, the synthesizer piece 'Antarctica' by Vangelis was commissioned to accompany a film directed by Koreyoshi Kurahara. Most other Antarctic-related music was also composed as film soundtracks, notably John Cale's 'Antarctica' for the 1995 Manuel Huerga film *Antarctica as a State of Mind*, and much of these are 'New Age' compositions.

n

Nansen, Fridjof (1861–1930) Norwegian scientist, politician and explorer. Born near Oslo, Nansen studied at Oslo and Naples universities before travelling to the Arctic in 1882. In 1888–90 he crossed the interior of Greenland with five others.

From 1893 to 1896, in the purpose-built ship FRAM, Nansen explored the Arctic, investigating the drift of polar sea ice and, on an overland journey, reaching 86°14'N—the highest latitude achieved at that time

From 1896 Nansen was professor of zoology, then oceanography, at Oslo University and carried out extensive oceanographical studies in northern waters. He was Norway's ambassador to BRITAIN, led the Norwegian delegation at the League of Nations, and worked for the repatriation of prisoners of war, famine relief and refugees. He was awarded the Nobel Peace Prize in 1922.

Nansen's direct association with Antarctica dates from 1909 when he lent the *Fram* to Roald AMUNDSEN. Two of his 'inventions' were invaluable to all polar explorers: the Nansen SLEDGE on ski-shaped runners, lightweight but strong and flexible over uneven surfaces, and the Nansen stove, which had a double-boiler used for melting drinking water. He was supportive of Antarctic explorers, and generous with first-hand advice on polar conditions.

Nares, George Strong (1831–1915) British naval officer and explorer. The leader of the CHALLENGER EXPEDITION was born in Aberdeen, Scotland, and entered the British Navy in 1845. From 1854 to 1863 he was involved with cadet training and in survey work around the British Isles. Nares was a steady and well-respected commander of the *Challenger*, and was fully supportive of the scientific aims of the expedition. The trawling system he designed proved efficient in collecting specimens from the sea floor. He was recalled from the *Challenger* expedition to command an Arctic expedition, the chief aim of which was to reach the North Pole. Although not attaining this goal, the expedition completed significant scientific work. In 1876 he received a knighthood, and in 1878 he was involved in surveying the MAGELLAN STRAIT.

NASA Established in 1958, the USA's National Aeronautics and Space Administration (NASA) has explored the inter-relationship Antarctica has with the Earth's climate and global weather systems and its importance as an early barometer of global warming and the GREENHOUSE EFFECT, as well as exploring outer space.

In the process, NASA has provided a 'bird's eye' view of Antarctica. SATELLITES have allowed scientists to view Earth from the outside and study the ATMOSPHERE, continents and oceans as a whole. Weather monitoring is an important part of NASA's work, and encompasses studies of the total Earth system and the effects on that system of both natural and human-induced change. As part of this research, an instrument developed by NASA is being used to map the OZONE HOLE.

SEA ICE and ICE SHEET investigations, the study and archiving of METEORITES and Antarctic technology developments are also carried out by NASA scientists. MICROBIAL RESEARCH is of interest for its extraterrestrial applications, and NASA scientists consider the DRY VALLEYS provide the closest environmental conditions to those found on Mars. Research was carried out in the Dry Valleys prior to the launch of the Viking Mars mission.

National Antarctic Expedition (1901–04) British expedition to Antarctica, led by Robert SCOTT. The expedition made the first extensive land explorations of Antarctica and carried out a scientific programme under the direction of Edward WILSON. The initiative for the expedition came from Sir Clements MARKHAM, who raised funds from private and public sources. The expedition ship, the DISCOVERY, was purpose built in Dundee, Scotland.

On 3 February 1901 the DISCOVERY anchored in the BAY OF WHALES. The next day Scott made the first BALLOON ascent in Antarctic history, ascending 244 m (800 ft) in a hydrogen balloon to survey the ice shelf. The expedition went on to winter at Hut Point, MCMURDO SOUND. To help pass the expedition's first winter, Ernest SHACKLE-

Below: Starting out on their great southern journey are (left to right) Ernest Shackleton, Robert Scott and Edward Wilson.

Below: Robert Scott (seventh from left), leader of the 1901–04 National Antarctic Expedition, on board the Discovery.

TON produced monthly editions of the *SOUTH POLAR TIMES*. This newsletter contained articles, humorous pieces, caricatures and puzzles.

On 2 November 1902 Scott, Wilson and Shackleton set out for the SOUTH POLE. They reached approximately 82°17'S before turning back, closer to the Pole than any previous party. Other groups made sledging trips from the *Discovery* base. Albert Armitage led a party to explore the western mountains in an attempt to find a route up to the POLAR PLATEAU and into the interior.

When the relief ship, *Morning*, arrived in January 1903, the *Discovery* was frozen into the ice and Scott decided to spend another winter in Antarctica. Some men, including the ailing Shackleton, were sent north on the *Morning*. In September Scott led a 10-man party on another journey, man-hauling sledges to the FERRAR GLACIER. In January 1904 the *Morning* returned with another relief ship, the TERRA NOVA. Using explosives and the swell, the *Discovery* was finally freed from ice in February, just as plans for abandoning the ship were being prepared.

The expedition's experiences of winter yielded some valuable lessons for future Antarctic exploration: in particular, suitable clothing, the working pace of dogs and the inadequacies of some food supplies. Six volumes of scientific reports, predominantly biological descriptions by Wilson and Thomas Hodgson and including meterological observations and detailed magnetic records, were published by the British Museum.

National Institute of Polar Research Japan's inter-university research institute, established in 1973. The institute maintains four RESEARCH STATIONS in Antarctica: SYOWA STATION, Mizuho Station, Asuka Station and Dome Fuji Station, and operates Japanese Antarctic research expeditions on upper atmosphere physics, meteorology, glaciology, earth sciences, biology and polar region engineering.

National Ozone Expeditions A 1986–87 project involving scientists from many nations to investigate the OZONE HOLE. Ground and SATELLITE measurements of ozone concentration were made. Aircraft, flying from CHILE, made repeated sweeps at an altitude of about 18 km (11 miles) to gather data directly from the ATMOSPHERE above Antarctica. They flew about 175,000 km (109,000 miles), and confirmed that there was indeed a 'hole' in the OZONE, and that this was mirrored by a rise in the levels of chlorine monoxide. This confirmed the theory that CHLOROFLUOROCARBONS were finding their way into the upper atmosphere and releasing chlorine, which is partly responsible for the breakdown of ozone over Antarctica.

National Science Foundation An independent USA government agency responsible for promoting science and engineering. The National Science Foundation (NSF) was given responsibility for the UNITED STATES ANTARCTIC RESEARCH PROGRAM in 1959, and in 1971 assumed overall responsibility for managing all American activity in Antarctica. NSF's UNITED STATES ANTARCTIC PROGRAM encom-

passes both research and operational activity, involves around 3000 people each year, and has an annual budget of several hundred million dollars.

The NSF operates three year-round RESEARCH STATIONS in Antarctica: MCMURDO STATION, AMUNDSEN-SCOTT SOUTH POLE STATION and PALMER STATION. It also manages the deployment of research ships and icebreakers, aircraft, remote-sensing equipment such as satellites and research balloons and a number of automated weather stations and geophysical observatories.

Navigation Each summer a number of ICEBREAKERS venture south, carrying SCIENTISTS to Antarctic research stations and also providing support to projects located in and around the SEA ICE. Occasionally such vessels travel in winter, as this gives scientists a rare view of the environment outside the usual research season.

Nearly all other travel by sea to Antarctica is done by YACHT or on commercial CRUISES. The window for such TOURISM is from December to March, when the sea ice extent is at a minimum. Most trips leave from southern parts of South America and travel to the ANTARCTIC PENINSULA. On average, 35 yachts will head to the Peninsula every year, and a number of cruise ships.

All these vessels must negotiate the notorious DRAKE PASSAGE, the ROARING FORTIES, FURIOUS FIFTIES and SCREAMING SIXTIES. There are a sequence of high and low pressure systems embedded within the prevailing westerly WINDS which make for exciting and dramatic weather. However, navigating the Passage and surrounding waters is safer now than in the days of early explorers because of SATELLITE weather information, good charts which show sheltered anchorages, and radio contact with other ships in the area. Most boats will wait in shelter for a gap in the weather before racing across the Passage as quickly as possible—usually a three to six day trip for yachts.

Once near the continent, yachties must keep constant watch for ICEBERGS and BRASH ICE. Most yachts that go south are constructed of steel or aluminium, which cope better than wood or fibreglass with the abrasion of ice. Retractable keels are another advantage, as these enable anchorage in very shallow but sheltered bays. Anchorages must be chosen with care as there is always the possibility of a bay becoming blocked by an iceberg grounding in the entrance. One charter yacht crew has estimated in 10 years' Antarctic sailing an average of half of the time stormbound in anchorages. When not stormy, the summer conditions can be sunny, calm, even balmy.

While travelling there are many hazards from ICE which is sometimes almost completely submerged and very hard to see. The break-up of GLACIER faces also poses a danger, as falling ice creates tidal waves, or pieces from the submerged glacier foot, extending a deceptively long way out from the visible wall, surface under a yacht. Furthermore, the temptation to cruise close around bergs must be tempered by their potential to roll.

An expedition must be totally self sufficient in food, and FUEL, although fresh water can be sourced by melting ice. A good warm stove and insulated boat is a big advantage.

Most yachts do not travel further south than Marguerite Bay—there is little shelter beyond that point and the sea remains heavily iced year-round. Cruise ships make trips south from NEW ZEALAND to the ROSS SEA, then around the coast to the Peninsula and up to South America. A few excursions journey to the WEDDELL SEA also, but much of the rest of the continental shore lacks any shelter for small boats and cruise ships.

Nematodes (Nematoda) A class of unsegmented worms, also called roundworms, of which 43 terrestrial free-living species are recorded from the Antarctic continent and nearby islands. There are

Below: There are 43 species of free-living terrestrial nematodes in Antarctica, and in 1998 one, Panagrolaimus davidi, *was found to survive intracellular freezing.*

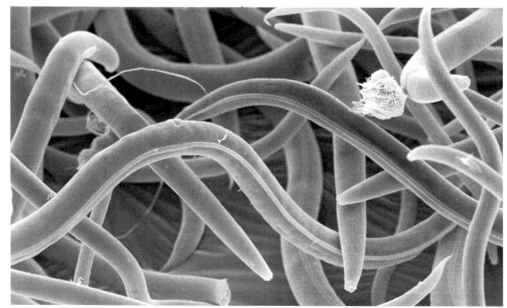

also marine free-living species and parasitic species associated with Antarctic BIRDS, MAMMALS, FISH and marine INVERTEBRATES. Nematodes occur in soils, plants, animals and dead organic material and, although most species are found in terrestrial or freshwater habitats, nematodes also live in the SOUTHERN OCEAN. Studies of population density on Signy Island reveal that between 1 and 5 million nematodes may be found per sq m (1.2 sq yd) of MOSS carpet habitat.

Nematodes are microbial feeders, mainly HERBIVORES, that feed on plant material such as ALGAE growing in the moss carpets as well as BACTERIA, yeast and filamentous FUNGI. Some nematodes may also eat PROTOZOA. At least one Antarctic nematode is carnivorous, and nematodes themselves fall prey to predatory fungi.

Neumayer Station The first Neumayer Station was built by West GERMANY in 1981 on the EKSTRÖM ICE SHELF, in the northeast WEDDELL SEA area. Georg von Neumayer, after whom the station is named, was chairman of the INTERNATIONAL POLAR COMMISSION and an important promoter of Germany's early research activities in Antarctica. The year-round Neumayer Station is a research observatory for geophysical, meteorological and air chemistry measurements.

New Zealand New Zealand's first involvements in Antarctica were through SEALING and WHALING voyages, as a staging post for European expeditions, and as a participant in the 1911–14 AUSTRALASIAN ANTARCTIC EXPEDITION. In 1923, BRITAIN, on behalf of New Zealand, made a TERRITORIAL CLAIM to the ROSS DEPENDENCY. In contrast to other Antarctic claimants, New Zealand has not actively supported its claim; rather, it promoted early calls for the internationalization of Antarctica under the UNITED NATIONS and for the establishment of a WORLD PARK. For New Zealand, one of the 12 original CONSULTATIVE PARTIES, the importance of the ANTARCTIC TREATY was the declaration of a demilitarized, nuclear-free zone of peace and science in Antarctica. However, New Zealand supported the ANTARCTIC TREATY SYSTEM against the criticism of the supporters of the COMMON HERITAGE PRINCIPLE in the 1970s and 1980s.

New Zealand participated in the INTERNATIONAL GEOPHYSICAL YEAR, and built SCOTT BASE as part of the 1955–58 COMMONWEALTH TRANSANTARCTIC EXPEDITION. It made the first inspection conducted under the auspices of Article VIII of the Antarctic Treaty, in late 1963, when two observers visited MCMURDO BASE, AMUNDSEN-SCOTT SOUTH POLE BASE and BYRD STATION.

Mainly reliant on the logistics support provided by the USA, New Zealand allows the Americans to use facilities in Christchurch, from where the Italian and most of the large USA Antarctic operations depart. Although New Zealand has some interest in promoting itself as a GATEWAY for Antarctic TOURISM, there have been no sightseeing flights over the continent since the 1979 AIR CRASH on Mount EREBUS, which killed all 279 people on board. CRUISE ships to Antarctica depart from New Zealand ports.

New Zealand is a member of the Commission

Above The Nimrod *leaves Lyttelton Harbour, New Zealand, 1 January 1908.*

for the CONVENTION ON THE CONSERVATION OF ANTARCTIC MARINE LIVING RESOURCES (CCAMLR) and usually adopts a pro-conservation stance at CCAMLR meetings. It is interested in FISHING in the SOUTHERN OCEAN and, although it asserted the right to an Exclusive Economic Zone (EEZ) off the Ross Dependency in 1979, an EEZ has never been implemented. It is also a signatory to the MADRID PROTOCOL.

New Zealand Antarctic Programme Initially run by government departments, NEW ZEALAND's research programmes are now administered by the New Zealand Antarctic Institute, operating as ANTARCTICA NEW ZEALAND, which was established in 1996.

Nimrod The expedition vessel on the 1907–09 BRITISH ANTARCTIC EXPEDITION, the *Nimrod*, was 40 years old, but was sturdy and reliable. To conserve coal, the ship was towed from Lyttelton, NEW ZEALAND, to the edge of the PACK ICE by the *Koonya*—the longest towing operation ever carried out to that time.

Nimrod Glacier The 135 km (84 mile) long Nimrod flows from the POLAR PLATEAU into Shackleton Inlet and the ROSS ICE SHELF. It is named after the ship used in the 1907–09 BRITISH ANTARCTIC EXPEDITION.

'Ninety Degrees South' Expedition (1986–87) Private British-Norwegian expedition, led by Monica Kristensen. This attempted to follow Roald AMUNDSEN's route to the SOUTH POLE, using DOG teams and air-dropped supplies. Although they failed to reach the Pole and were forced to turn around at 85°59'S (about 320 km/200 miles short), the expedition made some valuable scientific observations on the ROSS ICE SHELF.

Non-consultative party A state that has signed and ratified the ANTARCTIC TREATY, but which has not fulfilled all the requirements necessary to be granted the status of a CONSULTATIVE PARTY at an ANTARCTIC TREATY CONSULTATIVE MEETING (ATCM). Non-consultative parties do not have votes at an ATCM, and they have only been

granted the right to observe the meetings since 1983. In 2001, there were 17 non-consultative parties to the Antarctic Treaty.

Nordenskjöld, Otto Gustaf (1869–1928) Swedish explorer and geologist. Nordenskjöld was professor of geology and mineralogy at the University of Uppsala. From 1895 to 1897 he led a geological expedition to Patagonia and TIERRA DEL FUEGO, and in the following year undertook an expedition to the Yukon.

On 16 October 1901, as leader of the SWEDISH SOUTH POLAR EXPEDITION, he set sail from Göteborg aboard the *Antarctic* commanded by Carl LARSEN. They reached the SOUTH SHETLAND ISLANDS in January 1902 but were unable to enter very far into the WEDDELL SEA. In February, Nordenskjöld and five others established a research station on SNOW HILL ISLAND off the ANTARCTIC PENINSULA and were stranded there for two winters when the *Antarctic*, returning to pick up the party in February 1903, became ice-bound and was crushed. They were rescued in November 1903 by the Argentinian naval vessel *Uruguay*. During their time there, much valuable scientific data had been collected, including FOSSILS from the Jurassic era. Like Erich von DRYGALSKI, who was working at the same time but on the opposite side of the continent, the expedition never crossed the ANTARCTIC CIRCLE.

Nordenskjöld undertook two further expeditions to eastern Greenland, but plans for another expedition to Antarctica were thwarted by the outbreak of World War I.

Northern giant petrel (*Macronectes halli***)** Along with the SOUTHERN GIANT PETREL, these are the largest of the petrels. They have wing-spans of about 1.8 m (6 ft) and are 87 cm (34 in) long. Although similar in appearance to southern giants, they have greenish bill tips rather than red ones. Once thought to be the same species as the southern giant petrel, they also differ from southern giants by breeding in small scattered groups or singly, and usually north of the ANTARCTIC CONVERGENCE.

Northern giants forage at sea as well as on land, feeding on SQUID, FISH and CRUSTACEANS at sea and other birds and carrion on land. Northern giants are found all around the subtropical and subantarctic zones. They breed on a number of islands, including SOUTH GEORGIA, MACQUARIE and CAMPBELL, and begin breeding at five to seven years of age. Northern giant petrels tend to avoid areas frequented by humans and are highly vulnerable to disturbance; however, populations are rising in some areas, possibly because of increased SEAL numbers and therefore increased scavenging opportunities.

Norway A polar country with interests in both the Arctic and Antarctic regions, Norway's main historic involvements in Antarctica have been in WHALING and early EXPLORATION. The first confirmed landing in Antarctica, at Cape ADARE on 24 January 1895, was by the NORWEGIAN ANTARCTIC EXPEDITION. Notable among later expeditions was that of 1910–12 led by Roald AMUNDSEN, which became the first to reach the SOUTH POLE.

In 1928 Norway made TERRITORIAL CLAIMS for BOUVETØYA in 1928, and for PETER I ØY in 1931. A claim to DRONNING MAUD LAND was formalized in 1939.

Norway participated in the INTERNATIONAL GEOPHYSICAL YEAR and was one of the original 12 CONSULTATIVE PARTIES to the ANTARCTIC TREATY. Although it has not maintained a year-round base in Antarctica since 1960, it operates two summer-only stations.

A member of the Commission for the CONVENTION ON THE CONSERVATION OF ANTARCTIC MARINE LIVING RESOURCES (CCAMLR), Norway has usually adopted a pro-conservation stance at CCAMLR meetings. It is also a signatory to the MADRID PROTOCOL.

Norwegian Antarctic Expedition (1910–12) Private Norwegian expedition, the first to reach the SOUTH POLE, led by Roald AMUNDSEN. It was meticulously organized. Amundsen borrowed the ship, the FRAM, from Fridtjof NANSEN initially for an expedition to the North Pole. However, when he heard that the North Pole had been reached, he changed plans and decided to try for the South

Pole instead. He kept the destination secret from the crew until the *Fram* had reached Madeira. He sent a telegram to Robert SCOTT, who was also attempting to be the first to reach the South Pole. It read 'Beg leave to inform you *Fram* proceeding Antarctic. Amundsen.'

The *Fram*, commanded by Lieutenant Thorvald Nilsen, had set sail from Norway on 9 August 1910, with 19 men, 97 dogs, four pigs, six pigeons and a canary. On 14 January 1911 it arrived at the BAY OF WHALES, where Amundsen set up camp, 111 km (69 miles) closer to the Pole than Scott's base camp at MCMURDO SOUND. A hut was erected 3.7 km (2⅓ miles) inland at a site the Norwegians named FRAMHEIM. This was the first expedition to overwinter on the ROSS ICE SHELF.

The expedition was visited by the *TERRA NOVA*, Scott's ship, on 4 February. The British were entertained on board the *Fram* and afterwards every Norwegian caught a cold. Amundsen set up a chain of supply depots between Framheim and the South Pole before the start of winter. Provisions were laid up to 80°S and the route was marked with bamboo poles and flags. Throughout the winter the men followed strict routines, preparing provisions and refining tents and other supplies.

After one abandoned attempt in September, Amundsen set off for the Pole on 19 October with four others: Helmer HANSSEN, Olav BJAALAND, Oscar WISTING and Sverre HASSEL. They took four sledges, each drawn by 13 dogs. Every day, they built a beacon to mark their progress and left a record with the distance and bearings to the next destination. After climbing what Amundsen named QUEEN MAUD MOUNTAINS, they shot 24 dogs, which were no longer needed—the dog meat provided fresh food for the men and other dogs.

The five men reached the South Pole on 14 December 1911. They planted the Norwegian flag, and the next day Amundsen circled the camp in a radius of about 22 km (13½ miles). He left behind a black tent, called Poleheim, a note for Scott, and a letter addressed to King Haakon VII for Scott to deliver should the Norwegian team

not survive the return journey. On their last night at the Pole, they feasted on seal meat and smoked cigars. The return journey went smoothly and they returned to Framheim on 25 January with just 11 dogs. The expedition had taken 99 days and had covered 2594 km (1611 miles). The *Fram*, which had been visited on 16 January by members of the JAPANESE ANTARCTIC EXPEDITION, set sail from Framheim on 30 January to Hobart, AUSTRALIA, from where Amundsen sent a cable around the world with the news.

Norwegian Antarctic Territory In 1928 NORWAY made TERRITORIAL CLAIMS for BOUVETØYA, and for PETER I ØY in 1931. A claim to DRONNING MAUD LAND between 20°W and 45°W longitude was formalized in 1939, prompted by the activities of the 1938–39 GERMAN ANTARCTIC EXPEDITION in that area. The main reason in making these claims was to protect Norwegian WHALING interests in the SOUTHERN OCEAN. The claimed area has been a Norwegian dependency since 1948.

Norwegian-British-Swedish Antarctic Expedition (1949–52) The first large-scale international scientific expedition. It carried out ground-breaking geophysical, geological and glaciological work in DRONNING MAUD LAND, making the first detailed study of the Antarctic interior using ECHO SOUNDERS and taking the first significant ICE CORE samples.

Norwegian Polar Institute Responsible for Norway's scientific research, environmental monitoring and mapping in both polar regions. The Norwegian Polar Institute had its beginnings in the Svalbard and Arctic Sea Research Body, established in 1928, which was renamed in 1948 in order to incorporate Norway's Antarctic TERRITORIAL CLAIMS. The formation of the Institute marked the beginning of Norway's scientific involvement in Antarctica. Today Norway also operates two field stations, Troll and Tor, and concentrates on environmental management with research on CLIMATE and POLLUTION and topographical mapping.

Norwegian Sealing and Whaling Exploration (1893–95) See ANTARCTIC EXPEDITION.

Notothenioidei The dominant suborder of FISH in Antarctica. This group of bony, perch-like fish is confined exclusively to the SOUTHERN OCEAN and accounts for about 53 percent of species present and approximately 90 to 95 percent of the FAUNA abundance and biomass on the Antarctic CONTINENTAL SHELF. The origins of the group are unknown, but the fauna of which they are a part probably evolved 40 to 60 million years ago.

The Antarctic Notothenioidei is comprised of four main families: the Nototheniidae (ANTARCTIC COD), the Harpagiferidae (spiny PLUNDER FISH), the Bathydraconidae (dragon fish), the Channichthyidae (ICEFISH).

Most notothenioids are bottom-dwelling, or DEMERSAL FISH, which lack swim-bladders. Some are adapted to temporary or permanent pelagic life so that they can use food resources such as

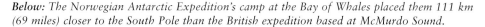

Below: The Norwegian Antarctic Expedition's camp at the Bay of Whales placed them 111 km (69 miles) closer to the South Pole than the British expedition based at McMurdo Sound.

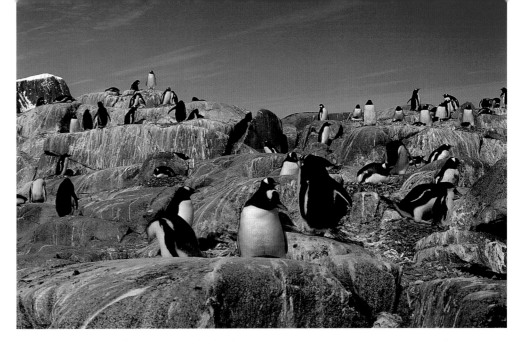

Above: The nitrogen from guano at bird colonies comprises an important component of Antarctica's nutrient cycle.

KRILL. Many species also spawn so that their larvae can take advantage of summer PHYTOPLANKTON blooms for feeding.

Adaptations for the extremes of their environment include the presence of GLYCOPROTEIN compounds in their blood, which act like ANTIFREEZE, preventing ice crystals from forming in body fluids at subzero temperatures. Their kidneys are also adapted so as not to excrete the antifreeze. Although notothenioids can tolerate very low temperatures, they can only survive a very narrow range of –2.5°C to 6.0°C (27.5°F–42.8°F).

Novolazarevskaya Station Established by the USSR in February 1961, on the edge of the Lazarev Ice Shelf in DRONNING MAUD LAND. On 30 April 1961 at Novolazarevskaya, the physician Leonid Rogozof successfully performed SURGERY on himself to remove his appendix. Novolazarevskaya operates as a year-round base, and conducts RESEARCH into geosciences, meteorology and glaciology. The blue ice AIRSTRIP at the station has been investigated for use in TOURISM operations. Five people were killed in a HELICOPTER accident near the station on 1 June 1998.

Nuclear contamination There has been one reported incidence of nuclear contamination in Antarctica. In March 1962 the USA began operating a 1.8 megawatt pressurized-water NUCLEAR POWER plant, nicknamed 'Nukey Poo', at MCMURDO STATION. The reactor was shutdown 10 years later after it leaked coolant water. The plant's site and the surrounding area were found to be contaminated. After a clean up, which took three years and involved the US Navy shipping back 800 tonnes (800 tons) of radioactive junk and another 12,200 tonnes (12,200 tons) of earth and gravel to the USA for disposal, the site was declared 'decontaminated to levels as low as reasonably achievable'.

Nuclear-free zone The ANTARCTIC TREATY, signed in 1961, was the first nuclear test ban treaty and established the world's first nuclear-free region, although it permitted the peaceful use of nuclear energy.

Nuclear power The only nuclear power plant to have existed in Antarctica began operating in March 1962 at the USA's MCMURDO STATION. It was a 1.8 megawatt pressurized-water nuclear power plant providing heat and power to the station. However, the reactor was prone to breakdowns, which made it uneconomic, and it required back-up oil-powered generators. In 1972, after it leaked coolant water, it was shut down. The plant was dismantled and shipped back to the USA along with NUCLEAR CONTAMINATION from the site. This was the first time a nuclear reactor had been disassembled and transported.

Nuclear waste Article V of the ANTARCTIC TREATY prohibits the disposal of radioactive waste material in Antarctica, and the MADRID PROTOCOL requires all radioactive waste generated in Antarctica to be removed by the state that created it.

At one time, Antarctica's remoteness was thought to make it an ideal place to bury radioactive wastes. However, disposal sites need to be geologically as well as geographically suitable—sufficiently stable that leaks are unlikely. Even without the provisions of the Antarctic Treaty, burying waste in the Antarctic ice cap is unlikely to work, as the ICE eventually drifts out to sea; burying it in BEDROCK would be technically difficult and extremely expensive. For the foreseeable future it is unlikely that Antarctica will be used as a dumping ground for radioactive waste.

Nudibranchs Nudibranchs, or sea slugs as they are sometimes known, are an order of gastropod MOLLUSCS without protective shells. Nudibranchs are mostly carnivorous. They have rosettes of retractile naked gill fronds surrounding their anus, hence the name 'nudibranch', meaning 'naked gill'. Aeolid nudibranchs differ markedly from others in that the gill rosette is replaced by clusters of delicate, branching cerata, which form rows down each side of the body and contain extensions of the liver.

The *Notaeolidia* genus of aeolid nudibranchs is endemic to Antarctica. *Notaeolidia gigas* has been found at depths from 3–50 m (10–164 ft) and can measure up to 8 cm (3 in) in length. A predator of large hydroids and soft CORAL, it has at least three rows of cerata, numbering at least 200, on each side.

To compensate for the lack of protective shells, some nudibranchs use chemical deterrents obtained from their prey to safeguard themselves from potential predators such as SEASTARS, SEA URCHINS and FISH. Aeolid nudibranchs are known to incorporate the living, undigested stinging cells of their cnidarian prey into their cerata, thus adding another layer of protection.

Nunataks Isolated, spire-shaped peaks or MOUNTAINS of rock projecting above the surface of the ice or snow. The word 'nunatak' is Eskimo in origin.

Nutrient cycling The terrestrial and marine environments of Antarctica are at opposite extremes

in terms of nutrient availability for growth. As well as sunlight, photosynthesis and growth require nutrient salts to form enzymes and structural molecules—a shortage of even one essential nutrient will limit the production of new living matter. There is an abundance of such nutrients in the SOUTHERN OCEAN, which is one of the richest oceans in the world. Upwelling of the deep oceanic water at the ANTARCTIC DIVERGENCE, the EAST WIND DRIFT and the stormy turbulence of the Southern Ocean generally all help to bring nutrient-rich water to the surface, where it is available for photosynthesizing PHYTOPLANKTON and thus enters the marine FOOD WEB. South of the ANTARCTIC CONVERGENCE, nutrient levels in the surface waters equal or exceed the maximum levels found in other oceans.

By contrast, the lack of soil and its reservoir of essential nutrients means that PLANT growth in the terrestrial environment is severely limited; it is further limited by freezing TEMPERATURES and a scarcity of WATER. MOSSES and LICHENS subsist on nutrients leached from rocks and those deposited by wind-blown dust, snow and sea-spray. The marine system also contributes essential nutrients through animal waste—some of the most luxurious terrestrial plant growth is found near sea bird nesting colonies, where guano nourishes ALGAE, fixing nitrogen from the air into a usable form for other plants. Eggs, moulted feathers and dead chicks are all ways through which birds contribute to the terrestrial nutrient cycle. On MARION ISLAND it has been estimated that sea birds contribute 90 percent of all nitrogen introduced annually into the vegetated areas. SEALS also contribute nutrients into the terrestrial system, mainly through excreta and hair cast at moulting.

Nutrients enter the FOOD CHAIN through primary producers, are consumed by HERBIVORES and subsequently their predators, and are cycled within the ecosystem by decomposing organisms such as BACTERIA and FUNGI. In the terrestrial environment the spring thaw means that freeze thaw damage to older plant tissues releases nutrients to MICROORGANISMS, which then show a burst in population growth.

O

Oases Around 20 mainly ice-free oases exist in Antarctica. They cover less than 2 percent of the continent, but are scientifically significant areas. Exposed ROCK has a significant impact on the local CLIMATE; over summer the rock absorbs heat which evaporates any snow falling on the oasis. Along with strong KATABATIC WINDS, this allows the oases to remain ice-free.

The most studied and extensive oases are the DRY VALLEYS in southern VICTORIA LAND. Other smaller coastal oases are found at Cape Hallett in north Victoria Land; MAC.ROBERTSON LAND; at the BUNGER HILLS in WILKES LAND and in the VESTFOLD HILLS in PRINCESS ELIZABETH LAND.

Oates, Lawrence Edward Grace (1880–1912) British army captain and explorer, born in Putney, London. He was educated at Eton and joined the Sixth Inniskilling Dragoons of the British Army. He was seriously wounded in the Boer War, leaving him with a limp. He later served with the army in Ireland, Egypt and India. Horses were his passion and he wrote that 'hunting is the only thing that makes life endurable'.

Oates joined the 1910–13 BRITISH ANTARCTIC EXPEDITION, and was charged with taking care of the expedition's ponies. Known for his silence—Edward WILSON described him as 'extraordinarily silent and laconic'—he seemed to be only enthusiastic about horses. He reached the SOUTH POLE with Robert SCOTT on 17 January 1912. On the return journey to base camp he suffered from severe frostbite and his feet turned gangrenous. On 15 March 1912, conscious that he was slowing down his companions, he walked out into the snow to his death, saying 'I am just going outside and may be some time.' In a letter to friends written a few days before arriving in NEW ZEALAND with news of the polar party's fate, Scott's second-in-command Edward EVANS wrote, 'Capt. Oates, the Inniskilling Dragoon came out of it best of all ... By Jove he was a fine man.'

Oates Land A mountainous section of EAST ANTARCTICA, sandwiched between the ROSS SEA to the east and GEORGE V LAND to the west. It was discovered on 22 February 1911 by Lieutenant H Pennell from the TERRA NOVA and is named after Lawrence OATES.

Observation Hill A hill that overlooks Cape EVANS on ROSS ISLAND. Just before they left Antarctica aboard the TERRA NOVA, the surviving members of the BRITISH ANTARCTIC EXPEDITION erected a memorial cross on the top of Observation Hill in memory of Robert SCOTT and his four companions who died on the return journey from the SOUTH POLE. As well as their names, the final line of Lord Tennyson's *Ulysses*—'To strive, to seek, to find, and not to yield.'—is inscribed on the cross.

Observatories The first land-based observatories in Antarctica were established by the 1882–83 GERMAN INTERNATIONAL POLAR YEAR EXPEDITION at SOUTH GEORGIA. Hourly meteorological observations were recorded manually, and magnetic, glacial, astronomical and tidal records kept. Today, permanent observatories, including at AMUNDSEN-SCOTT SOUTH POLE STATION, and networks of automatic observatories in remote regions collect year-round data from geospace. They house magnetometers, which measure changes in the Earth's magnetic field caused by electrical disturbances in the upper ATMOSPHERE; 'ionosondes', first used in Antarctica in the 1950s, which beam into space radar signals that are reflected back by charged particles in the ionosphere; antennae to transmit radio waves, and often an all-sky camera to record images of the AURORA AUSTRALIS and other OPTICAL PHENOMENA.

Ocean Camp The name given to the floating camp that Ernest SHACKLETON and the members of the IMPERIAL TRANSANTARCTIC EXPEDITION established on an ice floe on the WEDDELL SEA after they abandoned the *ENDURANCE* on 27 October 1915. They left Ocean Camp on 22 December and set out carrying three lifeboats across the ice floes on the journey that eventually enabled them to reach open water and sail to remote ELEPHANT ISLAND.

Ocean currents About 90 percent of the SOUTHERN OCEAN's volume moves in deep currents produced by the interaction of large water masses (see ANTARCTIC CONVERGENCE, CIRCUMPOLAR DEEP WATER and ANTARCTIC BOTTOM WATER). The remaining 10 percent of the Southern Ocean moves in surface currents that are produced by WINDS blowing across the ocean, and transferring energy to the surface (see WEST WIND DRIFT and EAST WIND DRIFT and WEDDELL DRIFT).

Ocean ridges Since the opening of the DRAKE PASSAGE about 23 million years ago—the final separation of Antarctica from GONDWANA—the continent has been surrounded by mid-ocean ridge systems. These mark the spreading centres of large crustal plates, into which new sea-floor volcanic material is introduced. The PACIFIC ANTARCTIC RIDGE is the largest and most active. It can be followed from a point midway between NEW ZEALAND and Antarctica, northeast to where it joins the East Pacific Rise at the margin of South America.

Oceanographic research The harsh environmental conditions of the SOUTHERN OCEAN and the extensive SEA ICE cover make oceanographic research in Antarctic waters not only fascinating but also very challenging. This science, which is concerned with the oceans and seas and their interaction with other global systems, includes research into physical and chemical properties of WATER, the geologic framework of the ocean floor, and marine ecosystems and species.

Early explorers conducted the first research with basic equipment. James Clark ROSS dredged the seabed on his 1841 expedition: '... I have no doubt that from however great a depth we may be able to bring the mud and stones of the bed of the ocean, we shall find them teeming with animal life,' he said. The first expedition focusing prima-

Below: Captain Lawrence Oates was in charge of the ponies on the ill-fated 1910–13 British Antarctic Expedition.

Above: Observation Hill on Ross Island is an important memorial site. The surviving members of the 1910–1913 British Antarctic Expedition erected this cross to their colleagues in the pole party here. It is inscribed 'To strive, to seek, to find, and not to yield'.

rily on marine science was 30 years later when the 1872–75 CHALLENGER EXPEDITION led by George NARES crossed the ANTARCTIC CIRCLE; records of the worldwide scientific voyage are contained in the 50-volume *Report on the Scientific Results of the Voyage of the H.M.S.* Challenger, which was published between 1881 and 1895.

Now scientists use a vast array of equipment and technology to study the Southern Ocean. ICE-BREAKERS are floating laboratories with specialized drilling and sampling equipment. Submersibles and robotic vehicles are used to sample the sea floor. Sound waves reflected from the bottom of the ocean allow scientists to build a picture of a 'slice', or 'seismic profile', of the sea floor. Mapping the sea floor also helps scientists to interpret processes such as glacial erosion that have shaped it.

SATELLITES are used extensively in oceanography, to gather data about surface WIND and WAVE conditions, water TEMPERATURE and PHYTOPLANK-TON distribution.

Oceanic Research Foundation An independent, non-profit organization based in AUSTRALIA, established in 1977 by David LEWIS and others. The Foundation conducts research into marine resources, ecology, environment and heritage of AUSTRALIA and the SOUTHERN OCEAN, and has organized a series of oceanic research expeditions, including five to Antarctica, between its establishment and 2001.

Oceans The warmer waters of three major oceans—the PACIFIC, ATLANTIC and INDIAN—merge with the SOUTHERN OCEAN at the ANTARCTIC CONVERGENCE at a latitude of around 55°S to 60°S.

Octopus Little is known of the Antarctic species of the CEPHALOPOD class of MOLLUSCS, which includes octopus and SQUID. The *Pareledone* species of octopus are believed to be a key component of the Antarctic benthic FOOD WEB and, along with FISH, are top-level predators.

Office of Polar Programs (OPP) Responsible for the Antarctic and Arctic activities of the NATIONAL SCIENCE FOUNDATION which, in 1971, was given overall responsibility for the USA's activities in Antarctica.

Oil In 1972 the American research vessel *Glomar Challenger* drilled four holes in the ROSS ICE SHELF and found traces of hydrocarbons. The first serious exploration began in 1981 with SEISMIC RESEARCH undertaken by the Japan National Oil Corporation in the BELLINGSHAUSEN SEA. Since then, a number of countries and companies have conducted surveys in Antarctic waters, and estimates of possible recoverable oil reserves range from 3000 million to 100,000 million barrels. However, with the adoption of the MADRID PROTOCOL in 1991, all mining was banned in Antarctica for 50 years. The dangers of oil drilling in Antarctica's inhospitable conditions are considerable. If a blowout were to occur and not be controlled by the end of the summer season, it would rage unchecked for at least another six months. The construction of pipelines, storage tanks and on-shore crew facilities would have significant impact on BIRDS and MAMMALS occupying the coastal areas. In addition, the risks of spillages and shipping accidents would be considerable.

Below: The Southern Ocean has an unforgiving nature and has shaped the history of Antarctic exploration.

Above: Appearing to defy nature, the upper Oynx River flows inland from the coast.

The disastrous consequences of a major oil spill were demonstrated when the Argentinian supply ship *Bahia Paraiso* sank near PALMER STATION in 1989. More than 681,894 litres (150,000 gallons) of fuel seeped from the ship and within four days an oil slick of about 100 sq km (62 sq miles) had spread across the ocean surface. The Argentinian and Chilean navies and the NATIONAL SCIENCE FOUNDATION cleaned up the spillage, skimming off most of the surface oil and removing the remaining fuel from the wreck. Although there were few marine mammals in the area, it is estimated that about 30,000 PENGUINS came into contact with the spill, as the accident occurred when most parent birds were foraging at sea to feed their young.

Onyx River Located in the DRY VALLEYS region, the perennial meltwater Onyx River arises in the Lower Wright Glacier and flows about 30 km (18½ miles) to Lake VANDA. It is believed to be Antarctica's longest RIVER, and is one of the few rivers in the world to flow inland from the coast.

Operation Deepfreeze (1955–57) Official American expedition, part of the INTERNATIONAL GEOPHYSICAL YEAR, led by Richard BYRD and George DUFEK. There were two parts to the expedition. The first, in the summer of 1955–56, was under Byrd's command. It established an AIRSTRIP at MCMURDO SOUND to be used as a base to support construction during the second operation, and a second base near Byrd's former headquarters at LITTLE AMERICA in KAINAN BAY. From here, land parties were to attempt to establish an inland station in MARIE BYRD LAND, although this was not accomplished. Seven SHIPS, 19 AIRCRAFT, TRACTORS and other VEHICLES and 1800 men were involved in this part of the expedition.

The second operation in the 1956–57 summer led by Dufek involved the construction of a permanent base at the SOUTH POLE—the AMUNDSEN-SCOTT SOUTH POLE STATION—along with BYRD STATION in Marie Byrd Land, WILKES STATION in Vincennes Bay, ELLSWORTH STATION on the FILCHNER ICE SHELF, and the joint US-New Zealand HALLETT STATION. In total, 12 ships and 3400 men were involved. On the first inspection trip to the Amundsen–Scott Station site on 31 October 1956, Dufek became the first person to set foot on the SOUTH POLE since Robert SCOTT's party in 1912. 'It was like stepping out into a new world,' he wrote. In a wasteful operation, the station was established entirely by air: aircraft dropped materials and SUPPLIES, many of which were smashed on impact with the ground or blown away in the wind. Dr Paul SIPLE, chief of scientific staff, complained that 'these young titans of modern Antarctica, accustomed to the opulence of the military services, operated by a philosophy that was "to get the job done and to hell with conserving supplies".' In all, in the four months it took to complete the station, 84 flights dropped 760 tonnes (760 tons) of supplies.

Upon completion in March 1957, 17 scientists arrived to join Paul Siple in being the first people to spend winter at the South Pole. They received a posthumous message from Byrd (who had died earlier that month): 'Your work ... may well mark the beginning of permanent occupancy of the Antarctic continent.'

Operation Highjump (1946–47) American naval expedition, officially known as the United States Antarctic Developments Project; the largest ever mounted in the Antarctic, led by Richard BYRD. Mainly a military training exercise, it was also aimed at establishing a USA presence in Antarctica. The expedition comprised 13 SHIPS under the control of Richard Cruzen, along with a 35,000 tonne (35,000 ton) aircraft carrier, 19 AIRCRAFT, four HELICOPTERS and 4700 men. It was the first time that helicopters and ICEBREAKERS were used in the Antarctic.

The expedition was divided into three groups. Ships from the first group entered PACK ICE in the ROSS SEA on New Year's Eve 1946 and arrived at the BAY OF WHALES on 15 January 1947. A base was established about 3.2 km (2 miles) north of LITTLE AMERICA, Byrd's base from the 1939–41 UNITED STATES ANTARCTIC SERVICE EXPEDITION. A landing strip was built for the twin-engined R4D aircraft carried by the *Philippine Sea*, which arrived at the end of January with Byrd on board.

Over the next month the planes flew a total of 36,530 km (22,700 miles), taking aerial photographs of the interior. At least three major mountain ranges were discovered. The longest flight was made on 15 February 1947 and reached 160 km (100 miles) beyond the SOUTH POLE—18 years after Byrd's historic flight that was long believed to have crossed over the Pole. Meanwhile, two other groups sailed east and west around the continent, carrying out surveys as they went. The most publicized discovery was a snow-free OASIS in the BUNGER HILLS covering an area of over 259 sq km (100 sq miles). In all, nearly 3.9 million sq km (1.5 million sq miles) of Antarctica were sighted and over 70,000 photographs were taken, covering 60 percent of the Antarctic coastline.

Operation Tabarin (1943–45) Initially a secret naval expedition by BRITAIN, Operation Tabarin was intended to establish a permanent British presence on the ANTARCTIC PENINSULA to counter TERRITORIAL CLAIMS made by ARGENTINA and CHILE. Led by biologist James Marr, who had Antarctic experience with the DISCOVERY EXPEDITION, and organized by a committee that included notable scientists such as James Wordie, the expedition set up three RESEARCH STATIONS and conducted a series of SCIENTIFIC INVESTIGATIONS. At the end of World War II Operation Tabarin was renamed the Falkland Islands Dependencies Survey, which later became the BRITISH ANTARCTIC SURVEY.

Operation Windmill (1947–48) American naval expedition, follow-up to OPERATION HIGHJUMP, used HELICOPTERS extensively (hence its name). The purpose of Windmill was to chart ground control points from which to construct maps using the aerial survey data obtained during Operation Highjump.

Optical phenomena Geomagnetic conditions in Antarctica are such that optical phenomena such as AURORA AUSTRALIS, HALOS and MIRAGES occur regularly. The sun produces a continuous supersonic stream of electrically charged, high-energy particles (plasma), or 'solar wind'. The optical phenomena are produced when the SOLAR WIND interacts with the magnetic field of the Earth, which deflects the particles so that they are directed down into the Earth's ATMOSPHERE over the Antarctic (and Arctic) regions.

Orca whale (*Orcinus orca*) See KILLER WHALES.

Above: The sun and a sun dog: an example of common optical phenomena over the Polar Plateau.

Orcadas Station Originally built for the 1902–04 scottish national antarctic expedition on Laurie Island in the south orkneys, the meteorological base was handed over to argentina in 1904 and renamed Orcadas. It is the oldest base in use in Antarctica. Research conducted here has included meteorology, atmospheric physics, geomagnetism, glaciology and biology.

Ornithological research See zoological research.

Ousland, Børge (1963–) Norwegian skiier. Ousland made the first solo crossing of the Antarctic continent on his second attempt. Previously, in 1995, Ousland made a solo journey from berkner island to the south pole in 44 days. In 1996–97, he successfully made the first solo transcontinental crossing, unassisted, from Berkner Island to scott base. He travelled 2845 km (1764 miles) in 64 days. On his fastest day on skis, he covered 226 km (140 miles) using a parasail. Ousland published his experiences in *Alone Across Antarctica*.

Overwintering For staff spending the relentlessly dark, cold polar winter at Antarctic bases, life is centred on dealing with the extreme environmental conditions, and coping with the psychology of living in confined spaces among a small group of people. According to a psychologist who has specialized in selecting overwinterers: 'The normal ways we deal with things when we're fed up—either withdrawing and shutting the door or going out to seek other people—are not available.' Modern communications make daily contact with home easy, but the physical distance can be aggravating. Overwinterers at the south pole are cut off from mid-February to late October; the Canadian flight to the Pole in April 2001 to pick up a seriously ill base camp member was exceptional. One feature of the South Pole winter is the 300 club, which involves streakers running across the ice wearing only shoes. The biggest milestone at most bases is midwinter, which from early

Left: Orcadas Station lies on an isthmus on the Orkney Islands, just above sea level.

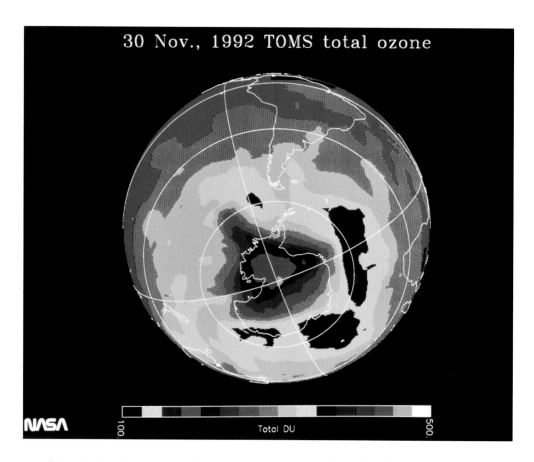

30 Nov., 1992 TOMS total ozone

NASA

100. Total DU 500.

Left: A false colour NASA image showing the ozone hole on 30 November 1992, *with the Antarctic continent outlined just below the centre of the globe. The colour scale shows total ozone values, which are measured in Dobson units; reds and greens indicate high ozone concentrations; the blues and purples indicate low ozone concentrations; and black areas indicate missing data. At the time represented in this image, the size of the ozone hole was 1.7 million square miles.*

expedition days has been celebrated by banquets and parties.

Aside from fun and camaraderie, staff must be able to handle isolation and the stresses associated with living in the dark for five months. The darkest months leading up to SUNRISE—July and August—are usually the most difficult: tempers may begin to fray and some people suffer depression. Frederick COOK, medical officer on the 1897–99 BELGIAN ANTARCTIC EXPEDITION recorded that, in the depths of winter, 'The curtain of blackness which has fallen over the outer world of icy desolation has also descended upon the inner world of our souls. Around the tables, in the laboratory, and in the forecastle men are sitting about sad and dejected, lost in dreams of melancholy.'

Being trapped inside for months with the same work routine can be psychologically demanding. James Clark ROSS described being '... so terribly weary and dejected for lack of work, lack of variety, lack of mental occupation, lack of thought ...' To relieve boredom, members of the 1901–04 NATIONAL ANTARCTIC EXPEDITION produced a monthly journal, called the *SOUTH POLAR TIMES*. When he returned to Antarctica in 1907, Ernest SHACKLETON bought a printing press and expedition members produced the *AURORA AUSTRALIS*.

Humour is one weapon against the repetiveness of the winter months. Professor A J W Taylor reported a message an overwinterer had posted on a noticeboard in one of the bases: 'With just 180 days to go, most people are going to do a surprising number of things. "Chomper" will fill 1600 man days of food boxes, George will send 908,300 dits and 804,150 dahs. Dave's ionosonde will send up 46,780,000 individual pulses of 10 kw each, not counting World Days, John will

photograph the sky 24,000 times in colour and 30,000 times in black and white ... and [somebody] will wash his socks once more.'

Ozone Molecules of ozone contain three atoms of oxygen instead of the normal two. Although it comprises only 0.0001 percent of the ATMOSPHERE, ozone shields life on Earth from the damaging effect of ULTRAVIOLET RADIATION, like a protective cocoon. Ozone is present throughout the ATMOSPHERE, but the bulk of it is in the stratosphere (see ATMOSPHERE) in a layer about as deep as Mount Everest is tall. When ultraviolet (UV) radiation strikes ordinary molecular oxygen (O_2), it is split into two single oxygen atoms (O_2 + UV light = 2O). These free atoms can join with oxygen molecules to produce ozone ($O + O_2 = O_3$).

Ozone is broken down in the lower stratosphere by complex photochemical reactions involving compounds of nitrogen, chlorine and hydrogen, which are emitted on the Earth's surface from biological sources, such as bacteria and VOLCANOES, and artificial sources (see CHLOROFLUOROCARBONS). When a free atom of oxygen collides with the chlorine monoxide, for example, molecular oxygen (O_2) is produced. This releases the chlorine atom to attack ozone and break it back down into molecular oxygen. In this way, one atom of chlorine can destroy about 100,000 molecules of ozone.

Ozone is depleted in Antarctica at a much higher rate than in the rest of the world. This is mainly due to the existence of polar stratospheric clouds (PSCs), which form in very cold TEMPERATURES inside the Polar vortex, a giant whirlpool of WIND that seals off the Antarctic stratosphere in WINTER. In spring, when the ideal

ozone-destroying combination of light and low stratospheric temperatures occurs, the PSCs provide a surface for chemical reactions, for free chlorine atoms from stable molecules to begin the cycle of ozone destruction.

As the vortex disintegrates in spring, the slow downward circulation drives pools of ozone-poor air over the Southern Hemisphere. In this way, the vortex seems to act as a 'chemical processor', with ozone-rich air going in, and ozone-poor air coming out. This process reaches its maximum in September (although often the lowest ozone readings are in early October), and erodes more than half of the ozone over Antarctica, resulting in the infamous OZONE HOLE. By November, the stratosphere begins to warm sufficiently to evaporate the PSCs. Ozone-rich air seeps back from higher LATITUDES, the chlorine is dormant for another year, and ozone destruction ceases.

Ozone hole Term used to describe areas of the ATMOSPHERE where OZONE has been reduced by at least 50 percent. This is a reoccurring scenario at the end of the polar WINTER over Antarctica. The returning SUN activates chemical reactions in the stratosphere that eat the core out of the ozone layer, leaving only a thin layer in the upper and lower stratospheres, where temperatures are not low enough for ozone destruction to take place.

Monitoring of Antarctic ozone began at HALLEY STATION during the INTERNATIONAL GEOPHYSICAL YEAR in 1957. By 1984, observations at Halley Station revealed ozone values 30 percent below those observed in the previous decades of measurement, and the researchers announced their remarkable discovery of an ozone 'hole.' Their findings, published in *Nature* in 1985, provided alarming evidence of the rate of ozone destruction. This became the most famous science project ever conducted in Antarctica.

Since its discovery, the hole has continued to grow. By 1993, ozone reached a record springtime low, when volcanic acid particles from the 1991 eruption of Mount Pinatubo in the Philippines accelerated ozone loss. At this time, 70 percent less ozone was present in the stratosphere than in 1960. The hole has since grown to twice the size of the continent, affecting an area of more than 29 million sq km (11.2 million sq miles) in the year 2000. International agreements, such as the MONTREAL PROTOCOL, have limited the release of ozone-destroying chemicals and reduced levels in the lower atmosphere. However, it is unlikely that ozone concentrations will recover by 2050, as once estimated.

p

Pacific Antarctic Ridge The largest and most active of the mid-OCEAN RIDGE systems in the Antarctic area. It runs from a point midway between NEW ZEALAND and Antarctica, northeast to where it joins the East Pacific Rise at the margin of South America.

Pacific Ocean The ocean, which is bordered by the Asian continent to the northwest and the Americas on the east, merges with the SOUTHERN OCEAN at the ANTARCTIC CONVERGENCE, north of WEST ANTARCTICA and the ROSS SEA. The southern Pacific Ocean includes New Zealand's SUBANTARCTIC ISLANDS and MACQUARIE ISLAND.

Pack ice Plates of thin, frozen SEA ICE that surround Antarctica in winter months. Pack ice drifts mainly with the wind—for example, northwest across the ROSS SEA and around the continent from east to west. In a few places it meets ICE left from previous seasons, forming areas of denser and thicker sea ice. James COOK was the first to name this feature, sighted in 1774: 'We discovered field or pack ice, and we had so many loose pieces about the ship that we were obliged to luff for one and bear up for the other,' he recorded in his journal. In turning back to higher latitudes he wrote, 'I will not say that it was impossible anywhere to get in among this ice, but I will assert that the bare attempting of it would be a very dangerous enterprise.'

Pack-ice formation is one of the most magnificent natural phenomena. As the air temperature plummets with the approach of winter, the sea surface surrounding Antarctica begins to freeze. The sea ice advances at the extraordinary average rate of 57 sq km (35 sq miles) per minute. As Apsley Cherry-Garrard commented, 'so great is the wish of the sea to freeze, and so cold is the air that the wind has only to lull for one instant and the surface is covered with a thin film of ice as though by magic.'

Although pack ice may form a year-long barrier around the continent, in summer there are usually pockets of clear water immediately surrounding the coast. The Antarctic 'freeze-up' commences in March when pack ice forms in sheltered bays: first, in the southerly parts of the WEDDELL, BELLINGSHAUSEN and ROSS SEAS. It extends northward as the sea surface temperature decreases.

By the end of winter, the pack ice attains its maximum coverage of about 19 million sq km (7.3 million sq miles)—about 8 percent of the Southern Hemisphere—and extends as far as 2200 km (1364 miles) from the coast. This doubles the area of Antarctica, and is double the quantity of pack ice that forms in the Arctic.

As the ice spreads, it also thickens by freezing of the water underneath and via snowfall on the surface. SNOW has a cushioning effect and increas-

Above: *Pack ice forms as the sea freezes, slowly incorporating brash and frazil ice into a solid expanse that is also thickened by snowfall.*

es friction against ships' hulls, making ice-breaking by ships more difficult. The action of WINDS, OCEAN CURRENTS, particularly in September and October, and some melting results in extremely active ice movement. It grinds, cracks and buckles as giant ice floes break apart and ride up over each other.

Sailing ships had little navigational control once they entered the pack ice. Six expeditions were stranded in the ice over winter, drifting aimlessly with the flow of the ocean. In 1915, Ernest SHACKLETON's ships, ENDURANCE and AURORA, were confined in pack ice for the winter. 'Where would the vagrant winds and currents carry the ship during the long winter months that lay ahead of us,' Shackleton worried. As the ice broke up in October, 'huge blocks of ice, weighing many tons, were lifted into the air and tossed aside as other masses rose beneath them.' The crew could only stand by and watch as *Endurance* was crushed by pack ice and sank in the WEDDELL SEA. Shackleton 'could not describe the impression of relentless destruction that was forced upon me as I looked down and around. The floes ... were simply annihilating the ship.'

The first images of the yearly growth and decay cycle of the pack ice were from the spacecraft *Nimbus 5*, launched in 1970.

Palmer, Nathaniel Brown (1799–1877) American sealer. Born at Stonington, Connecticut, where his father was a ship owner. As a 19-year-old he joined a SEALING expedition to the SOUTH SHETLAND ISLANDS.

The following season, commanding the *Hero*, he returned to the area as part of a large fleet. On

16 November 1820 he is reputed to have sighted the ANTARCTIC PENINSULA. On 25 January 1821 Palmer encountered the expedition led by Thaddeus BELLINGSHAUSEN in the South Shetlands. According to the Russian explorer, 'I lay to, despatched a boat, and waited for the Captain of the American boat. ... Soon after, Mr Palmer arrived in our boat and informed us that he had been here for four months sealing. ... They were engaged in killing and skinning seals, whose numbers were perceptibly diminishing. There were as many as eighteen vessels about at various points ... the whole fleet of sealers had killed 80,000.' In mid-October of the following season, Palmer met James WEDDELL in the FALKLAND ISLANDS and another British sealer, George Powell, with whom he discovered the SOUTH ORKNEY ISLANDS.

After a few years trading in the Caribbean he returned to the Antarctic in 1829, but the sealing was poor and he was raided by pirates. He made a fortune in the Atlantic in the 1830s and was later involved in the clipper trade to the Far East. He died at sea during a return voyage to San Francisco in 1877.

Palmer Archipelago The northern arch of the ANTARCTIC PENINSULA, extending to the DRAKE PASSAGE near the SOUTH SHETLAND ISLANDS. The Archipelago is composed of a series of islands that are covered in ice.

Palmer Land The name usually given to the southern half of the ANTARCTIC PENINSULA; named after Nathaniel PALMER, who claimed to have sighted the Antarctic Peninsula on 16 November 1820.

Above: The USA's only Antarctic base north of the Antarctic Circle, Palmer Station is primarily a research facility focused on ecological processes and pack ice.

Palmer Station Named after Nathaniel PALMER, Palmer Station was built by the USA on ANVERS ISLAND, off the ANTARCTIC PENINSULA. The first structure was built in 1965 and the remainder of the station completed in 1968. RESEARCH into marine ecology is undertaken here, as well as work on meteorology, upper atmosphere physics, glaciology and geology. Palmer can accommodate 10 people OVERWINTERING and about 40 in summer.

Pancake ice Circular, flat pieces of floating SEA ICE that resemble pancakes or lily pads. Pancake ice forms when FRAZIL ICE consolidates in turbulent water; or when GREASE ICE thickens and then is broken up by WIND and WAVE action.

Pancakes often collide, which causes them to have upturned edges. 'Ramping' occurs when ICE slabs ride up over each other, a process that tends to lock-in pockets of seawater, where marine organisms can survive winter months. When groups occur in sheltered waters, pancakes may meld and bond together: this layer is often thickened by frazil ice that forms at the base and from SNOW falling on the surface, and is eventually broken into ICE FLOES by waves.

Parasites Organisms that live on another—the host—from which they obtain nourishment. The majority of Antarctic INSECTS are parasites on warm-blooded animals. Many Antarctic MITES are also parasitic. By living on a warm-blooded host, parasites are protected from the harsh Antarctic CLIMATE.

Of the 67 insect species recorded from the Antarctic continent, including the ANTARCTIC PENINSULA, 45 species are parasitic. Biting LICE are the most numerous, living in the feathers of BIRDS. SEALS are host to a few species of sucking lice. The Antarctic flea (*Glaciopsyllus antarcticus*) is the world's largest flea and is found on ANTARCTIC FULMARS and PETRELS.

Many of Antarctica's 528 mite species are also parasitic, with feather mites living on most Antarctic BIRDS and other mite species inhabiting the nasal passages of seals.

Parhelia Type of OPTICAL PHENOMENON known as HALOS, parhelia are bright luminous spots that often form on either side of the sun. They occur when long edges of ice crystals hang vertically in the sky, caught in wispy cirrus clouds moved by high WINDS. Ernest SHACKLETON described one as having 'the form of a wide halo with two mock suns at either extremity of the equator of the halo parallel to the horizon and passing through the real sun.' The word 'parhelia' comes from the Greek 'para' (meaning beside) and 'helios' (sun). Also called SUN DOGS or MOCK SUNS.

Parry Mountains Non-existent mountain range 'discovered' by James Clark ROSS in 1841, and now believed to have been a MIRAGE of another range about 400 km (248 miles) further into the distance than Ross's estimation of 40 to 50 km (25–30 miles).

Patagonian toothfish (*Dissostichus eleginoides*) A close relative of the ANTARCTIC TOOTHFISH, it lives in subantarctic waters on shelves around islands and submarine banks. Since the 1970s, illegal, unregulated and unreported TOOTHFISH FISHING has endangered the species.

Patterned ground Polygonal patterns on PERMAFROST, permanently frozen ground. The process of freezing stirs the soil on the permafrost surface into polygon patterns. Coarse rocks gather on the outer edge of the polygons and finer sediment migrates into the centre.

Paulet Island A circular volcanic island, about 1.5 km (1 mile) wide, Paulet Island was discovered by James Clark ROSS's 1839–1843 expedition. Located off the northeast tip of the ANTARCTIC PENINSULA, its large ADÉLIE PENGUIN rookeries attract tourists. The Adélies have also made their homes in the remains of a hut built by Otto NORDENSKJÖLD's party in 1903, after their ship *Antarctic* was crushed by PACK ICE and sank 40 km (25 miles) east of the island. The men sledged for 14 days to reach solid ground, and all but one survived WINTER in their makeshift 10 m (33 ft) by 7 m (23 ft) shelter.

Pearlwort (*Colobanthos quitenis*) See ANTARCTIC PEARLWORT.

Pelagic fish FISH that occur in the upper waters of the open sea. There are no shoaling pelagic fish species in the SOUTHERN OCEAN and no families of fish that are confined to surface waters throughout their lifecycle. Only the Antarctic silverfish (*Pleurogramma antarcticum*) is well adapted to operate as a permanent pelagic planktivore (plankton-eater).

Some NOTOTHENIOIDEI species inhabit the Antarctic BENTHOS (sea floor) as adults, but are pelagic as juveniles so that they can use food resources such as KRILL. In particular, many spawn so that their larvae can take advantage of summer PHYTOPLANKTON blooms for feeding.

The eggs and larvae of the ANTARCTIC TOOTH-

Below: Pancake ice is readily identified by its ridged edges and flat, circular shape.

Above: A parhelion over Scott Base.

FISH (*Dissostichus mawsoni*) are pelagic, floating near the sea surface where the larvae feed on ZOOPLANKTON. Adult Antarctic toothfish, although they lack a swim-bladder for buoyancy, are considered to be neutrally buoyant and are permanent members of the mid-water community, living at depths of 300 to 500 m (984–1640 ft).

Huge schools of lantern fish (Myctophidae) are found in the Southern Ocean north of the PACK ICE. These non-notothenioid fish are named for their luminescent (light-emitting) organs (photophores). Myctophids spend the day at depths of about 800 m (2624 ft), but rise to within 50 m (164 ft) of the surface at night.

Penguins *See following pages.*

Pensacola Mountains A major 400 km (250 mile) long mountain chain that runs between the Foundation Ice Stream and SUPPORT FORCE GLACIER SHELF. The range was discovered and photographed on 13 January 1956 from the air by OPERATION DEEPFREEZE and mapped in detail by the US GEOLOGICAL SURVEY in the 1956–57 season.

The ancient foundations of the Pensacola Mountains were formed millions of years ago, before the break-up of GONDWANA. Geologists have been investigating the origins of the mountains' BEDROCK, and believe that it was pushed to the surface in a giant 'super plume' of magma from within the Earth's mantle. Part of the rock foundation, the Dufek Massif, may be rich in MINERAL RESOURCES. The massif is underlain by an intrusion layer, which forms when volcanic magma collects in an underground chamber; during the cooling process, minerals separate into layers and solidify. Platinum and other elements, such as cobalt, nickel, vanadium, copper and iron, may be present.

Permafrost Areas of permanently frozen ground, or rocks and soil particles consolidated by ICE. In Antarctica, permafrost is up to 1 km (½ mile) thick. On the surface of the frozen ground the soil becomes stirred and sorted by freezing into polygons, known as PATTERNED GROUND.

Pesticides Pesticides such as DDT that were used on the world's farms and as insecticides through-out the 1950s and 1960s have made their way via OCEAN CURRENTS and airborne particles to Antarctica, where the simple marine FOOD CHAIN is particularly vulnerable to pollution. DDT and other chlorinated hydrocarbon pesticides have been found in the tissue of EMPEROR PENGUINS and in other Antarctic organisms.

Peter I Øy A COASTAL ISLAND, offshore from ELLSWORTH LAND, Peter I Øy was discovered by Thaddeus BELLINGSHAUSEN in 1821 and named after Peter the Great. Encased in PACK ICE for most of the year, the island covers an area of 158 sq km (61 sq miles), 95 percent of which is glaciated.

Petermann Island One of Antarctica's 10 most visited areas, and site of the southernmost GENTOO PENGUIN rookeries. Only about 2 km (1⅓ miles) long, the island is situated south of ANVERS ISLAND, off the west coast of the ANTARCTIC PENINSULA.

Petrels Of the Procellariiformes, a large order of 80 to 100 pelagic birds, divided into four distinct families: ALBATROSSES; the medium-sized petrels made up of GADFLY PETRELS, PRIONS, SHEARWATERS and FULMARS; the tiny STORM PETRELS; and the distinctive DIVING PETRELS. Petrels are characterized by their hooked beaks covered with horny plates, and their conspicuous raised nostrils. They comprise the greatest biomass and number of bird species in Antarctic waters. At least 40 petrel species breed on the islands further to the north— off the ANTARCTIC PENINSULA and on the widely scattered Antarctic and subantarctic islands.

Phantom Islands Islands that have been charted, then subsequently found not to exist. Many early phantom islands are believed to be the result of miscalculations—particularly before LONGITUDE could be determined accurately—or of copying errors by cartographers. Not until the advent of SATELLITE surveillance has it been possible to conclusively disprove the existence of many of these islands, such as EMERALD ISLAND, reported by early 19th-century whalers to lie in the region of 57°S and 162°E and searched for by Charles WILKES in 1840 and by the 1893–95 ANTARCTIC EXPEDITION. Although such islands have been reported in every ocean, in the Antarctic region phantom islands have been particularly prevalent for several reasons: MIRAGES and other OPTICAL PHENOMOENA occur frequently and produce confusing distortions; from a distance and in bad weather conditions ICEBERGS can be easily mistaken for islands; and, because of volcanic activity, a number of islands have disappeared (or been created). The earliest phantom island reported in the Antarctic region, Elizabethides, was charted in 1578 by Francis DRAKE between South America and the ANTARCTIC PENINSULA: if it existed at all, it may have been destroyed in a volcanic eruption.

Pharmaceuticals The great biodiversity of Antarctica's marine life is valuable in pharmaceutical research. For instance, scientists have found that some of the molecules used as defense mechanisms in Antarctic marine SPONGES can fight diseases in humans, and have made synthetic copies of these molecules.

Philately The issuing of stamps is an historic way of proving TERRITORIAL CLAIMS. In January 1908, 24,000 copies of the NEW ZEALAND 'Penny Universal' stamp overprinted with the words 'King Edward VII Land' were issued for the 1907–09 BRITISH ANTARCTIC EXPEDITION and the New Zealand government appointed Ernest SHACKLETON as postmaster. In November 1910, two further New Zealand stamps were overprinted with 'Victoria Land' for Robert SCOTT's 1910–13 expedition. The first FALKLAND ISLANDS stamp was made in 1878, but the centennial issue of 1933 was the first to depict scenes relating to Antarctica. Many countries that have had bases in Antarctica or sent expeditions there have either used regular stamps on their mail or issued special commemorative stamps. Mail sent from LITTLE AMERICA II during BYRD'S SECOND EXPEDITION made use of a 1933 US general-issue stamp commemorating Byrd's polar flights. Stamps have been issued since the 1950s by Australian postal authorities for the AUSTRALIAN ANTARCTIC TERRITORY, by New Zealand for ROSS DEPENDENCY, and by FRANCE for TERRES AUSTRALES ET ANTARCTIQUES FRANÇAISES.

Photography Frederick COOK, on the 1897–99 BELGIAN ANTARCTIC EXPEDITION, and Louis BERNACCHI, on the 1898–1900 *SOUTHERN CROSS* EXPEDITION, were among the first photographers in Antarctica. They photographed daily life on the continent, the wildlife and the landscape. By the early 20th century, photographers were replacing ARTISTS as official recorders on expeditions.

The 1907–09 BRITISH ANTARCTIC EXPEDITION took 10 cameras, including one for movie FILM, and a variety of members acted as photographers. The photograph of Roald AMUNDSEN's party at the SOUTH POLE with a tent and Norwegian flag was an iconic image used extensively in what is now regarded as the first European travel film, *Scott of the Antarctic*. Herbert PONTING and Frank HURLEY (who continued taking photographs on his 'Vest Pocket Kodak' after the *ENDURANCE* sank) are the two best known early

Continued page 144

Penguins (Spheniscidae)

Flightless, aquatic birds of the Southern Hemisphere, penguins are among the most specialized and well-adapted of all creatures. They are true ocean-dwellers and extremely accomplished at swimming and diving. Their distinctive black-and-white coats, streamlined torpedo-shaped bodies and upright swagger have made penguins one of the world's favourite birds.

Penguins in general spend about 80 percent of their time in the ocean and come ashore for extended periods only to breed and moult. Of the 18 living penguin species, seven can be classified as Antarctic or subantarctic: two—the EMPEROR and the ADÉLIE—breed on the Antarctic continent below the ANTARCTIC CIRCLE. Three species—the CHINSTRAP, GENTOO and MACARONI—breed on the ANTARCTIC PENINSULA as well as on some SUBANTARCTIC ISLANDS. The KING and ROCKHOPPER breed only in the subantarctic zone. Another penguin—the ROYAL—is confined to MACQUARIE ISLAND, and is similar in appearance to the macaroni peguin.

The remaining penguin species live in warmer zones, as far north as the Equator (Galapagos Islands). Penguins are not found in the Northern Hemisphere—photographs of penguins with polar bears are faked.

Penguins have heavier wing bones than those of flighted birds; they use their modified wings as highly efficient flippers to propel themselves through water. Whereas other aquatic birds use their webbed feet for propulsion, penguins use their webbed feet, and their short tails, for steerage only. They intersperse swimming with por-

Below: A gentoo penguin (Pygoscelis papua) *chick moults its down feathers in readiness for adult plumage.*

poising (leaping above the surface of the water to breathe and to escape predators) and can reach speeds of up to 10–12 km (6–7½ miles) an hour. Penguins are extremely proficient divers, although the smaller species generally only dive for a few minutes at a time; the larger emperor penguins have been recorded staying underwater for over 18 minutes, and diving as deep as 534 m (1752 ft). Although they are short-sighted, penguins have retinas that are similar to those of FISH, and are especially sensitive to green, blue and violet, allowing them to readily detect prey underwater.

On land, penguins 'walk' on their toes, and use their powerful clawed feet (and often their bills) for grip. They have surprising strength and endurance, and are able to climb cliffs and cover considerable distances—100 km (62 miles) or more—over ICE on their bellies, propelling themselves with their feet and flippers.

Like all birds, penguins are warm-blooded, with a body temperature that is close to that of humans'. As insulation against the cold, they have a layer of fat beneath the skin and layers of densely packed, flattened feathers above. The feathers overlap, creating a watertight covering beneath which, in most species, is a layer of down that traps warm air close to the skin. When penguins are in cold water the blood flow in their feet and flippers is reduced to lessen heat loss.

Penguins have salt glands located in the skull above their eyes; these glands enable them to drink sea water and eat salty foods such as the CRUSTACEANS, SQUID and fish that are their main diet. Their hardened tongues are barbed to help them hold their slippery prey. At the ANTARCTIC CONVERGENCE, where the cold, dense Antarctic waters meet warmer subantarctic seas, minerals and nutrients are flushed to the surface from deep below. This creates favourable conditions for the abundance of PHYTOPLANKTON, fish and crustaceans—food for the penguin species living near and south of the Convergence.

Penguins themselves are preyed upon by

Above: Emperor penguins (Aptenodytes forsteri) *are the largest of the penguins, and the only ones to spend winter on the continent.*

marine MAMMALS, including LEOPARD SEALS, KILLER WHALES, HOOKER'S SEA LIONS, and, in the warmer areas, by sharks. On some subantarctic islands, introduced species such as dogs, cats and ferrets also endanger breeding penguins.

Coming onshore to breed, the penguins gather in large colonies—some Adélie rookeries have been estimated at close to a million birds. Generally they choose coastal sites, although some emperor colonies are up to 100 km (62 miles) from the ocean. The inherent sociability of these enormous breeding colonies, along with the birds' distinctive and quirky behaviour patterns, reinforce the popular perception of penguins as human-like. The 19th-century French novelist Anatole France set his best-selling morality tale *Penguin Island* in a penguin colony and endowed the birds with human status; his colourful pen-

Below: Adélie penguins (Pygoscelis adeliae) *adrift on an ice floe. Chicks may perish if they drift too far from the colony.*

Above: *Penguins are prey species for killer whales or orca* (Orcinus orca), *which attempt to ambush the birds at the ice edge.*

guin-human personalities acted out a powerful polemic on contemporary French life.

Most penguins nest in the area in which they themselves were hatched and, depending on the species, will raise one or two chicks in a breeding season. Mates tend to remain together from season to season, although some birds do change mates. Egg size, egg laying, incubation and fledging times all vary from species to species. One constant is the red colour of the egg yolks, due to the diet of crustaceans.

Although they would have been known to the indigenous inhabitants of the northern coastal lands, such as Maori in New Zealand and Aborigine in Tasmania, Australia, the first recorded sighting of penguins was in 1499, when a member of Vasco da Gama's voyage in search of a passage to the East described some large flightless birds with cries 'resembling the braying of asses' that were sighted near the Cape of Good Hope. Two decades later, Antonio Pigafetta, a member of Ferdinand MAGELLAN's expedition which sailed round TIERRA DEL FUEGO, recorded large flocks of black, flightless, fish-eating 'geese'.

As the sailing routes and SEALING and WHALING stations became established, penguin colonies became convenient food sources—both meat and eggs were taken. In the late 19th century penguins began to be exploited commercially. Although not as profitable as whales or seals, oil was extracted from the fatty layer beneath the penguin's skin and used for heating and cooking. For a time, the skins became fashionable in Europe, particularly

when made into muffs. A penguin-oil factory operated for about 20 years on Macquarie Island, but was shut down in 1919, by which time penguins were becoming popular exhibits in zoos.

The first scientific study of penguin behaviour was by K Von den Steinen, who was medical officer and zoologist on the GERMAN INTERNATIONAL POLAR YEAR EXPEDITION of 1882–83. During the expedition he made a census of the penguin rookeries on SOUTH GEORGIA and captured king penguins for study (tethering them in leather corsets).

FOSSIL evidence shows there were once more species of penguin than exist today. Ten to 15 million years ago, penguins were much larger than emperor penguins, with the largest, from SEYMOUR ISLAND, probably standing about 1.7 m (5½ ft) tall. Most scientists believe that penguins evolved from a flying ancestor about 80 million years ago, and that both present-day penguins and PETRELS are derived from it.

Below: *King penguins* (Aptenodytes patagonicus) *resemble the larger emperors, but are confined to the subantarctic islands.*

photographers of Antarctica, but other expedition members—such as James Paton, who visited the ice with Robert SCOTT and Ernest SHACKLETON, and Henry BOWERS, whose South Pole images were recovered with his body—took cameras south and produced invaluable photographic records. By the mid-20th century, Antarctica was photographed by almost every visitor.

Because of its stark beauty and unique wildlife Antarctica is a popular destination for amateur and professional photographers alike, but the environment demands a few extra precautions due to the extreme cold and dry atmosphere. Generally cameras must be kept warm, usually inside clothing, except for the brief time necessary to take a photograph, as batteries will fail in cold temperatures. This is particularly true for those with automatic winders, as the cold makes film brittle and it tends to break when wound quickly. In addition, many photographers keep their cameras in sealed plastic bags to prevent condensation occurring when moving from the cold, dry outside, to the warm, humid indoor environments and back outdoors again.

Phytoplankton Phytoplankton are the 'pasture' of the SOUTHERN OCEAN—the main source of primary production in the Antarctic FOOD CHAIN. All marine animals ultimately depend on these simple PLANT organisms. Phytoplankton are linked to the higher animal species through ZOOPLANKTON and DEMERSAL FISH, and through KRILL, SQUID and PELAGIC FISH.

Phytoplankton are free-floating ALGAE, drifting in the upper 200 m (656 ft) of the open ocean, where sunlight is available. In Antarctic waters nearly all phytoplankton (99 percent) are DIATOMS, tiny single-celled plants contained in an elaborately sculptured box of silica. There are believed to be about 100 species of diatoms around Antarctica, some of which form long connected chains. In summer, diatoms give the waters a greenish tinge and stain ICE FLOES red-brown.

The constant upwelling of mineral-laden water makes the ANTARCTIC SURFACE WATER rich in nutrients, and in spring and summer phytoplankton numbers increase rapidly, forming blooms that can cover thousands of square kilometres of ocean.

The overall productivity of the Southern Ocean may be as much as four times that of other OCEANS: not only are the waters rich in minerals, but low temperatures mean the cold water dissolves more oxygen and CARBON DIOXIDE, which is then readily available for respiration and photosynthesis, and long summer daylight hours mean almost continuous photosynthesis is possible.

Phytoplankton levels and productivity are seasonal. At their lowest, in winter, plant and animal plankton descend into the sub-surface layers of the ocean. Levels increase dramatically again from October, when the annual blooming begins in open water north of the PACK ICE and spreads southward as the ICE melts. Between February and April, phytoplankton levels fall rapidly again.

Microscopic organisms such as phytoplankton remove carbon dioxide from the ATMOSPHERE. They also release chemicals that have a role in CLOUD FORMATION. It is believed that some species of phytoplankton also produce their own 'sun screen', which prevents them being damaged by harmful ULTRAVIOLET RADIATION.

Pilot whale See LONG-FINNED PILOT WHALE.

Pintado petrel See CAPE PETREL.

Plants Cryptogamous plants (those which do not produce seed) dominate the modern-day FLORA of Antarctica, with DIATOMS comprising approximately 99 percent of the marine PHYTOPLANKTON species. The terrestrial flora is mainly composed of ALGAE, LICHENS, FUNGI, MOSSES and LIVERWORTS. Two species of FLOWERING PLANTS—ANTARCTIC HAIRGRASS and ANTARCTIC PEARLWORT—are found on the ANTARCTIC PENINSULA but do not survive further south in the harsher environment of the rest of the continent.

Plate tectonics Continental plates make up the Earth's outer shell, called the lithosphere. According to the theory of plate tectonics, the plates move slowly across the ocean floor and shear past each other at their boundaries. When they are compressed together, the plates buckle into MOUNTAIN ranges, and, when they rift apart, they produce VOLCANOES and EARTHQUAKES.

Past crustal movement beneath Antarctica uplifted the TRANSANTARCTIC MOUNTAINS, where a deep fault system lies between the thinner crust of WEST ANTARCTICA and the deep ancient shield beneath EAST ANTARCTICA. At present, however, the continent is the most earthquake-free area in the world.

In 2001, nine seismic stations were set up to measure tremors: those recorded are usually caused by either volcanic activity or the fracturing of the ICE SHEET. But some offshore islands are extremely geologically active. A massive earthquake that occurred off the BALLENY ISLANDS in March 1998 measured 8.1 on the Richter scale (the Kobe earthquake in Japan, which killed 6000 people in 1995, measured 7.1).

In 1994 the Antarctic Crustal Profile Project (ACRUP) was conducted by a multinational group of scientists to gather information about deep crustal structures in the ROSS SEA region and build up a 'time-line' of plate movement beneath Antarctica. Giant rift basins beneath the Ross Sea, which formed with the break-up of GONDWANA and are now filled with deep marine sediments, were investigated. From the information obtained, scientists concluded that the Ross Sea region is still being 'stretched', which explains the volcanic activity in this region.

Efforts have also been made to match plate boundaries in Antarctica with those on other continents. Similarities between ancient crustal belts suggest that Antarctica may have been connected to southwestern North America more than 600 million years ago, in the late Precambrian era. This theory has come to be known as the SWEAT (Southwest US-East Antarctica) hypothesis.

Plunder fish Two families of plunder fish—Harpagiferidae (spiny plunder fish) and Artedidraconidae, which includes the barbeled plunder fish—are found in Antarctic waters.

Five of the six species of spiny plunder fish found are endemic to Antarctica. They belong to the suborder NOTOTHENIOIDEI and are DEMERSAL FISH living near the bottom of the CONTINENTAL SHELF. They tend to be small—usually from 10–30 cm (4–12 in)—and are scaleless. One species, *Harpagifer antarcticus*, is common in tide pools in SOUTH GEORGIA and is found in shallow water around the northern end of the ANTARCTIC PENINSULA.

The Artedidraconidae include 23 species of the barbeled plunder fish, all exclusively Antarctic. New fish species continue to be found in Antarctica, including the first of the genus *Artedidraco* to be discovered in 80 years, and the brainbeard plunder fish, *Pogonophryne cerebropogon*, which was collected at a depth of 300 m (984 ft) in the ROSS SEA. It is almost 38 cm (15 in) long and uses its long chin barbel as a lure to attract prey.

Pointe Géologie A coastal point on Débarquement Rock, one of a number of islands just off the coast of TERRE ADÉLIE, named by the expedition led by Jules-Sebastién DUMONT D'URVILLE, which collected rock samples there when they landed on 21 January 1840.

A rich variety of wildlife lives on and around Pointe Géologie, which has been designated a SPECIALLY PROTECTED AREA. This special status was highlighted in 1983 when FRANCE began construction of a rock AIRSTRIP at the Pointe, drawing PROTESTS from environmental groups.

Below: A religious shrine at Poland's Arçtowski Base in the South Shetland Islands.

Poland Poland is active in Arctic and Antarctic research. Polish meteorologist Henryk ARÇTOWSKI, who took part in the 1897–99 BELGIAN ANTARCTIC EXPEDITION, made the first complete meteorological observations in Antarctica.

Above: 'An unbroken desert of snow', was how Lincoln Ellsworth described the Polar Plateau, which covers the majority of East Antarctica.

Poland sent several expeditions to Antarctica from 1958, took over the Russian Oazis Base in 1959, and took part in USSR programmes in the 1960s and 1970s. Poland applied to be included in the ANTARCTIC TREATY but was denied; however, it became the first country to accede to the treaty in 1961. In 1976 Poland established its first permanent year-round station, ARÇTOWSKI BASE, in the SOUTH SHETLAND ISLANDS.

Poland was the first acceding state to become a CONSULTATIVE PARTY to the Antarctic Treaty. It is a member of the Commission for the CONVENTION ON THE CONSERVATION OF ANTARCTIC MARINE LIVING RESOURCES (CCAMLR), and usually adopts a pro-fishing stance in its meetings; it is involved in research and commercial FISHING for KRILL and other species in the SOUTHERN OCEAN.

Polar desert Although covered in freshwater ICE, Antarctica is arid; it is the world's largest desert. Low TEMPERATURES over the POLAR PLATEAU mean that air masses can hold very little water vapour. The humidity and, consequently, PRECIPITATION are very low. No rain falls in the interior, and the small amount of SNOW that is carried this far inland by cyclonic systems (see WEATHER) is equivalent to about 5 cm (2 in) of melted water annually—only slightly more than the Sahara Desert receives. In the DRY VALLEYS, it is believed that significant precipitation has been absent for about 2 million years.

The aridity of the CLIMATE preserves any evidence of human presence in Antarctica for hundreds of years. 'Nothing has changed at all but the company,' said one of Ernest SHACKLETON's men on revisiting their old base at Cape ROYDS. Three years after their 1907–09 expedition the remains of lunch, left when the men rushed to board their ship, were still intact.

Polar easterlies A narrow ring of low-altitude easterly winds occur between the Antarctica coast and the CIRCUMPOLAR TROUGH. They form in a zone of higher air pressure around the coast and are driven by KATABATIC WINDS, which are diverted to the west by the spin of the Earth—known as

the 'Coriolis Force'. Polar easterlies are relatively mild and bring clear WEATHER to coastal areas.

Polar Front The ANTARCTIC CONVERGENCE was commonly referred to as the Polar Front. The first scientific definition of the Front was by Wilhelm Meinardus, meteorologist on the 1901 GAUSS EXPEDITION, who charted the course of the transition from warmer to low sea temperatures at between 105°W and 80°E.

The terms 'Polar Front' and 'Meinardus Line' were replaced by the name 'Convergence' in 1928, when Dr G Wüst, scientist on the METEOR EXPEDITION published his theories on ocean circulation. Many of these theories have subsequently been proved.

Polar Night Jet In MIDWINTER, when the CIRCUMPOLAR VORTEX—a whirlpool of WIND that spirals around the SOUTH POLE over winter—is at its strongest, reaching speeds of more than 100 m (328 ft) per second, it is known as the 'Polar Night Jet'. See ATMOSPHERE.

Polar Plateau A seemingly never-ending expanse of thousands of miles of apparently flat, snow-covered ice, the Polar Plateau covers most of the East Antarctic interior, extending from the TRANSANTARCTIC MOUNTAINS in the west outwards to the mountains that ring most of the coast.

To an observer at the SOUTH POLE, the plateau is completely featureless, and the horizon is unbroken in every direction. Robert SCOTT found the environment oppressive: 'The scene about us is the same as we have seen for many a day, and shall see for many a day to come—a scene so wildly and awfully desolate that it cannot fail to impress one with gloomy thoughts.'

The low TEMPERATURES and low humidity combine to make the plateau a POLAR DESERT. On his flight across the continent, 'Droning along nearly a mile above an unbroken desert of snow,' Lincoln ELLSWORTH described the plateau as a 'single, huge, conical glacier 7000 feet thick, flowing down on all sides from the Pole itself over a descending series of plateaus.'

Polar Years See FIRST INTERNATIONAL POLAR YEAR and INTERNATIONAL GEOPHYSICAL YEAR.

Pole of Inaccessibility Base In 1958, during INTERNATIONAL GEOPHYSICAL YEAR, the USSR established the Polyus Nedostupnosti Base at the POLE OF RELATIVE INACCESSIBILITY.

Pole of Relative Inaccessibility Also known as the 'Pole of Maximum Inaccessibility' and the 'Pole of Greater Inaccessibility', it is the centre of the Antarctic continent, the point that is the greatest distance from all the coasts, and is at an altitude of 3718 m (12,200 ft). It is located at around 85°S and 65°E.

Politics The path of Antarctic politics follows the development of human interest in the region and its discovery, exploration and potential for exploitation. Early explorers in the SOUTHERN OCEAN were often seeking land for potential colonies or safe harbours. Private firms sent ships into the Southern Ocean in search of new SEALING and WHALING grounds to exploit commercially. During the 'HEROIC AGE' of Antarctic EXPLORATION, expeditions were attracted by considerations of personal and national prestige, as much as by interests in scientific investigations. Some expeditions were designed as 'Pole hunts' to capture the public imagination and attract funding for further exploration. Others were concerned with possession: from the 19th century on, different states made, or considered making, TERRITORIAL CLAIMS in Antarctica.

The most significant turning point in Antarctic politics was the signing of the ANTARCTIC TREATY, which was agreed in 1959 at the close of INTERNATIONAL GEOPHYSICAL YEAR. Since then, several CONVENTIONS concerned with Antarctic resources and environmental protection have been signed.

Issues relating to marine living resources are handled by the 1982 CONVENTION ON THE CONSERVATION OF ANTARCTIC MARINE LIVING RESOURCES (CCAMLR), with the exception of seals and whales, which are covered by the CONVENTION FOR THE CONSERVATION OF ANTARCTIC SEALS and the INTERNATIONAL CONVENTION FOR THE REGULATION OF WHALING. The impetus for CCAMLR was the growing interest in the harvesting of KRILL.

In the late 1980s the key issue was the negotiations leading to the 1988 CONVENTION ON THE REGULATION OF ANTARCTIC MINERAL RESOURCE ACTIVITIES (CRAMRA). When PROTESTS against mineral exploitation in Antarctica prevented the ratification of CRAMRA, the treaty CONSULTATIVE PARTIES responded by negotiating the 1991 MADRID PROTOCOL, which placed a total ban on MINING in Antarctica for 50 years.

Antarctic political developments can be slow. It may take several years for a decision to be made after discussions have been initiated. In recent years non-governmental organizations, such as GREENPEACE and the ANTARCTIC AND SOUTHERN OCEAN COALITION, have played a critical role, debating the issues publicly and forcing the decision-making of the ANTARCTIC TREATY SYSTEM (ATS) to become more open and visible. Supporters of the COMMON HERITAGE PRINCIPLE

Above: This spill of dumped waste oil below Russkaya Base was photographed in the summer of 1990–91. This site has since been cleaned up.

have been contaminated by stable pesticides such as DDT, polychlorinated biphenyl (PCBs) and heavy metals such as mercury. The use of many of these substances has been banned for several decades in some countries: it is believed that residues come mainly from human activity in the world's poorer countries, where these chemicals are still used extensively.

Polynesian legends See LEGENDS.

Polynyas Areas of open water among the PACK ICE in the SOUTHERN OCEAN. They tend to recur in the same locations year after year and are areas of enhanced biological activity. Providing 'skylights' to the ATMOSPHERE for marine mammals such as WHALES and SEALS, they are sources of PLANKTON, KRILL and FISH, food for mammals and large colonies of BIRDS as well. Most EMPEROR PENGUIN colonies are located close to reoccurring polynyas.

Coastal polynyas are thought to be caused chiefly by persistent local WINDS such as the KATABATIC, which cascades from the SOUTH POLE down to the coast, driving SEA ICE offshore and causing warmer water to well up. Storms may also contribute to the formation of polynyas by drawing ICE away from the coast.

The shape of the local coastline also affects their formation: in TERRA NOVA BAY, for example, the Drygalski Ice Tongue blocks the northward movement of sea ice from the ROSS SEA, resulting in a reoccurring 1000 sq km (620 sq mile) polynya. Local gyres, eddies and OCEAN CURRENTS

have also helped to reduce secrecy in the ATS by using the UNITED NATIONS as a forum to discuss Antarctic issues.

However, ongoing political problems exist. Examples include slow progress toward meeting the obligations set out in the Madrid Protocol, discussion proceeding slowly on the need for a permanent secretariat for the Antarctic Treaty, and the effects of a dramatic increase in illegal, unregulated and unreported (IUU) fishing in the

Southern Ocean that CCAMLR has been unable to adequately control.

Pollution As well as the environmental impact of bases—including those caused by construction, power generation, fuel and toxic spills, and problems of WASTE DISPOSAL—there is evidence of pollution in Antarctica carried by OCEAN CURRENTS, atmospheric circulation and FAUNA. Since the 1970s, Antarctic organisms, including PENGUINS,

Below: Possession Island is one of a group of two islands lying in the Ross Sea.

may keep coastal polynyas open for long periods of time. Open-ocean polynyas are larger and last longer. Caused by the deep upwelling of warmer water, some remain open for years at a time. The best example is the vast Weddell Polynya in the WEDDELL SEA, which persisted during the winter months from 1974 to 1976 and measured around 350 by 1000 km (217 by 620 miles). The heat production from an ocean polynya upwelling, measured by scientists in 1994, equated to 20 times the energy required to power every USA household over a winter season.

Early explorers would anticipate breaking into open waters when they observed grey 'water sky'—caused by the low reflectivity of open water compared with that of the surrounding pack ice. They often thought they had discovered new oceans when they broke into large polynyas. Polar-orbiting satellites now provide important navigational information about the behaviour of these features in the Southern Ocean.

See also LEADS.

Ponies Robert SCOTT opted to use both DOGS and Manchurian ponies to haul his SLEDGES. The ponies, under the care of Lawrence OATES, were not as well adapted to the Antarctic environment as the dogs, and floundered in deep, soft snow. Exposed to brutal WEATHER conditions on sledging journeys, they either froze to death or had to be destroyed.

Ponting, Herbert (1870–1935) British photographer and explorer, born in Salisbury, England. He travelled widely during the first half of his life, working as a war correspondent in eastern Asia and during the Spanish-American war, and he

Below: Herbert Ponting's still and moving images of Antarctica taken on the 1910–13 British Antarctic Expedition helped to define the world's idea of the Heroic Age.

lived on a ranch in California. He took up professional photography in 1900.

By 1909, when Ponting met Robert SCOTT, he had developed a reputation as a photographer. Ponting convinced the explorer to appoint him 'camera artist' to the 1910–13 BRITISH ANTARCTIC EXPEDITION. Ponting was the first professional photographer to go to the Antarctic.

Although his reputation was much enhanced by his Antarctic work, the results brought him little financial reward. His book, *The Great White South*, and the film, *Ninety Degrees South*, were acclaimed, but he profited little from them and few of the business ventures he tried during the later part of his life were successful.

Pourquoi Pas? A three-masted schooner, built with French government funds by Jean-Baptiste CHARCOT for his 1908–10 expedition. The ship sailed from St Malo in Normandy, France, in May 1908 and spent two austral summers charting the ANTARCTIC PENINSULA coastline, returning to France in June 1910. The *Pourquoi Pas?* made a number of voyages to the northern polar regions, and in 1936 sank with Charcot on board.

Port Martin Base This base was built in TERRE ADÉLIE in 1950 by FRANCE and was destroyed by fire in 1952. It is now unoccupied.

Portugal Like SPAIN, Portugal signed the 1494 TREATY OF TORDESILLAS, under which it laid claim to lands 'from the Arctic to the Antarctic Poles' east of a boundary between 46°W and 49°W longitude. Portugal has never made a claim to Antarctic territory.

Portugal is not a member of the ANTARCTIC TREATY SYSTEM or the CONVENTION ON THE CONSERVATION OF ANTARCTIC MARINE LIVING RESOURCES, but has been interested in commercial FISHING in the SOUTHERN OCEAN, and some Portuguese-flagged fishing vessels have been involved in illegal, unregulated and unreported TOOTHFISH FISHING.

Possession Islands Discovered by James Clark ROSS and claimed for Britain in January 1841, the group consists of two islands, Possession and Foyn. They are located in the Ross Sea, at the northern tip of VICTORIA LAND.

Precipitation Any form of water—liquid or solid—that precipitates, or falls, from the ATMOSPHERE and reaches the ground. In Antarctica, precipitation originates from water that evaporates from ice-free areas of the SOUTHERN OCEAN. It is carried in cyclonic systems that spread water vapour over the continent, mostly in a band from 80°E to 140°W—WILKES LAND to MARIE BYRD LAND. Because of low TEMPERATURES, and the fact that the continent is isolated from the ocean by SEA ICE for much of the year, the Antarctic atmosphere contains only about one-tenth of the water vapour found in temperate (northern) latitudes. This accounts for very low precipitation over Antarctica, particularly on the POLAR PLATEAU, where water vapour freezes and falls as SNOW (see POLAR DESERT). Along the coast, precipitation is

heavier, amounting to about 38 cm (about 15 in) of water per year. The ANTARCTIC PENINSULA is the only area where rainfall is common.

Some researchers believe that global warming may increase the amount of precipitation over Antarctica—that as the Southern Ocean warms, more water vapour may be released, which would be circulated over the continent. Because snow supplies mass to the ICE SHEET, this process may balance out the loss of ice from ICE SHELVES breaking up at the coast.

Recent research near the SOUTH POLE has found that accumulation has increased more than 20 percent since the 1980s. However, it has yet to be determined whether this increase is due to global warming or if other factors, such as periodic variations in atmospheric circulation, have played a role.

Priestley, Raymond (1886–1974) British geologist, born in Tewksbury, England. He was appointed by Ernest SHACKLETON as assistant geologist to Edgeworth DAVID on the 1907–09 BRITISH ANTARCTIC EXPEDITION, on which he formed part of the support party for Shackleton's attempt to reach the SOUTH POLE. Together with Bertram Armytage and Philip Brocklehurst he ascended the FERRAR GLACIER, looking for FOSSILS and carrying out geological and survey work. At one point the three men were carried out to sea with only two days' provisions on a small ICE FLOE surrounded by KILLER WHALES.

He was recruited by Robert SCOTT for the 1910–13 BRITISH ANTARCTIC EXPEDITION. Priestley was a member of the six-man northern party, led by Victor Campbell, which carried out exploratory surveys of the VICTORIA LAND coastline. Because the *TERRA NOVA* was unable to pick them up in mid-February 1912 as arranged, they were forced to OVERWINTER in an ICE CAVE, living off SEALS and PENGUINS. The following spring they walked to Cape EVANS, a journey that took 40 days.

Priestley won a Military Cross in World War I, and was vice-chancellor of both Melbourne and Birmingham universities. He made two later visits to Antarctica.

Prince Edward Islands Two large islands, MARION and Prince Edward, lie just north of the ANTARCTIC CONVERGENCE in the south Indian Ocean. Part of SOUTH AFRICA's Cape of Good Hope Province, they have a combined area of 316 sq km (122 sq miles). Prince Edward has a coastline of steep vertical cliffs, and Marion is covered by small hills and lakes. Bleak and barren over winter, the islands are transformed into lush, fertile pastures during summer. TEMPERATURES, however, remain low, and the WEATHER is usually overcast with strong westerly WINDS.

The islands were sighted in 1663 by a Dutch navigator and again in the mid-18th century by Marion du FRESNE and James COOK. French sealers probably made the first landing about 30 years later. SEALING companies operated on the islands up until 1930. BRITAIN's 1908 SOVEREIGNTY claim was challenged in 1947–48 by SOUTH AFRICA, which set up a permanent meteorological station on Marion Island. This has been developed into a

Above: Fairy prions (Pachyptila turtur) are relatively common in the subantarctic, although they fall prey to skuas and feral cats while nesting.

scientific station that houses about 70 people over summer.

FUR and ELEPHANT SEALS have re-established breeding colonies on the islands. The long history of human occupancy has inevitably resulted in INTRODUCED SPECIES, notably rats and cats, gaining a foothold and ravaging bird populations. From five cats brought to the Marion Island station in 1949, the population had grown to about 3400 by 1977. These have now been eradicated, and populations of PETRELS, ALBATROSSES and PENGUINS that breed on the islands are now thriving.

Princess Elizabeth Land Also known as Princess Elizabeth Coast. Located in EAST ANTARCTICA between MAC.ROBERTSON LAND and WILHELM II LAND. Discovered by the BRITISH-AUSTRALIAN-NEW ZEALAND ANTARCTIC RESEARCH EXPEDITION in 1930–31, it was named after the British princess who became Queen Elizabeth II.

Prions (*Pachyptila* spp.) Six species of prions—small PETRELS—exist solely in Southern Hemisphere waters. Of these, four regularly visit Antarctic waters: the ANTARCTIC PRION, FULMAR PRION, THIN-BILLED PRION and FAIRY PRION; a fifth, SALVIN'S PRION, breeds on SUBANTARCTIC ISLANDS. The species differ in bill size and function, but are otherwise very similar in appearance.

Prions are small, blue-grey and white in colour, with a black open 'M' across the upperwings and black-tipped tails—a colour scheme that provides effective camouflage against predation and makes them difficult to detect at sea. From above, their plumage blends with the sea; from below, the white underparts merge with the open sky.

A fast, twisting flight is characteristic of all prions. Another defensive strategy most prions employ is to return to their nests after nightfall. Also collectively known as whalebirds, the larger-billed species have bills fringed with lamellae that

allow them to sieve seawater in the manner of BALEEN WHALES. Prions feed from the ocean's surface, taking CRUSTACEANS, SQUID and some small FISH. They rarely follow ships.

Protests Political protests about Antarctica and the SOUTHERN OCEAN are mostly organized by environmental groups and usually involve CONSERVATION and environmental issues. The earliest protests were against WHALING, and were in part responsible for the INTERNATIONAL WHALING COMMISSION adopting a moratorium on commercial whaling in 1985. Other Antarctic issues on which environmentalists have campaigned include: MINING, illegal and driftnet FISHING, the GREENHOUSE EFFECT, the OZONE HOLE, POLLUTION, and the call for a WORLD PARK in Antarctica.

Protests are often organized on an international basis by groups such as GREENPEACE. Rarely carried out in Antarctica itself because of difficulties travelling there, the protests often take place at venues where Antarctic issues are being discussed, such as ANTARCTIC TREATY meetings, or where agencies responsible for Antarctic activities are headquartered. Greenpeace has also campaigned in the Southern Ocean, using their own ships to disrupt whaling and fishing operations.

Environmentalists have protested at the secrecy in which the CONSULTATIVE PARTIES to the Antarctic Treaty conduct their meetings, and have lobbied for a greater public role in Antarctica. Access to these meetings for environmental groups has improved—from no representation in the 1970s to having individuals attached to national delegations. The ANTARCTIC AND SOUTHERN OCEAN COALITION is now invited as an observer in its own right to consultative party meetings and meetings of the Commission for the CONVENTION ON THE CONSERVATION OF ANTARCTIC MARINE LIVING RESOURCES.

The success of protests has been mixed. Some

consultative parties are less amenable to public pressure than others. When FRANCE began building an AIRSTRIP at DUMONT D'URVILLE BASE in 1983, it ignored widespread international criticism, as well as the recommendations of its own THALER COMMITTEE to abandon the project. On the mining issue, environmental protests have been more successful and contributed to the abandonment of the CONVENTION ON THE REGULATION OF ANTARCTIC MINERAL RESOURCE ACTIVITIES and subsequent negotiation of the MADRID PROTOCOL. On some issues, the interests of the consultative parties may be incompatible with those of the protestors: for example, the calls for Antarctica to be declared a WORLD PARK.

Protocol on Environmental Protection to the Antarctic Treaty See MADRID PROTOCOL.

Psychological effects The unusual experience of living in Antarctica involves spending long periods of time in confined spaces with small groups, separated from friends and family, and coping with extreme WEATHER conditions. Frederick COOK, medical officer on the 1897–99 BELGIAN ANTARCTIC EXPEDITION, described the mental state of Tollefsen, one of the team members: 'Poor fellow! His brain has for a long time been unsteady as a result of the unbroken daylight and hopeless isolation ... his mind is now permanently deranged.' Cook himself reported seeing 'marvellous delusional pictures'—including 'huge creatures misshapen and grotesque [that] writhed on the horizon'—that sometimes drove him to the 'verge of madness'.

Richard BYRD described the experience of isolation: '... there is no escape anywhere. You are hemmed in on every side by your own inadequacies and the crowding pressure of your associates. The ones who survive with a measure of happiness are those who live profoundly off their intellectual resources, as hibernation animals live off their fat.'

Polar explorers usually had specific characteristics in mind when selecting their team members. Roald AMUNDSEN wanted men with 'a flamboyant fatalism to which risk appeals'; Douglas MAWSON selected graduates 'with the dash and recuperative power of youth', and Ernest SHACKLETON sought men who were '... able to live together in harmony ... [and had] generally marked individuality.'

After World War II, the USA began to psychologically screen potential Antarctic staff. Criteria were developed to assess people's performances and the results used to improve future selections. Now staff are selected to go to Antarctica not just for their work skills but also for their psychological stability, which can have a huge impact on the success of a season's research. Professor A J W Taylor, who has carried out extensive research into Antarctic psychology, concluded that those selected for OVERWINTERING should be 'the most stable and experienced people ... because they are least likely to become self-centered, dissatisfied, preoccupied and inefficient. Unfortunately, such people are not always over represented in society at large, much less people who want to work in polar regions.'

Effort is also put into designing bases that are not only functional, but are easy to live in. Staff are provided with areas in which to socialize, and private spaces such as libraries. However, even with the comforts of modern bases, overwintering can take its psychological toll on staff. Depression can result from social isolation from family and society; from emotional pressures, which can give rise to insecurity and anxiety; from confinement with the same group of people, and from boredom. Many people report that they slow down mentally and physically over winter. They may have abnormal mood swings, become irritable and sleep disorders are common. A condition called 'big eye', for example, includes disorientation and sleeplessness due to the lack of a regular light-dark cycle. Sensory deprivation is another problem, caused by the lack of normal sights, smells and sounds associated with living on Earth. These symptoms are commonly referred to as 'cabin fever,' or 'winter-over syndrome'.

Psychological studies in Antarctica are about more than staff selection and base design. Living in Antarctica has been likened to living on a space station: the continent is seen as an ideal place in which to study the effects of isolation. Twin studies in space and in Antarctica may be beneficial to people living in both environments—for instance, by studying staff psychology before and after their experiences in Antarctica, psychologists are able to assess which personality characteristics affect the ability of staff to cope with isolation, and integrate into a group. Researchers are also investigating how individuals from different cultures respond to living in Antarctica.

q

Queen Alexandra Range Located on the western edge of the ROSS ICE SHELF, its highest point is Mount Kirkpatrick, at 4528 m (14,856 ft). The range is separated from the QUEEN MAUD MOUNTAINS by the BEARDMORE GLACIER and is flanked on the other side by the DRY VALLEYS.

Queen Mary Land An area of EAST ANTARCTICA lying east of WILHELM II LAND. Members of the AUSTRALASIAN ANTARCTIC EXPEDITION landed there in 1911, near the SHACKLETON ICE SHELF. The coast was explored further during the BRITISH-AUSTRALIAN-NEW ZEALAND ANTARCTIC RESEARCH EXPEDITION. The USSR maintained a number of research stations in the Queen Mary Land area, the most prominent of which, MIRNYY BASE, is still in operation year-round.

Queen Maud Mountains A rugged group of MOUNTAINS in central Antarctica that extend southeastwards for 800 km (500 miles) from the head of the ROSS ICE SHELF. A major subdivision of the TRANSANTARCTIC MOUNTAINS, they were discovered in 1911 by the party led by Roald AMUNDSEN, who traversed them on their successful journey to the SOUTH POLE and named the

mountains after the queen of Norway. The mountains are heavily glaciated and several peaks reach more than 4000 m (13,000 ft) above sea level. Geologists mapped the mountains from the 1930s through to the 1970s and found significant coal deposits in some places.

Quest **Expedition (1921–22)** British expedition, led initially by Ernest SHACKLETON then by Frank WILD. Shackleton's intention was to circumnavigate the Antarctic continent and carry out meteorological and geological research. The expedition was funded primarily by John Rowett, an old school friend of Shackleton's, who donated £70,000. Shackleton chose the *Quest*, a Norwegian sealing ship, for the expedition with Frank WORSLEY as captain.

The *Quest* sailed on 18 September 1921. Delays caused by engine repairs in Lisbon and Rio de Janiero meant the expedition was one month late. Many of the expedition members had accompanied Shackleton on previous expeditions, and to them he appeared to lack his former energy and enthusiasm. In Rio de Janiero he suffered a heart attack, then in SOUTH GEORGIA died from another heart attack on 5 January 1922. Wild took command and followed Shackleton's original plan. They sailed to the Antarctic coastline, making observations and taking soundings until 21 March 1922.

However the *Quest* was short of supplies, the engines were unreliable, and the ship leaked. The expedition returned to South Georgia, where they constructed a cairn over Shackleton's grave, then sailed back to England.

Below: Protests in Antarctica tend to focus on environmental issues, such as the French blasting at Dumont d'Urville Base to construct an airstrip.

r

Radar Originally used on long overland journeys, methods of using radar in Antarctica were developed in the 1960s. Sledges carrying radar antennae were dragged across the surface of the continent to give the first continuous plots of the shape, or topography, of the BEDROCK and the elevation of the ICE SHEET. This is still an effective way of mapping small-scale landscapes, such as a glacial basin.

However, radars on aerial flights and SATELLITES are now used to build a picture of larger areas. Early flights revealed the extraordinary hidden topography of MOUNTAIN ranges, such as the Gamburtsev Mountains that rise 2600 m (8500 ft) above sea level, but are completely covered by ICE; SUBGLACIAL LAKES and basins. Radar beams are continually scattered over Antarctica from a satellite known as RADARSAT, which can measure the surface motion of GLACIERS and has revo-lutionized MAPPING in Antarctica. It is able to operate through winter darkness, penetrate through thick cloud cover, and rotate in orbit, changing its perspective on the continent.

Radio The first radio link from Antarctica was during the 1911–14 AUSTRALASIAN ANTARCTIC EXPEDITION, via a relay on MACQUARIE ISLAND. The use of radio in Antarctica was greatly expanded by Richard BYRD, who used radio to communicate with field parties and different bases, and to link up with the rest of the world. When the plane *Floyd Bennet* made what, at the time, was believed to be the first flight over the SOUTH POLE in November 1929, its radio signal was picked up by a ham radio operator in New York and broadcast in Times Square (ham radio operations were a crucial communication tool in Antarctica for many years).

During BYRD'S SECOND EXPEDITION, a licensed radio station with an American network announcer was installed at LITTLE AMERICA. Byrd was not without regrets that the sense of total isolation was gone: 'When too much talk seems to be the cause of much of the grief in the world, no man could break the isolation of the Last Continent of Silence without a twinge of remorse.'

Below: Satellite telephone technology powered by solar panels is a far cry from the heavy and unreliable radio equipment used on earlier polar expeditions.

At one stage it was believed that Antarctica might be a useful location for ground relay stations for global radio transmissions, but communication SATELLITES have filled this role more effectively. The southern auroral belt made communication between some parts of Antarctica by radio difficult, but this problem has also been resolved with the use of satellites.

Radioactivity In Antarctica, ICE CORES measure global levels of radioactivity, and have shown that increases in nuclear radiation since 1945 can be attributed to nuclear weapons tests. Scallops and FISH in Antarctic waters have also been contaminated by radioactivity. In 1972, radioactive material was found on the site of a NUCLEAR POWER plant operated by the USA at MCMURDO STATION; the plant has since been dismantled and shipped back to the United States.

Rea, Henry (b. c. 1804) British naval officer. Rea joined the British Navy in 1820, and in 1833 was chosen by the Admiralty to follow up the discoveries of various SUBANTARCTIC ISLANDS made by John BISCOE. Two ships were made available by the ENDERBY BROTHERS company: the *Hopefull* and the *Rose*. Rea was an uninspired leader and a clash with Captain Prior of the *Rose* caused the latter to resign when the expedition was at the FALKLAND ISLANDS. At 60°S, the *Rose* was trapped between two ICEBERGS and crushed. This mishap and the rebellious crew—Rea was later to refer to them as a 'mutinous set of dogs'—forced the expedition back to England.

Red snow A phenomenon found on snowfields near the Antarctic coast, particularly on the ANTARCTIC PENINSULA. The surface is coloured by ALGAE that contain green and red pigments. The algae are introduced by windblown dust, and can cling to the surface of SNOW and ICE. The snowfields where this occurs are usually close to PENGUIN colonies, which supply enough nutrients for these algal populations to thrive. Apsley Cherry-Garrard wrote that '... snow seldom looks white, and if carefully looked at will be found to be shaded with many colours but chiefly with cobalt blue or rose madder, and all the gradations of lilac and mauve which the mixture of these colours will produce.'

Rescue The history of Antarctic exploration is studded with dramatic rescues: the recovery of the CASTAWAYS Gustav and Frederick Stoltenhoff from INACCESSIBLE ISLAND by the *CHALLENGER* EXPEDITION in 1873; the failed attempt to pull a sailor out of icy Antarctic waters by Lieutenant Lecointe of the *BELGICA* in 1897; two relief ships pulling the *DISCOVERY* out of the PACK ICE in February 1904; the Argentinian-led rescue of the men from the 1901–04 SWEDISH SOUTH POLAR EXPEDITION who had been stranded on PAULET ISLAND over the 1903 winter and, of course, the desperate epic journey Ernest SHACKLETON made with five expedition members to SOUTH GEORGIA in order to mount a rescue voyage for the rest of his crew.

Rescues are always hazardous in Antarctic

Above: *New Zealand scientists undertake preliminary drilling on the Ross Ice Shelf between White and Ross Islands. The project, entitled ADDRILL, is aimed at mapping the Earth's profile beneath the sea floor.*

conditions, and timing is crucial. Search and Rescue (SAR) teams often have a very short period of good weather in which to work. The most extensive SAR capability in Antarctica is a joint effort between the USA and NEW ZEALAND programmes. Using personnel and equipment from MCMURDO STATION and SCOTT BASE, the SAR teams can access most areas of the Antarctic continent in EMERGENCIES. If a party does not communicate by RADIO with a station for 72 hours, SAR teams will be alerted and a search may be carried out, usually using a fixed-wing AIRCRAFT or HELICOPTER. The 1979 Air New Zealand AIR CRASH on Mount EREBUS was the biggest combined search and rescue effort that has been coordinated in Antarctica, and extra personnel were sent in from New Zealand to assist.

From time to time over summer, seriously ill staff are evacuated from Antarctic stations; it may be impossible for the rest of the year. In April 2001—when the continent was in the throes of winter—a Royal New Zealand Air Force plane evacuated four ill USA personnel from McMurdo Station, and two Canadian planes flew to the SOUTH POLE to airlift out American doctor Ronald Shemenski, who was suffering from pancreatitis.

Most private expeditions carry SATELLITE beacons (and have insurance cover), so they can request a rescue if necessary. In 1995, 41 days into his attempt at the first solo unsupported crossing

of Antarctica, equipment problems forced British explorer Roger Mear to call his base for rescue. Using a satellite beacon, Mear sent a message that was received by the Rescue Coordination Centre in Plymouth, England, and relayed to ADVENTURE NETWORK INTERNATIONAL (ANI) in Punta Arenas, Chile. ANI deployed a rescue plane from its base at PATRIOT HILLS and the explorer's tent was spotted in the snow within two hours.

Research From the earliest days of EXPLORATION, scientists have accompanied expeditions to Antarctica and conducted SCIENTIFIC RESEARCH there in a wide variety of disciplines. The ANTARCTIC TREATY noted the contribution to knowledge derived from research in Antarctica and provided for the freedom of scientific investigation that developed in the INTERNATIONAL GEOPHYSICAL YEAR of 1957–58 to continue, and it promoted the exchange of information and scientists among treaty members. The MADRID PROTOCOL further recognized the importance of the Antarctic environment to scientific research.

The SCIENTIFIC COMMITTEE ON ANTARCTIC RESEARCH helps to foster research efforts among different international and national groups. The CONVENTION ON THE CONSERVATION OF ANTARCTIC MARINE LIVING RESOURCES also established a scientific committee to coordinate research into the ECOSYSTEM of the SOUTHERN OCEAN as well as generate advice for its decision-making commission.

Research in Antarctica is usually carried out at RESEARCH STATIONS and by field expeditions. Research in the Southern Ocean uses data gathered by scientists and by commercial FISHING operations. SATELLITES are also useful for mapping and other projects, such as ATMOSPHERIC RESEARCH. Common areas of scientific investigation in Antarctica include astronomy, biology, climatology, glaciology, geology, geomagnetism, human physiology, meteorology, oceanography and ornithology. They include specialized research projects that are difficult to do in other parts of the world, such as studying PENGUINS, drilling ICE CORES and ICE SHELF research, collecting METEORITES and studying the OZONE HOLE.

See also, ASTRONOMICAL RESEARCH, CLIMATO-LOGICAL RESEARCH, CONSULTATIVE PARTY, GEOLOGICAL RESEARCH, GLACIOLOGICAL RESEARCH, METEOROLOGICAL RESEARCH and SEISMIC RESEARCH.

Research stations Research stations are bases containing facilities to assist scientists who are living in Antarctica to conduct scientific research. Establishing a research station is usually regarded as an important step in becoming a CONSULTATIVE PARTY to the ANTARCTIC TREATY.

Resources The natural riches on the continent, its offshore islands and the SOUTHERN OCEAN include biological and MINERAL RESOURCES. Broadly speaking, these resources also include the continent itself: the ICE, CLIMATE and research potential of Antarctica. The exploitation of resources initially centred on the Southern Ocean and SUB-ANTARCTIC ISLANDS, as expeditions exploited SEALING and WHALING grounds in the 18th and 19th centuries.

Today, the FISHING industry is a significant exploiter of Antarctic resources. Attempts have been made to regulate harvesting to avoid over-exploitation through the INTERNATIONAL WHALING COMMISSION and the establishment of management regimes: the CONVENTION ON THE CONSERVATION OF ANTARCTIC SEALS and the CONVENTION ON THE CONSERVATION OF ANTARCTIC MARINE LIVING RESOURCES. Nevertheless there are problems with illegal, unregulated and unreported TOOTHFISH FISHING and BY-CATCH fishing. So far there has been little resource exploitation on the continent itself because of the harsh environmental conditions and the 50-year ban on mining of MINERAL RESOURCES under the MADRID PROTOCOL. However, as technology and the pattern of global demand for resources change, so will the importance of the perceived resources of Antarctica.

Right whale See SOUTHERN RIGHT WHALE.

Riiser-Larsen, Hjalmar (1890–1965) Norwegian explorer. Riiser-Larsen accompanied Roald AMUNDSEN through the north-west passage aboard the *Gjöa* in 1903, he fitted out the two hydroplanes that Amundsen used in his failed attempt to fly over the North Pole on 21 May 1925, and he accompanied Amundsen on the flight of the dirigible *Norge* to the North Pole on 9 May 1926.

In 1929 Riiser-Larsen captained the whaling ship *Norvegia* to the SOUTHERN OCEAN on an expedition to explore east DRONNING MAUD LAND using a seaplane. On 14 January 1930 he met Douglas MAWSON at Proclamation Island and they agreed to limit their respective TERRITORIAL CLAIMS to either side of 45°E, with the west reserved for NORWAY. In 1933 he planned an ambitious winter expedition, with three men and 50 DOGS, to sledge over the SEA ICE around the coast from Princess Olga Coast to the ANTARCTIC PENINSULA—a distance of around 14,500 km (3000 miles). However, the plan was aborted when, soon after they landed, the ice broke away, taking dogs and stores with it. In 1940, erroneously reported killed during the German invasion of Norway, he escaped to England.

Rivers Flowing water with enough discharge to form a river is rare in Antarctica. One of the longest rivers flows down the LAMBERT GLACIER into the AMERY ICE SHELF.

The longest non-glacial river is the ONYX, which most summers flows 48 km (30 miles) westwards from the Lower Wright Glacier to Lake VANDA, a SALINE LAKE in the DRY VALLEYS. It is one of the relatively few rivers in the world to flow inland from the coast.

The ALPH RIVER, which flows along the MORAINES of the Koettlitz Glacier, was discovered by a sledging party led by Griffith TAYLOR in 1911. After hauling their sledges over the steep icy terrain of the Glacier, they were surprised by a 'steep gully about 100 ft deep at the bottom of which was a strongly flowing stream.' Discharge measurements have since shown that the Alph River, although shorter, may have a higher discharge than the Onyx.

Roaring Forties Concentrated STORM formation over the stretch of SOUTHERN OCEAN between latitudes 40°S and 50°S. In combination with the FURIOUS FIFTIES (between 50°S to 60°S) and the SCREAMING SIXTIES (between 60°S to 70°S), this legendary wind belt is associated with many adventures and catastrophes at sea. At its most furious—between Antarctica and the tip of South America—it is considered the most hostile WIND in the world and creates violent seas that have claimed numerous ships over the centuries. After members of the AUSTRALASIAN ANTARCTIC EXPEDITION disembarked from the *Aurora* on the SHACKLETON ICE SHELF in 1912, Captain John DAVIS experienced fierce winds that lasted for a week. While steaming at full speed into the wind, the ship was pushed backwards in the gale, so that the wake of the ship flowed back past the bow.

Robots As scientists try to minimize the impact their projects have on the Antarctic environment, robots are being developed to collect samples by remote control. They are likely to be used in any exploration of Lake VOSTOK, one of the SUBGLACIAL LAKES beneath the Antarctic ICE SHEET. The USA space agency NASA has conducted Antarctic trials to determine whether robotic equipment can withstand demanding environmental conditions on other planets. Robots that were sent to MARS in 1998 were tested in the DRY VALLEYS.

Meteorite-hunting robots, or 'meteorbots', have been equipped with specialized features to help them detect METEORITES in Antarctica; RADAR allows them to identify objects buried beneath the icy surface. There are plans to power the robots by WIND turbine, thus enabling them to operate over the long polar winter.

de la Roche, Anthony (*c.* 17th century) English merchant. Sailing from the Pacific round Cape Horn in 1675, de la Roche was blown off-course

Above: Rockhopper penguins (Eudyptes crestatus) *are skilled climbers and can move at surprising speed over land.*

to the east and south of the MAGELLAN STRAITS and spent 14 days sheltering in a bay of an unknown land, at a latitude of 55°S. From his description, he was probably in SOUTH GEORGIA. This is the first voyage recorded south of the ANTARCTIC CONVERGENCE.

Rockets The development of space technology has opened up new avenues for ATMOSPHERIC RESEARCH in Antarctica. Rockets have been used to a limited extent, along with observations from SATELLITES and ground-based data. During the INTERNATIONAL GEOPHYSICAL YEAR, physicist James VAN ALLEN used 'rockoons'—experimental rockets launched from high-altitude BALLOONS—to make soundings in the upper ATMOSPHERE over Antarctica. He detected a zone of intense RADIATION in the magnetosphere (see ATMOSPHERE), now known as the Van Allen Belt.

Rockhopper penguin (*Eudyptes crestatus*) The rockhopper is the smallest of the Antarctic and subantarctic penguins. It is also the most aggressive towards its companions—and is famously noisy. It averages 56 cm (22 in) in height and weighs about 3 kg (6½ lb). The rockhopper has red eyes and a distinctive crest of droopy yellow feathers behind its eyes. The estimated population numbers 3.7 million pairs. They are found throughout the subantarctic zone and as far north as TRISTAN DA CUNHA. As their name implies, rockhoppers are good at jumping and climbing among cliffs and rocks, helped by their strong claws and bills. They make their shallow nest in small rookeries on ledges, crevices and among boulders on the shoreline. Breeding occurs from September to November. Females lay two eggs but usually only one is incubated, for 35 days. The chick is then brooded for 26 days. Rockhopper penguins feed mainly on CRUSTACEANS and lantern fish.

Rocks The ages of Antarctic rocks span a vast period: the BEDROCK of EAST ANTARCTICA dates back more than 3000 million years; the youngest rocks are less than 10 million years old. Rocks are exposed over less than 2 percent of the continent: at the margins of the continent; on the ANTARCTIC PENINSULA, and in ice-free OASES such as the DRY VALLEYS. Where rocks do protrude above the surface, they stand out in stark contrast to the ICE.

Rongé Island Along with the neighbouring CUVERVILLE ISLAND, Rongé Island was discovered off the northwestern coast of the ANTARCTIC PENINSULA by Adrién de GERLACHE in 1897–99. There are several large CHINSTRAP and GENTOO PENGUIN rookeries on the island.

Ronne, Finn (*c.* 1900–80) Norwegian-American scientist and explorer. Ronne's father, Martin, accompanied Roald AMUNDSEN to Antarctica in 1910–12 (it was his silk tent that was left at the SOUTH POLE) and on other polar expeditions, including BYRD'S FIRST EXPEDITION.

By the early 1930s Finn Ronne was designer-

Below: Rockhopper penguins (Eudyptes crestatus) *are the smallest—and noisiest—of Antarctic penguins, and with their red eyes and crest of yellow feathers they are easily identified.*

engineer with Westinghouse in Pittsburgh. He travelled to Antarctica with Richard BYRD's 1933–35 expedition as expert skier, dog handler and radio operator. Back in the USA, Ronne resumed his career at Westinghouse, but was to return to Antarctica eight times.

In 1939–41, again with Byrd, he was a member of the UNITED STATES ANTARCTIC SERVICE EXPEDITION. He travelled on the *North Star* to lead the sledging teams on the western side of the ANTARCTIC PENINSULA, where the Ronne Entrance at the southern end of ALEXANDER ISLAND was named in his honour. During World War II he served in the United States Navy, from which he retired in 1962 as rear-admiral. In 1946 he organized Operation Nanook to the Arctic.

He led a privately sponsored scientific expedition to Antarctica: the RONNE ANTARCTIC RESEARCH EXPEDITION of 1946–48, during which the RONNE ICE SHELF was discovered. During the INTERNATIONAL GEOPHYSICAL YEAR, Ronne was back south again, this time sailing on the *Wyandot* to the edge of the FILCHNER ICE SHELF and establishing ELLSWORTH STATION. In 1961 and 1971 he was invited on flights to the SOUTH POLE to commemorate the 50th and 60th anniversaries of Amundsen's and Robert SCOTT's journeys to the Pole. 'On my nine expeditions to Antarctica I skied behind a dog-team more than six thousand miles,' he wrote in *Antarctica, My Destiny*. 'Bridging the pioneering and modern eras of polar exploration, I have seen the dogs replaced by airplanes, helicopters, and tracked vehicles. The full cycle of polar achievement has passed before my eyes.'

Ronne Antarctic Research Expedition (1946–48) Privately sponsored American scientific expedition led by Finn RONNE to undertake surveying and scientific research. From their base on STONINGTON ISLAND, land and air explorations were made of both coasts of the ANTARCTIC PENINSULA, along with an aerial reconnaissance of the RONNE ICE SHELF. This was originally named 'Edith Ronne Ice Shelf' after Ronne's wife (Edith and Jennie Darlington, the wife of one of the pilots, were the first WOMEN to live in Antarctica). The expedition explored and mapped over 174,000 sq km (450,000 sq miles) of newly discovered territory, flew over 28,000 km (45,000 miles), and took over 14,000 aerial mapping photographs. As the expedition sailed away from Stonington Island, Vivian FUCHS and his COMMONWEALTH TRANSANTARCTIC EXPEDITION sailed in.

Ronne Ice Shelf First sighted from the air by Finn RONNE, it is one of two ICE SHELVES located at the head of the WEDDELL SEA (the other is the FILCHNER ICE SHELF). The Ronne Ice Shelf is over 1300 m (4264 ft) thick at its grounding line, and extends inland for more than 840 km (521 miles). About the size of France, it is the second largest ice shelf in Antarctica, after the ROSS ICE SHELF. On its east and west margins it is fed by the Foundation and Evans ICE STREAMS. The shear created by the flow of these ice streams causes a very thin central area, where seawater is frozen on to the base of the shelf. A massive ICEBERG broke away from the shelf in October 1998. The event, observed via SATELLITE, carried away a German summer research station, which was uninhabited at the time. The iceberg—more than 64 km (40 miles) wide—was floating in the South Atlantic in 2001.

Roosevelt Island Located in the eastern ROSS SEA, the island is 130 km (81 miles) long and 65 km (40 miles) wide and is completely submerged by ICE. Discovered by Richard BYRD in 1934, and named after then president of the USA, Franklin D Roosevelt, the northern tip of the island is 5 km (3 miles) south of the BAY OF WHALES. It can be identified from the surface by a central ridge of ice that lies about 550 m (1800 ft) above sea level.

Ross, James Clark (1800–62) British naval officer and polar explorer. Ross was born in London, entered the British Navy at the age of 12, and joined his uncle, Sir John Ross, on two Arctic voyages. Between 1819 and 1827 he went with Sir William Parry on four expeditions, one of which was in search of the north-west passage. In 1829 he joined his uncle again in the successful discovery of the North Magnetic Pole.

Following his work on a magnetic survey of Britain, he was appointed to lead the 1839–43 voyage of HMS *Erebus* and HMS *Terror* to the SOUTHERN OCEAN. The *Terror* was under the command of Francis CROZIER, who had been with Ross on one of the Parry expeditions. Its main purpose was to locate the SOUTH MAGNETIC POLE and carry out magnetic observations, and it included onboard the naturalist Joseph HOOKER. They made first for MARION ISLAND then Îles KERGUÉLEN, where they surveyed the islands. In August 1840 they reached Hobart, AUSTRALIA, where Ross was welcomed by Sir John Franklin and met Charles WILKES.

The expedition crossed the ANTARCTIC CIRCLE on 1 January 1841. Eleven days later the watch reported sighting land: snow-covered mountains. Their furthest point south was 71°15', where a landing was made at POSSESSION ISLANDS; the coastline beyond was named VICTORIA LAND. On 27 January a second landing was made on a volcanic island named for Sir John Franklin, and on the following day they discovered ROSS ISLAND, on which was an active volcano they named Mount EREBUS and a smaller extinct cone they called Mount TERROR. They also discovered the ROSS SEA and ROSS ICE SHELF and reached within 258 km (160 miles) of the South Magnetic Pole. After wintering in NEW ZEALAND, in December 1841 the two ships returned south. New Year's Day was celebrated on an ICE FLOE with various games and a 'grand ball'. After sailing along the edge of the ice shelf, they made for the FALKLAND ISLANDS.

On 17 December 1842 the expedition sailed south for the third time, this time to the WEDDELL SEA. The area between the Antarctic Peninsula and JOINVILLE ISLAND, now called Erebus and Terror Gulf, was charted and landings were made on two newly discovered volcanic islands, PAULET and Cockburn. On 5 March 1843 Ross reached his highest latitude—71°30'S—before heading for

Above: *James Clark Ross. The magnetic observations undertaken on his 1839–43 voyage and the detailed scientific reports he produced on his return were of immense value to 19th-century science.*

home. In 1848–49 he was placed in command of the *Enterprise* in search of the ill-fated Arctic expedition of Sir John Franklin. On his return he was promoted to rear-admiral, and edited his scientific reports on Antarctica.

Ross Dependency In 1923 BRITAIN made a TERRITORIAL CLAIM to the Ross Dependency on behalf of NEW ZEALAND. The action was motivated by the British desire to annex the entire Antarctic continent with the help of New Zealand and AUSTRALIA. The claim area lies between 150°W and 160°E, and includes ROSS ISLAND, the site of New Zealand's SCOTT BASE, and many HISTORIC SITES. It encompasses approximately 440,000 sq km (170,000 sq miles) of land and 330,000 sq km (127,400 sq miles) of the ROSS ICE SHELF. Other RESEARCH STATIONS—notably the USA's MCMURDO STATION and ITALY's TERRA NOVA BAY—are located in the Ross Dependency.

Ross Ice Shelf The largest ICE SHELF in the world, the Ross Ice Shelf is about the size of the American state of Texas and bigger then France. Almost all the Ice Shelf is afloat: it occupies the entire south ROSS SEA. It is up to 1000 m (3280 ft) thick at its grounding line and is fed by five ICE STREAMS and seven major GLACIERS.

The traditional route to the SOUTH POLE was over its smooth, flat surface. 'It is impossible to conceive the stupendous extent of this ice cap, its consistency, utter barrenness, and stillness, which sends an indefinable sense of dread to the heart,' wrote Louis BERNACCHI, physicist on the 1898–1900 SOUTHERN CROSS EXPEDITION. The level expanse of ice is now used intensively for landing aircraft and over-snow transport. Its cliffed front ranges from less than 15 to 50 m (50–165 ft) high, it is 50 to 350 m (165–1150 ft) thick, and is a major source of ICEBERGS.

Above: Royal penguins (Eudyptes schlegeli) are found only on Macquarie Island, where they were once hunted for oil.

Termed the 'Great Ice Barrier' by James Clark ROSS, who sighted it in 1841, the seemingly impassable ice cliffs along the edge of the shelf awed early explorers. Ross sailed along its edge for two successive seasons unable to find a landing place. He observed that the Shelf 'presented an extraordinary appearance, gradually increasing in height as we got nearer to it, and proving at length to be a perpendicular cliff of ice between 100 and 200 ft above sea level, perfectly flat and level at the top ... we might with equal chance of success try to sail through the Cliffs of Dover, as to penetrate such a mass.'

Ross Island On the northwest edge of the ROSS ICE SHELF, it was discovered by James Clark ROSS in 1841, who named two mountains on the island after his ships *Erebus* and *Terror*. 'With a favorable breeze, and very clear weather,' Ross wrote, 'we stood to the southward, close to some land which had been in sight since the preceding noon, and which we then called the "High Island"; it proved to be a mountain twelve thousand four hundred feet of elevation above the level of the sea, emitting flame and smoke in great profusion; at first the smoke appeared like snow drift, but as we drew nearer, its true character became manifest.' This was the largest recorded eruption of Mount EREBUS. From Hut Point, the site of SCOTT BASE and MCMURDO STATION, plumes of steam can be seen funnelling from the mountain and drifting across the skyline on most clear days.

Ross Sea Deep indentation in the Antarctic coastline, about 960,000 sq km (370,656 sq miles) in area. Along with the ROSS ICE SHELF, the embayment reaches within 500 km (310 miles) of the SOUTH POLE. The sea is relatively shallow, and extends north as far as the edge of the CONTINENTAL SHELF. The waters are nutrient-rich, supporting an abundant ecosystem. It is the most southern breeding ground of ADÉLIE and EMPEROR PENGUINS.

The west coast of the Ross Sea is the most predictably open area of the Antarctic coast over summer and most early expeditions worked their way to the continent through this route. The first ships to enter the sea were the HMS *Erebus* and *Terror*, commanded by James Clark ROSS, in 1842. The BAY OF WHALES was the most-used anchorage in the Ross Sea for about 50 years until it broke off the Ross Ice Shelf in the 1950s, and was where Roald AMUNDSEN sited his base for his assault on the South Pole.

Two permanent year-round bases—SCOTT BASE and MCMURDO STATION—are located in the Ross Sea area, and there are a number of HISTORIC SITES. CRUISE ships are frequent visitors.

Ross seal (*Ommatophoca rossii*) Of all the seals in Antarctica, least is known about Ross seals. They were first recorded by James Clark ROSS on his 1839–43 expedition. These phocid seals live deep within heavy PACK ICE and, until the 1970s, only about 100 people had sighted them. They have a circumpolar distribution.

Ross seals have dark backs and silvery-grey stomachs with dark longitudinal bands on chests and throats. The stripes are clearly visible when the seals are disturbed: unlike other seals, Ross seals will rear back, with mouths open and throats inflated. They have small heads and the largest eyes of any seal. It is believed that their large eyes may be an adaptation for hunting in the dark Antarctic waters. Their diet consists mainly of SQUID, supplemented by FISH and KRILL. Their rear flippers are long, almost one-quarter of their body length.

Solitary animals, they are on average 2.3 m (7½ ft) long, considerably smaller than LEOPARD and CRABEATER SEALS. Females are slightly larger than males. Ross seals make very distinctive sounds, which include trilling, warbling and 'chugging'. Little is known about their breeding cycle. The pups are born in November.

Rothera Station Rothera Station was built by BRITAIN in 1975 to replace Adelaide Station at Rothera Point on ADELAIDE ISLAND and it has operated continuously since. Rothera is an important logistics centre for the BRITISH ANTARCTIC SURVEY on the ANTARCTIC PENINSULA, and has a wharf and crushed rock AIRSTRIP. Flights link Rothera directly with the FALKLAND ISLANDS. Research is focused on biology, geoscience, glaciology and atmospheric sciences. The station can house 130, and has facilities for OVERWINTERING about 20 people.

Rotifers Tiny INVERTEBRATE animals, less than 0.5 mm (0.02 in) long, rotifers have a wheel-like, ciliated organ that is used both for feeding and locomotion. They live partly in fresh water—large numbers stain pond water red—but several species inhabit moist environments such as MOSS banks. Rotifers are mainly HERBIVORES, surviving on the microflora of BACTERIA, ALGAE, yeast and filamentous FUNGI in the moss community, and overwintering as eggs. Some rotifers are also predators, feeding on protozoa and, in turn, being preyed upon by several species of TARDIGRADES. Like tardigrades, rotifers are able to undergo CRYPTOBIOSIS (surviving without water in their body). When in a cryptobiotic state, they are able to tolerate harsh extremes of temperature and show none of the usual metabolic signs of life.

Royal albatross See SOUTHERN ROYAL ALBATROSS.

Royal Geographical Society Founded in 1830, the society was one of the key supporters of Britain's early Antarctic EXPLORATION, instructing James Clark ROSS that: 'The subject of most importance, beyond all question, to which the attention of Captain James Clark Ross and his officers can be turned—and that which must be considered as, in an emphatic matter, the great scientific object of the Expedition—is that of Terrestrial Magnetism ...'.

The society again took a prominent role in Antarctic exploration when, in 1896, the society's president, Sir Clements MARKHAM, initiated fundraising for the NATIONAL ANTARCTIC EXPEDITION. Markham persuaded the ROYAL SOCIETY to jointly sponsor the expedition, but the two societies came into conflict over who should lead the expedition. The Royal Geographical Society prevailed with its candidate Robert SCOTT, against the wishes of the Royal Society, which would have preferred a scientist to lead.

With a history of exploration that also includes such famous names as Stanley and Livingstone, the Royal Geographical Society today has a membership of around 13,000.

Royal penguin (*Eudyptes schlegeli*) Often classified as a subspecies of MACARONIS, royal penguins are similar in appearance, apart from their white cheeks, and have identical breeding and feeding patterns. Found only on MACQUARIE ISLAND, royals were, at one time, slaughtered for their oil.

Royal Society Founded in 1660, Britain's Royal Society is the world's oldest scientific academy. Projects supported by the society include James COOK's Pacific voyages, the CHALLENGER EXPEDITION, the 1839–42 expedition led by James Clark ROSS and the NATIONAL ANTARCTIC EXPEDITION headed by Robert SCOTT.

The Royal Society continues to promote science by funding research, publishing journals, organizing lectures and exhibitions and promoting science education and awareness.

Royal Swedish Academy of Sciences Modelled on the ROYAL SOCIETY in London and the ACADEMIE DES SCIENCES in FRANCE, the Royal Swedish Academy of Sciences was founded in 1739. It is an independent, non-governmental organization, the aims of which are to promote research in mathematics and the natural sciences, and it has awarded the Nobel Prizes in Physics and Chemistry since 1901.

Royds, Cape On the eastern coast of ROSS ISLAND. This was the site of Ernest SHACKELTON's base during his 1907–09 expedition; his HUT has been restored. The Cape has a large ADÉLIE PENGUIN rookery.

Royds, Charles (1876–1931). British naval officer. Born in Lancashire, England, Royds was the nephew of Sir Clements MARKHAM and Arctic explorer Albert Hastings Markham. He joined the British Navy and was appointed first lieutenant with the NATIONAL ANTARCTIC EXPEDITION of 1901–04. He led four journeys to Cape CROZIER; during the second, the EMPEROR PENGUIN colony was discovered. He was in the first party to climb Mount EREBUS, and, during an expedition to VICTORIA LAND, he was the first to find a route to the interior. An expedition was made about half-way across the ROSS ICE SHELF to determine if it was a floating ice mass or covered land. During this journey, Royds performed the first known SURGERY on the Antarctic continent when he operated on the hand of one of the party, using carbolic toothpaste as an antiseptic.

On his return, he continued his career in the navy, rising to the rank of vice-admiral. Following his retirement, he was appointed deputy commissioner of the Metropolitan Police in London.

Russia Tzar Alexander I sent Captain Thaddeus BELLINGSHAUSEN on an expedition to explore the SOUTHERN OCEAN in 1819–21. Bellingshausen discovered the Traversy Islands, PETER I ØY and ALEXANDER ISLANDS and circumnavigated the Antarctic continent.

There was little further Russian interest in Antarctica until just before World War II, when the USSR government protested the TERRITORIAL CLAIMS made by NORWAY, which included a claim to Peter I Øy. From 1946, the Soviet WHALING fleet began to operate in southern waters, and in 1950

Above: *Used primarily for meteorological and medical research, Russkaya Base was operational between 1980 and 1990.*

the USSR challenged attempts to resolve DISPUTED CLAIMS without its participation. It did not recognize any claims to SOVEREIGNTY in Antarctica, and, like the USA, reserved the right to make a claim in the future 'based on the discoveries and explorations of Russian navigators and scientists'.

The USSR participated in the INTERNATIONAL GEOPHYSICAL YEAR, establishing six bases, including one at the POLE OF RELATIVE INACCESSIBILITY and another at the SOUTH GEOMAGNETIC POLE. It was one of the original 12 CONSULTATIVE PARTIES to the ANTARCTIC TREATY. When the USSR broke up in 1991, Russia assumed the place of the USSR in the ANTARCTIC TREATY SYSTEM (ATS) (and the UKRAINE acceded to the Antarctic Treaty in its own right).

Russia has been a strong supporter of the ATS. Its major economic interest in Antarctica is in FISHING. The USSR was one of the last countries to cease commercial whaling, in 1988, and it began catching KRILL in the 1960s. This remained a large fishery until the end of the Cold War, when Russia was unable to continue subsidizing fishing operations. Russia is a member of the Commission for the CONVENTION ON THE CONSERVATION OF ANTARCTIC MARINE LIVING RESOURCES, and generally adopts a pro-fishing stance. It ratified the MADRID PROTOCOL in 1997.

Russian activities in Antarctica are organized through the ARCTIC AND ANTARCTIC RESEARCH INSTITUTE. In the post-Cold War era, although Russia has had some financial difficulties in maintaining its Antarctic programme, it still operates five year-round bases, including VOSTOK and MIRNYY, and three summer-only bases, and engages in a wide range of scientific research.

Russkaya Base Established by the USSR in 1980, the base operated for 10 years. It was located on Cape BURKS on the MARIE BYRD LAND coast and research into meteorology and medicine was conducted there.

Rymill, John (1905–68) Australian scientist and explorer. Born in Melbourne, Rymill went on an ethnological expedition to Canada in 1929 and two expeditions to Greenland in 1930–33 with Gino Watkins, taking over the leadership when the latter disappeared on 20 August 1932. On his return, he began planning the 1934–37 BRITISH GRAHAM LAND EXPEDITION, which sailed from England on 10 September 1934 and arrived at Port Lockerby on the west coast of the ANTARCTIC PENINSULA on 22 January 1935.

The expedition overwintered at one of the Debenham Islands and explored the WEDDELL SEA, travelling through a 'strait' charted by Hubert WILKINS but which they discovered was in fact a GLACIER. Although a small expedition, they produced a great amount of surveys and maps of previously unexplored areas, including aerial reconnaissance carried out in a single-engine de Havilland Fox Moth AIRCRAFT, and covered over 2100 km (1300 miles) by sledge.

Rymill returned home to become a pastoral farmer in South Australia and during World War II enrolled in the Naval Reserve. He was killed in a road accident.

S

Saline lakes The water in Lake Bonney in the Taylor Valley, Lake VANDA in the Wright Valley and in smaller LAKES and PONDS of the DRY VALLEYS region can be up to 12 times saltier than sea water. For every kilogram (2.2 lb) of water in DON JUAN POND in the Dry Valleys, there is half a kilogram (1.1 lb) of dissolved salt. Despite temperatures that drop to −51°C (−60°F), many saline lakes never freeze.

Saline lakes tend to be stratified, or layered, and each layer has different chemical and thermal (temperature) characteristics. The SALINITY tends to increase at greater depths because the deepest waters are the oldest, having been concentrated by evaporation and freezing. The resulting brine has a much higher density than any fresh water flowing into the lake; this, together with the ice covering the lake, prevents the wind from mixing the layers. Lake Vanda has at least 12 stratified layers, each one saltier than the layer above.

Temperature also increases with depth. The ice covering the lake's surface is made up of crystals that are aligned vertically as a result of ice constantly freezing underneath and evaporating from the top. The vertical crystals act as 'light pipes', and transmit summer SOLAR RADIATION to the water below, where it is absorbed by the dense, saline layers at lower levels. Lake Vanda can be 25°C (77°F) at its deepest layer, whereas the lake surface is at freezing point, and the air has a mean temperature of −20°C (−4°F).

Organisms that inhabit saline lakes can form populations within specific layers where conditions are ideally suited to their growth. Some ALGAE swim between layers using whip-like fins called flagells. Ablation Lake, on ALEXANDER ISLAND, is 100 km (62 miles) from open water but has tidal rhythms that indicate that it is directly connected to open sea. Within its complex ecosystem both freshwater CRUSTACEANS and marine species have been found.

Salinity A measure of salt concentration or 'saltiness'. Antarctica is thought to produce the world's saltiest ocean water. High-density, salty water seeps out from SEA ICE as it crystallizes and sinks. In inland LAKES, water can be up to 12 times saltier than sea water. Salt deposits are formed by evaporation, trapped sea water, wind-blown marine salts, volcanic or hot spring activity and weathering of salts from local bedrock; these salts then accumulate in the lakes. In the DRY VALLEYS, scientists have found that high salt concentrations are also biologically produced by sulphur-emitting ALGAE.

Salvin's prion (*Pachyptila salvini*) Closely resembling the ANTARCTIC PRION, the two species are not easily distinguished from each other at sea, although Salvin's has a slightly longer and wider bill and frequents warmer waters. They breed in millions within earth burrows on Îles CROZET and PRINCE EDWARD ISLANDS, range in the subantarctic zone during the breeding season, and move into subtropical waters in the winter.

Salvin's prions, which have larger lamellae than Antarctic prions, feed by sieving and filtering for tiny organisms, but can also catch and feed on SQUID and CRUSTACEANS.

SANAE Base Built between 1993 and 1998 on an inland rocky outcrop at Vesleskarvet in DRONNING MAUD LAND, this is the fourth base built by SOUTH AFRICA named SANAE, which is an acronym for SOUTH AFRICAN NATIONAL ANTARCTIC EXPEDITIONS. The first three SANAE bases had been built on an

Below: Because of their high salt levels, many saline lakes never freeze.

Above: The hoist used for lowering goods into South Africa's SANAE Base, which lies beneath snow.

ICE SHELF and became unusable. SANAE IV operates all year round, with 20 personnel OVERWINTERING and up to 60 in the summer. RESEARCH is conducted in the physical, earth, life and oceanographic sciences.

Sastrugi Sharp ridges in snow formed by wind erosion. They may be small-scale, similar to ripples in sand, or rough hard ridges several metres (feet) high. 'Fields' of sastrugi may be many kilometres (miles) wide and crossing them can be exhausting. Sometimes sets of sastrugi overlie each other, some having been formed by strong KATABATIC WINDS, for example, and others from blizzards blowing from a different direction.

Satellites Used in Antarctica since the 1970s for mapping, communications and research, a number of satellites continually sweep across the skies. Polar-orbiting satellites, which monitor changes in SEA ICE cover and the OZONE HOLE, are extremely valuable for research into CLIMATE change. Knowledge of the distribution of sea ice is also important for NAVIGATION. By 1980, global positioning systems (GPS) enabled scientists to use satellite signals to 'fix' a position on the continent to within a metre (3 ft). This technology has been used widely in Antarctica: for instance, stakes positioned on the surface of a GLACIER are fixed at different times in order to calculate the speed at which the ICE is flowing.

Radio COMMUNICATIONS at the SOUTH POLE are often disrupted over winter by magnetic storms (see ATMOSPHERE). Satellites that drift down to an angle of less than 9 degrees with the Equator can take over communications, relaying calls between bases.

Satellites provide a unique view of Antarctica from space. RADAR equipment mounted on a satellite called RADARSAT has recently enabled a joint Canadian-USA project to re-map many areas of Antarctica. Landsat satellites circle the Earth at an altitude of about 917 km (570 miles) taking aerial images. Although showing less detail than photographs taken from AIRCRAFT—the smallest object that can be identified is about 64 m (210 ft) across—because the satellites follow the same flight path and pass over Antarctica every 18 days, sequences of images can be compared, and changes in the landscape can be tracked over time—for example, the drift of a large ICEBERG can be followed as it floats into the SOUTHERN OCEAN. Satellites are also used for monitoring the location of FISHING vessels in the Southern Ocean.

Schouten, Willem Corneliszoon and **Schouten, Jan Corneliszoon** (*c.* 1580–1652) Dutch explorers. In search of an alternative trade route to the East Indies, on the *Eendracht* and *Hoorn*, along with Jacob LE MARIE, they sailed through DRAKE PASSAGE (where they rammed a 'sea monster'), and around the southern tip of TIERRA DEL FUEGO which, in January 1616, they named CAPE HORN after their ship. The *Hoorn* was subsequently lost by fire off the coast of Patagonia. The discovery meant that cartographers had to limit the proportions of TERRA AUSTRALIS INCOGNITA.

Science Vivian FUCHS described Antarctica as 'a continent for science', insulated from political pressures. In practice there has been an alliance between the interests of science and POLITICS from the earliest days of discovery and EXPLORATION, and on into the modern era. Before the INTERNATIONAL GEOPHYSICAL YEAR of 1957–58 and the signing of the ANTARCTIC TREATY in 1959, scientific RESEARCH in Antarctica was usually conducted through isolated national projects and scientists could be used to reinforce TERRITORIAL CLAIMS by occupation of RESEARCH STATIONS.

There were also political elements to the International Geophysical Year. Despite attempts by several governments to exclude it for political reasons, the USSR participated in the scientific programmes, thus establishing a foothold on the continent; and BASES were placed in strategic locations, such as the SOUTH POLE (by the USA) and in

Below: Blizzards and katabatic winds sculpt the ice into ridged patterns known as sastrugi.

the AUSTRALIAN ANTARCTIC TERRITORY (by the USSR). Such manoeuvrings prompted international law expert Francis Auburn to comment that scientific research was 'the currency of Antarctic politics'.

The Antarctic Treaty, signed in 1959, placed considerable emphasis on scientific activity in Antarctica; it provided for the exchange of information and scientists, freedom of scientific investigation, and the reservation of the continent for peaceful purposes. Scientific links among nations have been fostered by the SCIENTIFIC COMMITTEE ON ANTARCTIC RESEARCH.

The Antarctic Treaty's freeze on the SOVEREIGNTY issue generally reduced political pressures on the conduct of science in Antarctica. The exchange of scientists among nations acts as a *de facto* inspection system for compliance with the rules of the treaty. One area of friction has been the application of pure science, in fields such as geology or marine biology, to possible commercial EXPLOITATION of the resource wealth of Antarctica. Although the MADRID PROTOCOL has placed a moratorium on MINING activities for 50 years, a discovery of resources that could be profitably exploited would place pressure on the ANTARCTIC TREATY SYSTEM.

Antarctic science has re-emerged as a factor in global politics due to the potential of the GREENHOUSE EFFECT to increase temperatures, and the presence of the OZONE HOLE over Antarctica that can let more damaging ULTRAVIOLET RADIATION through the ATMOSPHERE. Scientific research programmes have also revealed the environmental impact of FISHING and by-catches on FISH and BIRD species in the SOUTHERN OCEAN.

Scientific Committee on Antarctic Research (SCAR) As part of the 1957–58 INTERNATIONAL GEOPHYSICAL YEAR, the INTERNATIONAL COUNCIL OF SCIENTIFIC UNIONS (ICSU) established the Special Committee on Antarctic Research in 1957 to coordinate the RESEARCH of the 12 nations working in Antarctica. This was turned into a permanent group and renamed the Scientific Committee on Antarctic Research (SCAR) in 1961. SCAR is an independent non-governmental organization that supplements the ANTARCTIC TREATY SYSTEM (ATS) by acting as its unofficial scientific body. SCAR can make suggestions and respond to requests for advice, and is represented by observers at ATS meetings.

SCAR has been a successful force for international cooperation, especially for RESEARCH in Antarctica and the SOUTHERN OCEAN. The Committee operates through permanently established working groups in particular disciplinary fields, such as biology and geology, and temporary groups of specialists charged with solving specific problems. One example of an effective RESEARCH programme coordinated by SCAR was the BIOLOGICAL INVESTIGATIONS OF MARINE ANTARCTIC SYSTEMS AND STOCKS in the 1980s. SCAR publishes a quarterly *SCAR Bulletin*, an irregular *SCAR Report* and occasional publications. Funding for SCAR is derived from the national governments of its members.

SCAR's members are drawn from scientific bodies rather than from state governments. SCAR has three types of membership: full membership, ICSU union membership and associate membership. Full members are countries with active scientific research programmes in Antarctica; union members have an interest in Antarctic research; and associate members are countries without an independent programme. In September 2000, there were 27 full members, seven union members and six associate members.

Scoresby, William (1789–1857) English explorer and scientist. The son of a whaling captain, Scoresby sailed in the Greenland seas with his father before attending Edinburgh University. Joseph BANKS, then president of Britain's ROYAL SOCIETY, encouraged Scoresby's scientific investigations, which he conducted when captaining whaling ships in the Arctic seas between 1811 and 1823. In 1820 he published the results of his meticulously recorded observations in the two-volume *An Account of the Arctic Regions*, which gives an account of Arctic geography, and charts his comprehensive investigations into polar seas, ice, meteorology and animals. In 1824 Scoresby left whaling to study theology, but continued to pursue his scientific interests. In 1856 he sailed to AUSTRALIA to study terrestrial magnetism.

It was not until the 20th century that the accuracy and importance of Scoresby's groundbreaking work on polar science was fully understood.

Scotia Arc The SOUTH SHETLAND ISLANDS at the tip of the ANTARCTIC PENINSULA form part of a large archipelago. Along with the SOUTH SANDWICH ISLANDS, SOUTH ORKNEY ISLANDS and SOUTH GEORGIA, this chain reaches 540 km (335 miles) into the SOUTHERN OCEAN. The Scotia Arc archipelago is the spine of a submarine MOUNTAIN range that links the mountains of Antarctica to the Andes Mountains in South America. The arc makes a hairpin bend to enclose the SCOTIA SEA and is terminated at its eastern end by the South Sandwich Trench.

The Scotia Arc islands are inhabited by 17 species of bird, including huge colonies of CHINSTRAP PENGUINS. There are also recovering populations of FUR and ELEPHANT SEALS that were hunted to near-extinction in the 18th and 19th centuries. In 1820–21, for example, 91 ships operated in the area, and two vessels alone took 60,000 seal skins.

Scotia Sea The southernmost section of the Atlantic Ocean, within which SOUTH GEORGIA, the SOUTH ORKNEYS and SOUTH SANDWICH ISLANDS lie.

Scott, Robert Falcon (1868–1912) British explorer, born in Devonport, England. In July 1881 he joined the British Navy as a cadet. In 1887 he was first noticed by Sir Clements MARKHAM, who was struck by his 'intelligence, information and the charm of his manner' and appointed him leader of the 1901–04 NATIONAL ANTARCTIC EXPEDITION.

The expedition made the first extensive land explorations of Antarctica and carried out scientific studies. Together with Edward WILSON and Ernest SHACKLETON, Scott reached 82°16'S in 1902, the southernmost point at that time. On his return to London, Scott was promoted to captain and published *The Voyage of the* Discovery in 1905.

Scott's BRITISH ANTARCTIC EXPEDITION left England in 1910 with the double purpose of being first to the SOUTH POLE and carrying out an extensive scientific programme. On the way to Antarctica, Scott received a cable from Roald AMUNDSEN, who announced he was also sailing to

Below: Captain Robert Falcon Scott with his wife Kathleen.

Above: New Zealand's Scott Base, established in 1957 by Edmund Hillary, in McMurdo Sound.

Antarctica. When the TERRA NOVA arrived at the BAY OF WHALES in January 1911, Amundsen had set up base already and was preparing his own attempt on the Pole. Scott refused to alter his plans. Aspley CHERRY-GARRARD wrote that 'Scott was the strongest combination of a strong mind in a strong body that I have ever known.'

Scott's party set out from Cape EVANS for the South Pole in October 1911, using a variety of transport methods. He disliked using dogs and eventually relied solely on teams of men on skis to haul the sledges. Together with Edward WILSON, Henry BOWERS, Lawrence OATES and Edgar EVANS, he arrived at the South Pole on 17 January 1912, to find that Amundsen's party had reached it a month earlier. 'Great God!' he wrote, 'This is an awful place and terrible enough for us to have laboured to it without the reward of priority.' On the return journey, lack of food made the party vulnerable to cold, injury and depression. Evans died first, then Oates, followed by Scott, Bowers and Wilson only 18 km (11 miles) from a store of food and fuel. The last words Scott wrote were on 29 March 1912: 'For God's sake look after our people.' A search party led by Edward Atkinson found this last camp the following November.

Scott Base Established in 1957 by NEW ZEALAND in the ROSS SEA region and named after Robert SCOTT. The first director of the base, which was built to support activities in the INTERNATIONAL GEOPHYSICAL YEAR and the COMMONWEALTH TRANSANTARCTIC EXPEDITION, was Edmund HILLARY. In 2000 New Zealand hosted the first 'on-ice' meeting of government representatives from 24 states at Scott Base.

Scott Base operates year-round, with an OVER-WINTERING population of 10, and up to 80 people in summer. Scientific research conducted here includes environmental monitoring, ionospheric, auroral and meteorological observations, marine and terrestrial biology and tide measurement.

Scott Glacier Over 32 km (20 miles) long and 11 km (7 miles) wide, the glacier flows to the coast between Cape Hoadley and Grace Rocks. Discovered by the 1911–14 AUSTRALASIAN ANTARCTIC EXPEDITION.

Scott Island A remote and barren MARITIME ISLAND lying northeast of Cape ADARE on the ANTARCTIC CIRCLE. Only 370 m (1214 ft) long, Scott Island is composed of the remnants of a volcanic cone, with high cliffs on the northern coast. It was discovered by Captain William COLBECK in 1902, en route to deliver supplies to Robert SCOTT in the relief ship *Morning*. The island is seldom visited.

Scott Polar Research Institute Established in Cambridge, England, in 1934 with Frank DEBEN-HAM as director. It was partly financed with sur-plus funds from the Scott Memorial Fund, set up for the dependants of those who died on the return journey from the SOUTH POLE. The aims of the Institute—part of Cambridge University since the 1960s—are to collect, collate and conduct research into the polar regions.

Scottish National Antarctic Expedition (1902–04) Privately funded Scottish scientific expedition. Financed by the Coats brothers of Paisley, its pur-pose was to study wildlife and complete an exten-sive survey of the SOUTH ORKNEY ISLANDS. The expedition included a bacteriologist, botanist, geologist, meteorologist, taxidermist, zoologist—and a bagpiper. The first to undertake a compre-hensive study of PENGUINS, the expedition includ-ed equipment to record the different calls of the birds onto gramophone discs. Led by William BRUCE, the expedition set out from the port of Troon on 2 November 1902 on the *Scotia*, under Captain Thomas Robertson.

They headed for the South Orkneys by way of the FALKLAND ISLANDS, reaching the South Orkneys on 3 February 1903. The ship was ice-bound during the first winter and, when freed (with the help of a little dynamite), sailed to Buenos Aires for repairs. Here, the expedition ceded the magnetic and meteorological observa-tion station that they had set up on Laurie Island to the Argentinian government and the station has been used almost every year since.

Above: Sea anenomes (Actiniaria) are common members of the Antarctic marine community, principally inhabiting the sea floor.

During the summer of 1904 the *Scotia* sailed along the western coast of the WEDDELL SEA and charted COATS LAND, named for the expedition's benefactors. Further west, off DRONNING MAUD LAND, oceanographical studies were carried out: the soundings taken showed that both James Clark ROSS and John MURRAY had been incorrect in their belief that a deep ocean trench ran along this part of the coast.

Screaming Sixties Concentrated STORM formation over the SOUTHERN OCEAN between latitudes 60°S and 70°S, part of the same wind belt as the ROARING FORTIES and the FURIOUS FIFTIES.

Sea anemones (Actiniaria) Sea anemones are very simple multicellular animals belonging to the cnidarian ('stinging threads') group of organisms, which also includes hydroids, jellyfish and CORAL. They have a central mouth, surrounded by tentacles with stinging cells that capture and paralyze small marine animals such as SEASTARS and SEA URCHINS. The five most conspicuous benthic (sea floor) species are *Isotealia antarctica*, *Stompia selaginella*, *Artemidactis victrix*, *Hormantha lacu-*

nifera and *Urticinopsis antarctica*.

Sea anemones usually attach themselves to rocks or coral near the sea floor. Although lacking any skeleton, they have highly developed muscles. Some *Stompia* species, including those found in Antarctica, are even believed to swim, flexing their column to escape after contact with potential predators, their pedal disc, or tail, forming a narrow cone as they move. *Urticinopsis antarctica*, found in Antarctica and the SOUTH SHETLAND ISLANDS, may bend over to engulf a seastar and also engages in territorial disputes with rival Antarctic species *Isotealia antarctica*.

Sea cucumbers (Holothuroidea) Sea cucumbers are fleshy cylinder-shaped ECHINODERMS, with the typical spines and plates found in echinoderms reduced to tiny ossicles (bones) embedded in the skin. In sea cucumbers, the anus doubles as a respiratory organ and also has a role in a bizarre defense mechanism. Although they have few natural enemies, when disturbed they may violently contract their bodies and expel most of their internal organs through the anal aperture. Surprisingly, the absence of internal organs is not

usually fatal in sea cucumbers—in most cases regeneration follows.

Sea cucumbers feed on organic material that falls to the sea floor, and Antarctic species can be found at a range of depths in the BENTHOS environment. They include *Ekmocucumis steineni*, which has been found at depths of 1200 m (4000 ft) and reaches up to 15 cm (6 in) in length. It may attach itself to other organisms such as hydroids and fan-shaped BRYOZOANS or, alternatively, live with the lower half of its body in the benthic sediment. The smaller *Abyssocucumis liouvillei*—up to 6 cm (2¼ in) long—lives attached to SPONGES and large stones. *Echinopsolus acanthocola*, which uses narrow, rod-like structures to elevate itself, is a suspension feeder and has been found attached to the spines of SEA URCHINS.

Sea ice An essential part of the SOUTHERN OCEAN system, it affects most physical and biological processes in Antarctica. Sea ice originates on the ocean surface and can take many forms, including PACK ICE, CONGELATION ICE, FRAZIL ICE, GREASE ICE and PANCAKE ICE. Distinctly different from GLACIER ice, it is saline and forms by crystal growth rather than from compressed SNOW. Unlike the Arctic, where sea ice grows to about 3 m (10 ft) thick, 'undeformed' Antarctic sea ice (without ridges or other deformities) usually only reaches about 1 m (3 ft) in thickness this is because of disturbances from WAVE action and OCEAN CURRENTS and because it rarely lasts for more than one year.

Sea ice acts as a barrier to energy exchange between the ATMOSPHERE and the ocean: in winter it insulates the ocean, and in summer it reflects 80 percent of SOLAR RADIATION, delaying the warming effect of the sun. When sea ice melts, it cools both the atmosphere and the ocean, which limits the water vapour supply to the atmosphere and is one reason why so little moisture reaches the continent. For seawater to freeze, the upper few metres (feet) must be supercooled: that is the TEMPERATURE must drop to, or below, freezing point (about −1.8°C/29°F). Unlike fresh water, as salt water cools it increases in density and sinks, bringing warmer deeper water to the surface.

When the sea does begin to freeze, the first crystals to form are tiny spheres of pure ice, which grow around impurities in the water—in the same way that snow forms around dust particles in the atmosphere. These crystals grow rapidly into thin discs, with a maximum diameter of 3 mm (⅛ in), which turn into hexagonal stars. The 'stars' grow until they overlap and freeze together: in calm conditions they become stacked vertically into long needles of congelation ice; in turbulent conditions the arms break off the stars and they are mixed in the ocean to form frazil ice or grease ice. These may consolidate to form ice pancakes, which can bond together with new crystal growth to form pack ice. 'Fast ice' is sea ice that forms and remains 'fast', or attached, to the shore. It tends to be the thickest form of undeformed Antarctic sea ice, reaching more than 3 m (10 ft) if it survives for several years.

Above: Most sea spiders measure between 1 to 10 mm (a fraction of an inch to ½ in), but in Antarctica Colossendeis *species grow much larger.*

Sea slugs Sea slugs, also known as NUDIBRANCHS, are an order of gastropod MOLLUSCS without protective shells. The *Notaeolidia* genus of aeolid nudibranch is endemic to Antarctica.

Sea spiders (Pycnogonida) Worldwide, most sea spiders are just 1 to 10 mm (a tiny fraction of an inch to ½ in) long, but in Antarctica *Colossendeis* sea spiders can reach far greater proportions, with some species having leg spans as wide as 50 cm (20 in). Like many other Antarctic INVERTEBRATES, not only do they grow exceptionally large, they live much longer than similar, warmer water species because the extreme cold means metabolisms are slow and there are relatively few predators in the Antarctic environment.

Sea spiders are not 'real' spiders (Arachnids), but marine, spider-like arthropods. Most have eight legs, but a number of Antarctic species have 10 or 12 walking legs. They have small bodies but their guts and reproductive organs extend almost to the tips of their legs. Although some are found in shallower water, most Antarctic *Colossendeis* species live in the BENTHOS (sea floor), where adults feed on SPONGES and BRYOZOANS or suck the juices from soft invertebrates such as soft CORALS.

In Antarctic sea spider species, the male cares not only for the eggs, but sometimes the newly hatched larvae as well. The male fertilizes and gathers the eggs as the female extrudes them, then cements the egg clutches to his body. Sometimes a male will gather and carry balls of eggs from as many as seven different females at a time.

Sea urchins The whole family of ECHINODERMS—meaning 'spiny skins'—takes its name from the spiny sea urchins (Echinoidea), of which at least 60 species are known to live in Antarctic and subantarctic waters. Brooding of young is prevalent among females, with 39 out of the 60 Antarctic and subantarctic species rearing their offspring internally within their test (shell).

The most common SOUTHERN OCEAN sea urchin, *Sterechinus neumayeri*, is extremely slow-growing, eventually achieving a maximum diameter of 7 cm (2⅔ in) at 40 years of age. It is main-

Left: During the winter freeze, the sea ice spreads at a rate of around 4 km (2½ miles) a day, until it covers an area equal to that of South America.

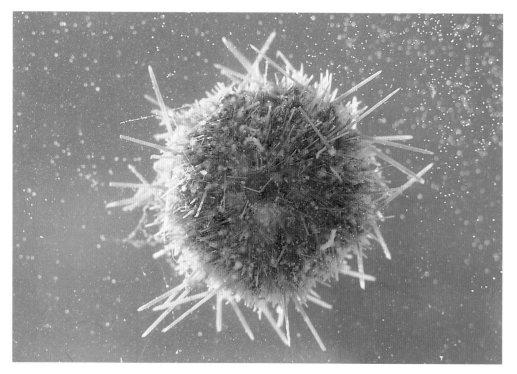

ly herbivorous, feeding on ALGAE, but will also eat small INVERTEBRATES such as BRYOZOANS. In shallow waters, numbers of *S. neumayeri* congregate below cracks in the ICE where WEDDELL SEALS defecate to feed on the seal faeces.

Sterechinus neumayeri, like many other urchins, is known to create a defensive shield out of debris. This camouflage often includes stinging hydroids that seem to protect the urchin from predation by SEA ANEMONES. It also uses ALGAE such as *Phyllophora antarctica*, which manufactures unpalatable defensive chemicals, as a detachable anemone deterrent. Other predators of *S. neumayeri* include OCTOPUS, FISH, SEASTARS and BRITTLE STARS.

Seals (Pinnipedia) Mammals that are closely related to bears and otters, Antarctic seals can be divided into two distinct families: 'earless' seals or Phocidae, and 'eared' seals or Otariidae. Both families have excellent hearing, but the phocids do not have any visible ear flaps.

Five of the six seal species found within the ANTARCTIC CONVERGENCE are phocid seals: the WEDDELL, ROSS, CRABEATER, LEOPARD and SOUTHERN ELEPHANT SEALS. Only the ANTARCTIC FUR SEAL and its close relative the KERGUELEN FUR SEAL are otariids, possessing pinnae or ear flaps.

On land, fur seals are much more agile than phocids. This is because their hind flippers can be rotated forward and can support some of their

Above: Sea urchins dwell on the ocean floor, where they graze on algae alongside other marine invertebrates.

Below: Weddell seals (Leptonychotes weddellii) *appear ungainly on the ice, but in the water they are graceful swimmers and expert divers.*

body weight, allowing them to stand—and run—on all four limbs. Phocid seals cannot raise their bodies off the ground, and move by undulating like giant insect larvae. On ice, phocids move with a slithering, snake-like motion, with leopard seals reaching speeds up to 25 km (15½ miles) an hour over short distances.

Once in the water, both groups are supreme movers, although there is another difference between the families: phocids generally use their hind flippers for propulsion, whereas eared seals 'fly' through the waters with their front flippers. The sea is their true habitat: Weddell seals even copulate underwater. The torpedo shape of their bodies allows for speed in water and their physiology is particularly adapted to the ocean world. Seals' nostrils are naturally closed: they have to contract muscles in order to take a breath. Before diving, they exhale air from their lungs, making them less buoyant and reducing the possibility of 'the bends'. Oxygen is stored in their blood stream, and their metabolism, including their heart rate, slows down. Weddell seals are capable of diving to 720 m (2361 ft), and can stay underwater for well over an hour (although most dives last less than 30 minutes). They feed on deepwater catches of FISH, KRILL, SQUID and CRUSTACEANS, and take occasional PENGUINS.

Breeding, which occurs in late spring to early summer, is the only time seals must come out of the water. They have a one-year reproductive cycle, and gestation lasts about 11 months. The suckling period is very short—varying from four to eight weeks for phocids and up to four months for the Antarctic fur seal.

Seals are well insulated with thick layers of blubber. As their name suggests, fur seals also have a dense pelt. The blubber and fur were targeted by sealers who hunted in the SOUTHERN OCEAN from the late 18th century onwards.

By the 1890s, seal populations on almost every island in the Southern Ocean were low enough to make the animals commercially extinct. After nearly a century of protection and, it is thought, an increase in the availability of their main food (krill), fur seal populations have recovered strongly. Southern elephant seal populations have also recovered under strict protection. The other species of Antarctic phocids were never targeted by sealers.

Sealing It is likely that 18th-century American sealers knew about seal stocks and had discovered various islands in the Southern Ocean but kept their locations secret and uncharted. Had it not been for William SMITH's accidental discovery of the SOUTH SHETLAND ISLANDS in the south Atlantic in 1820, the adjacent part of the continent may not have been charted until much later.

The first recorded big haul of skins was in January 1820 when Carlos Timblon aboard the *San Juan Nepomuceno* from Buenos Aires, Argentina, killed 14,000 animals in five weeks. In January 1821 Thaddeus BELLINGSHAUSEN predicted 'the numbers of these animals will rapidly decrease.' In that season alone, 55 American, British and Australian sealers were operating in the waters around the ANTARCTIC PENINSULA.

Above: Whalers sealing on Campbell Island between 1913–16.

The ENDERBY BROTHERS' enterprise on the opposite side of the continent had been active from 1805, when Captain Abraham Bristow in the *Ocean* discovered the AUCKLAND ISLANDS and established a sealing station on ENDERBY ISLAND two years later. The Enderbys, like the proprietors of several other sealing and whaling companies, encouraged their captains to combine exploration with commerce.

The same pattern of slaughter was played out on many SUBANTARCTIC ISLANDS: sealers 'worked' the beaches until they were devoid of life, and the industry collapsed. It has been estimated that a skilled sealer could kill and skin 60 fur seals an hour. By the mid-19th century, the seal populations were noticeably depleted and many of the sealing firms turned their attention to WHALING.

Seas Within the SOUTHERN OCEAN are eight seas, two of which—the ROSS and WEDDELL SEAS—are substantial embayments. The others, including BELLINGSHAUSEN, AMUNDSEN, DAVIS, HAAKON VII, SCOTIA and DUMONT D'URVILLE SEAS, are marginal seas off the continent's coast.

Seasons Antarctica has two distinct seasons, a short SUMMER and a long cold WINTER. The SUN is above the horizon for 24 hours a day in summer and below the horizon for 24 hours a day over winter. With increasing distance north of the SOUTH POLE the periods of sunlight during winter are longer, and summer nights also lengthen. Winter begins in March and finishes about August, when the SUNRISE floods the landscape with light, the SEA ICE surrounding the continent begins to break up and marine mammals return to Antarctic waters.

Seastars (Asteroidea) There are estimated to be at least 45 Antarctic species of seastars (also known as starfish), which belong to the family of ECHINODERMS. They can be long-lived: one species, *Anasterias rupicola* is known to reach an age of at least 39 years. Females of some Antarctic species brood their young, rather than spawning eggs into the water. Seastars prey on SPONGES, MOLLUSCS, other echinoderms and various zoophytes (plant-like animals such as CORAL). They have a unique method of digestion. Most have no teeth or feeding appendages; instead, they evert their whole stomach out of their body cavity and over their meal, the tissues of which are then dissolved by the seastar's digestive juices. When it is finished eating, the seastar simply withdraws its sac-like stomach, now full of food particles, back inside its body.

Acodontaster conspicuus is a predator of several sponge species, and is, in turn, preyed on by the smaller predatory seastar, *Odontaster validus*. When feeding, a single *O. validus* climbs on to an arm-ray of *A. conspicuus*, everts its stomach and digests a hole in the larger seastar. This alone is not fatal, but if other *O. validus* nearby respond by joining together in a deadly 'gang attack', eventually the larger *A. conspicuus* can become completely buried under high piles of attacking *O. validus*.

Seaweeds The world's southernmost populations of marine ALGAE are found in MCMURDO SOUND. Attached rather than free-floating like DIATOMS, seaweeds grow around the ANTARCTIC PENINSULA and coastal islands, with the richest flora found on the exposed West Antarctic coast. Seaweeds are far less important to the Antarctic FOOD CHAIN than diatoms, but do provide food for amphipods, osopods, FISH and other organisms.

Sei whale (*Balaenoptera borealis*) Sei (pronounced 'say') are the third largest of the BALEEN WHALES. Males reach a maximum length of 18 m (60 ft), and females grow to 21 m (70 ft). They are found throughout the oceans of the world and whalers considered them to be the fastest swimmers. It is believed that only larger individual seis venture south of the ANTARCTIC CONVERGENCE, where they arrive in January and leave by April. They gather in groups of three to five. Blue-grey in colour, they have lighter ventrals (throat grooves) and a prominent ridge along the top of the head, between the blowholes and snout.

Unlike other baleens, sei whales feed on the surface of the water, skimming along with their mouths open, catching food (predominantly COPEPODS and KRILL) and sieving it from the water. Because the sei whales feed in this manner, their backs and dorsal fins are visible for long periods. Their dives are rarely deeper than 300 m (984 ft) or over five to 10 minutes in length.

They breed in warmer northern waters during July and gestation lasts 11 to 12 months. They become sexually mature between the ages of eight and 11. Like other baleens, the females have a two- to three-year reproductive cycle and generally give birth to a single calf.

Their populations were devastated during the WHALING days and, although they are now considered vulnerable, sei populations have recovered faster than other exploited species.

Seismic research Antarctica is almost aseismic: that is, it has virtually no earthquakes. This is due to the stable crustal structure beneath the ice cap, the lack of plate boundaries within or close to Antarctica, and the continent's distance from any artificial sources of vibrations. These reasons make Antarctica a perfect 'baseline' for the measurement of tremors that occur in other parts of the globe.

Seismic surveys, which rely on sending shock waves through rock layers beneath the ice, provided important early information about the thickness of the ICE SHEET and the ROCK crust beneath the ice. Douglas MAWSON first suggested that the differences between the elasticity of rock and ice might be used, along with an ECHO SOUNDER, to measure ice depth. Soundings of the interior of Antarctica were first made by members of the NORWEGIAN-BRITISH-SWEDISH EXPEDITION, who travelled 650 km (403 miles) through DRONNING

MAUD LAND to the POLAR PLATEAU. After the INTERNATIONAL GEOPHYSICAL YEAR in 1957–58 many countries undertook seismic surveys. However, the equipment, which needs to be dragged over the surface, was cumbersome, and individual depth soundings often took two to three hours each. RADAR overcame these problems.

Sentinel Range Forming the northern half of the ELLSWORTH MOUNTAINS, the Sentinel Range includes most of Antarctica's highest peaks. Sheer ROCK walls and narrow ridges tower over the surrounding RONNE ICE SHELF. At 4897 m (16,066 ft), the VINSON MASSIF is the highest mountain in Antarctica. Only 52 m (170 ft) lower than Vinson Massif, Mount TYREE is regarded as Antarctica's most technically challenging climb.

Seymour Island Famous for its rich FOSSIL beds, Seymour Island does not have the permanent ice caps of the neighbouring islands at the tip of the ANTARCTIC PENINSULA. This is due to its location in the lee of MOUNTAINS on JAMES ROSS and SNOW HILL ISLANDS, which shelter it from prevailing storms. More than 20,000 ADÉLIE PENGUINS breed on the south coast. The Argentinian MARAMBIO STATION has a hard-rock AIRSTRIP and Hercules AIRCRAFT are able to land year-round on the island.

Shackleton, Ernest (1874–1922) Anglo-Irish explorer, born in County Clare, Ireland. His family moved to London when he was 10 and he entered the merchant navy at 16.

Ernest Shackleton as Victorian gentleman (below) and as veteran Antarctic explorer (right). Despite the fact that Shackleton never reached the South Pole, his name is synonymous with the Heroic Age of Antarctic exploration. He is possibly most renowned for his inspirational leadership.

Shackleton joined the 1901–04 NATIONAL ANTARCTIC EXPEDITION. Louis BERNACCHI, a fellow expedition member, wrote that, 'Shackleton was the life and soul of DISCOVERY. His mind was alert, his good humour inexhaustible ...' On 2 November 1902 Shackleton set out with Robert SCOTT and Edward WILSON to the SOUTH POLE. They reached 82°16'S on 30 December, but all three suffered from hunger, cold and scurvy. Shackleton collapsed twice on the return journey and Scott sent him back to NEW ZEALAND on the relief ship, *Morning*, in February 1903.

Back in England, Shackleton began organizing his own expedition. The BRITISH ANTARCTIC EXPEDITION left Torquay, England, in the *Nimrod* on 7 August 1907, with the goal of reaching the South Pole.

On this expedition the first ascent was made of Mount EREBUS and the SOUTH MAGNETIC POLE was reached for the first time. Shackleton, Frank WILD, Eric Marshall and Jameson Adams created a new 'farthest-south' record of 88°23'S, only 180 km (97 miles) from the South Pole; 'the only thing that stopped us from reaching the actual point was the lack of 50lbs of food,' Shackleton said on gramophone, recorded in New Zealand.

Above: Ernest Shackleton's Imperial Transantarctic Expedition camp was established when the Endurance *perished in pack ice. In the Heroic Age of Antarctic exploration tents such as Shackleton's were often modelled on traditional Eskimo structures.*

Shackleton returned to London a hero and was knighted by King Edward VII. Next he announced plans for a bold expedition to cross Antarctica from the WEDDELL SEA via the South Pole to MCMURDO SOUND in the ROSS SEA. The *ENDURANCE* left Plymouth, England, on 8 August 1914 under the command of Frank WORSLEY. As it turned out, Shackleton never set foot on the continent but the 1914–17 IMPERIAL TRANS-ANTARCTIC EXPEDITION became one of the most legendary Antarctic adventures. Shackleton's epic struggles across the ice, wild SOUTHERN OCEAN waters and SOUTH GEORGIA's glaciers to save the expedition members were immortalized in his book *South* and in Frank HURLEY's photographs and film.

He returned to England in spring 1917 after rescuing the expedition's Ross Sea party, who had been stranded on the other side of Antarctica when laying food depots for his intended crossing of the continent. Shackleton spent most of the remainder of World War I at the north Russian front organizing supplies for the British forces fighting there.

Shackleton's final expedition was the 1921–22 *QUEST* EXPEDITION to circumnavigate the continent and map the coastline. The *Quest* left London on 18 September 1921 with many former members of the *Endurance* expedition. Shackleton had a heart attack and died on 5 January 1922 in South Georgia.

Shackleton's death marked the end of the so-called 'HEROIC AGE' of Antarctic exploration. Years later, Raymond PRIESTLEY wrote that, 'Incomparable in adversity, he was the miracle worker who would save your life against all the odds and long after your number was up. The greatest leader that ever came on God's earth, bar none.'

Shackleton Glacier Over 97 km (60 miles) long and 8 to 16 km (5–10 miles) wide, the Shackleton Glacier descends from the POLAR PLATEAU to the ROSS ICE SHELF. It was discovered by the 1939–41 US ANTARCTIC SERVICE EXPEDITION.

Shackleton Ice Shelf Forming 386 km (239 miles) of coastal Antarctica, the Shackleton Ice Shelf projects 145 km (90 miles) into the SOUTHERN OCEAN near BUNGER HILLS at 100°E, and 64 km (40 miles) in the east. Charles WILKES charted the area in 1840 and named it Termination Land. Douglas MAWSON, who explored it during the 1911–14 AUSTRALASIAN ANTARCTIC EXPEDITION, named the ice shelf for Ernest SHACKLETON. Mawson's party erected huts, none of which have been sighted since.

Shackleton Range Lying between Slessor and Recovery Glaciers, the Shackleton Range reaches 1875 m (6150 ft) above sea level and stretches approximately 140 km (87 miles) south of COATS LAND near the FILCHNER ICE SHELF. Its highest peak is over 1800 m (5904 ft) above sea level. It was discovered from the air by the COMMONWEALTH TRANSANTARCTIC EXPEDITION and is also known as the Shackleton Mountains.

Shackleton Station Shackleton Station was built by BRITAIN in Vahsel Bay on the FILCHNER ICE SHELF to support the COMMONWEALTH TRANS-ANTARCTIC EXPEDITION. It was from here that Vivian FUCHS left on the first overland crossing of the continent.

Shags (Phalacrocoracidae) See CORMORANTS.

Shearwaters (Procellariidae) Shearwaters are slender-bodied and longish-billed PETRELS, most of which are superb divers. They use their long, narrow wings for swiftly wing-rowing deep beneath the sea's surface in pursuit of FISH, SQUID and CRUSTACEANS.

Twelve species of shearwater have been recorded in the SOUTHERN OCEAN region, but only three live in or regularly visit these waters: the SOOTY SHEARWATER, SHORT-TAILED SHEARWATER and the subantarctic LITTLE SHEARWATER. The first two travel into the Northern Hemisphere during the austral winter, a migratory pattern related to food supply rather than temperature. The name 'shearwater' originated from the flying habits of some species, which glide across the surface of the sea.

Like many PETRELS, shearwaters nest in burrows and lay single eggs. Different species have different breeding regimes: some breed in the summer months between November and February, and others in winter.

Shearwater FOSSILS similar to existing species have been found in marine sediments dated at 30 million years old, suggesting a slow rate of evolutionary change.

Shelter Any form of protection from the elements, from a basic ICE CAVE or IGLOO to a sophisticated scientific RESEARCH STATION. Field parties are instructed in how to build an EMERGENCY shelter in the SNOW, and may have to spend a night outside in a cave, or in tents, before being sent into the field. Most of the early permanent shelters in Antarctica were wooden HUTS, used as bases for

Below: Tents are the most common shelter used by field research parties. Those dubbed 'polar havens' have a strong frame and will often be 'home', the 'mess' or a research lab for an entire summer.

Above: Established by Britain in 1947 as a meteorological station, Signy Base was later redeveloped into a biological science facility.

sledge journeys or as a winter refuge. After World War II, bases became more sophisticated. Now even field camps are insulated, and some have internet and telephone links. Permanent bases are generally made from prefabricated panels that are sealed together and insulated with plastic foam and reflective surfaces. They generate their own power for heating, light and cooking, and are kept at a comfortable living temperature.

Ships About 100 ships visit Antarctic waters each year. These mostly consist of expedition vessels but CRUISE ships, research vessels, commercial FISHING boats and private YACHTS are also frequent visitors to the area. ICEBREAKERS, pioneered by OPERATION HIGHJUMP in 1946–47, carve paths through the ice to the coast when the PACK ICE begins to break up in summer.

The first ships to sail into Antarctica's ice-bound waters were Captain James COOK's *Resolution* and *Adventure* in 1772. The early ships had wooden hulls and were not equipped to travel through thick PACK ICE. Consequently, many were crushed and sank. In the 19th century, hull designs were adapted to polar conditions. Roald AMUNDSEN's *FRAM* had a saucer-shaped hull so that ice pressure would squeeze it upwards; in contrast, Ernest SHACKLETON's ill-fated *ENDURANCE* had a steep-sided hull that caved inwards as the moving ice pressed against the ship and crushed it in the WEDDELL SEA pack ice. Although steel hulls are now most common, wooden ships are still used in Antarctica. The 38 m (125 ft) wooden American ship, *Hero*, for example, operated until 1984, and was designed for offshore biological research.

Shipwrecks Furious storms, PACK ICE, ICEBERGS and enormous WAVES have claimed many ships in the SOUTHERN OCEAN. The most famous wreck was Ernest SHACKLETON's ship *ENDURANCE*, which was captured on film by Frank HURLEY as it was crushed by pack ice in the WEDDELL SEA in October 1915. In 2001, Robert Ballard who successfully located the wreck of the *Titanic*, announced plans of an expedition to locate the *Endurance*.

The AUCKLAND ISLANDS have claimed the most ships in Antarctic maritime history. When the *General Grant* sank on a voyage to London from the Australian goldfields in 1866, it was recorded as having 70 kg (154 lb) of gold aboard; however, it has always been believed that there may have been as much as 8 tonnes (8 tons). The ship ran into trouble when it was nudged onto the west coast of the Auckland Islands, where it became wedged in a huge cavern and sank. Of the 85 crew and passengers who abandoned the ship, only 15 survived an 18-month wait for rescue. Attempts to locate the wreck and recover the gold are ongoing.

When ARGENTINA's Antarctic resupply vessel *Bahia Paraiso* sank in 1989 near PALMER STATION on the ANTARCTIC PENINSULA, more than 681,900 litres (150,000 gallons) of FUEL spilt into the ocean, resulting in OIL slicks that covered approximately 100 sq km (39 sq miles). It is estimated that about 30,000 PENGUINS came into contact with the spill, as the accident occurred when most parent birds were foraging at sea for their young.

Shirase, Nobu (1861–1946) Japanese army officer and explorer. The child of a Buddhist monk, Shirase was born in Konoura in Akita, in north-

ern Japan. After a time in a school for priests in Tokyo, he joined the Japanese Army and in 1893 was part of an expedition to the Kuril Islands north of Japan. Initially, the unknown lieutenant's proposal to mount an Antarctic expedition to reach the SOUTH POLE was not well received. Eventually he gained influential support, and the JAPANESE ANTARCTIC EXPEDITION left Tokyo on 1 December 1910 in the three-masted, steam-powered fishing boat *Kainan Maru*.

On 26 February 1911 they encountered their first ICEBERG and on 6 March saw the coast of VICTORIA LAND and reached as far south as 74°16' before retreating from the ROSS SEA because of bad weather and heavy ice, and heading back to AUSTRALIA for the winter. There, receiving news of Roald AMUNDSEN's and Robert SCOTT's attempts on the Pole, Shirase shifted the focus of the expedition to science.

The expedition was back in Antarctic waters the following summer and on 16 January 1912 reached the ROSS ICE SHELF, then sailed further east to the BAY OF WHALES, where a base camp was established. From here a seven-man sledging party led by Shirase—the 'Dash Patrol'—reached 80°5'S, 415 km (258 miles) inland, before turning back. The expedition returned to Tokyo on 20 June 1912 and was welcomed by a crowd of around 50,000.

Shirase's immediate legacy was a very substantial debt, and he travelled widely in an attempt to raise money. He died penniless in his rented room in Toyota city at the age of 85. No one attended his funeral. He left a *waka*, a short poem: 'Study the treasures under the Antarctic and make use of them after my death.'

Short-tailed shearwater (*Puffinus tenuirostris*) Similar to SOOTY SHEARWATERS, but smaller in size and with a dark underwing. Regularly seen in Antarctic waters in February, the rest of the year they range over the Pacific region as far south as 65°S and breed on islands in BASS STRAIT, between TASMANIA and the Australian mainland. Like sooty shearwaters, the birds are known as 'muttonbirds', and about 300,000 nestlings are harvested in Tasmania each year. This species migrates in vast numbers to the northern Pacific for the southern winter. In polar waters they feed principally on KRILL but elsewhere often take FISH.

Signy Base Established by BRITAIN in 1947 on Signy Island in the SOUTH ORKNEYS as a meteorological station. In 1963 it was redeveloped into a biological science facility. In 1995 research in marine science was transferred to ROTHERA STATION, and Signy focused on terrestrial and freshwater biology. Since 1996, it has operated in the summer only, with a maximum of eight to 10 people.

Sinfonia Antartica The first major musical composition on an Antarctic theme, written by English composer Ralph Vaughan Williams. The five movement symphony, which captures the spirit of human endeavour in a hostile natural environment, was derived from a score Vaughan Williams originally wrote for the 1948 dramatic FILM directed by Charles Frend, *Scott of the Antarctic*. The work, characterised by a relentlessly climbing theme, is scored for a full orchestra and also utilises a soprano soloist and women's choir. It uses an organ to represent the crash of ice falling.

Siple, Paul Allman (1908–68) American scientist and explorer. From Ohio, Siple was 19 when, as an eagle scout, he won a Boy Scouts of America contest to accompany Richard BYRD on his 1928–30 expedition; he beat off thousands of applicants in the process. During the expedition he worked both as dog-handler and naturalist. Smitten with the Antarctic 'bug', he made another four journeys south.

A talented scientist, on BYRD'S SECOND EXPEDITION he was chief naturalist and led an 11-week SLEDGE journey of exploration across MARIE BYRD LAND—Byrd named the SIPLE COAST in his honour. After completing a doctorate in geography, he joined Byrd's 1939–41 UNITED STATES ANTARCTIC SERVICE EXPEDITION. As leader of the western base at LITTLE AMERICA, Siple undertook biological and geological survey work in the Edsel Ford Mountains in EDWARD VII LAND.

From 1946 until 1963 he worked as chief of research and development with the US Army and, during 1946–47, was involved with OPERATION HIGHJUMP. In Siple's view, '[scientists] were given so little opportunity to pursue their work that many of them vowed they would never return to the Antarctic with a Navy expedition.'

However, he returned with OPERATION DEEP-FREEZE in 1955–57 to supervise the construction of AMUNDSEN-SCOTT SOUTH POLE STATION and direct the INTERNATIONAL GEOPHYSICAL YEAR pro-

gramme. This was the first winter ever spent at the SOUTH POLE and on 18 September 1957, in the coldest temperature thus far recorded (–74°C/–104°F), he walked around the world accompanied by John Tuck and the dog Bravo—in two minutes: 'We realized that we had crossed the date line,' Siple reported. 'This would, of course, get us out of time with the rest of the camp personnel; so, rather than risk such a hypothetical mix-up, we took another walk around the world, "unwinding" in the opposite direction.' During this expedition Siple developed an index to measure WINDCHILL.

Siple Station Established by the USA in 1969 in ELLSWORTH LAND and named after Paul SIPLE. Siple Station operated continuously from 1973 until 1985, then as a summer-only base until 1988. RESEARCH into meteorology, upper atmosphere physics and glaciology was conducted here. For studies in the very low frequency range, Siple Station used a radio receiver and a crossed horizontal antenna that was 42 km (26 miles) long.

Sites of Specific Scientific Interest In 1964, general rules of conduct for scientific expeditions drawn up by the SCIENTIFIC COMMITTEE ON ANTARCTIC RESEARCH were incorporated into the AGREED MEASURES FOR THE CONSERVATION OF ANTARCTIC FLORA AND FAUNA. These included establishing the concept of SPECIALLY PROTECTED AREAS in which all activities were prohibited, and Sites of Specific Scientific Interest (SSSI), that required additional restrictions in order to protect their scientific value.

Skiing From the earliest forays into Antarctica's interior, skis have been used as a primary means of TRANSPORT. Roald AMUNDSEN took champion skier Oscar BJAALAND to the SOUTH POLE in 1911–12; Bjaaland had spent the previous winter preparing 10 sets of skis, as well as sledges and other equipment, for the assault on the Pole.

In the right weather conditions, some expeditions speed their progress by harnessing themselves to parasails, and cover remarkable distances. Reinhold Messner and Arved Fuchs crossed the continent, from the WEDDELL SEA to the ROSS SEA, on skis in 1989–90. On their best day, using parachutes, they traversed 1 degree of LATITUDE. In 2000–2001, Liv Arnesen, the first woman to ski solo to the South Pole, and Ann Bancroft skiied and 'sailed' across Antarctica from DRONNING MAUD LAND to the ROSS ICE SHELF, via the Pole—a journey that took 94 days. During the sixth crossing of Antarctica, Børge OUSLAND covered 2845 km (1764 miles) in 64 days; on his fastest day on skis he covered 226 km (140 miles) using a parasail.

Skuas (Stercorariidae) Skuas are rapacious, gull-like BIRDS superbly adapted to catch FISH. They also prey upon young chicks, eggs and small adults of other bird species, including PENGUINS, and also take carrion. Their black bills are strong and well hooked and, in contrast to GULLS, the females are slightly larger than the males; they weigh about 1.5 kg (3.3 lb). Skuas are strong fliers and fiercely defend their nest territories from all-comers (including humans). They are well known for their challenging 'long calls', given with the wings held high and fully extended, showing the

Below: Strong predators, skuas (Stercorariidae) prey on the chicks of other bird species, and also claim the carcasses of those that have died of natural causes.

Above: Sledging remains the traditional mode of transport for adventurers in Antarctica. However, modern technology has vastly improved the weight and functionability of sledges.

conspicuous white band at the base of the flight feathers.

Skuas consist of two genera and six species, with most having a bipolar distribution. There are two Antarctic species: SOUTHERN SKUA and ANTARCTIC SKUA. The females lay two eggs a season in their ground nests, and usually mate for life. Like most seabirds, skuas have salt glands allowing them to drink sea water when necessary, although they prefer fresh water when they can find it. Outside the breeding season, their migrations may range from one hemisphere to the other.

Sledges Since the early 20th century, sledges have been used for DOG- or man-hauling PROVISIONS and EQUIPMENT in Antarctica. Polar explorer Fridtjof NANSEN adapted Norwegian sledges for

Arctic use, and the Nansen sledge design remains basically the same. It has two runners, bent at the end like skis, and a framework of crossbars and slats to support the load. Originally, lashings were made of flexible leather and cord to allow the sledge to flex over uneven ground. Most of the materials used on new sledges are now synthetic and instead of bone or ivory runners they glide along on fibreglass or plastic edges.

Until quite recently, dog teams were the backbone of land transport. They have now all been removed from Antarctica, and sledges are usually pulled by motorized VEHICLES, such as SKIDOOS. When covered with a layer of soft snow, SEA ICE provides an ideal sledging surface. However, when the ice is pushed into hummocks and ridges in pressure zones, sledge hauling is exhausting. Travelling in deep snow or over SASTRUGI fields

with sledges is equally demanding. Raymond Priestley described sledging back to Cape EVANS in the spring of 1911: '... we were frequently floundering for several yards together up to the sockets of our thighs in snow, while the latter was hard and cloggy with a stiff crust, and every step was like drawing a tooth ...'

On crevassed ground, sledges distribute the load more widely, and can be safer than travelling on foot. In November 1912, Douglas MAWSON preceded Belgrave Ninnis over the same ground, but Ninnis and his dog team broke through a crevasse and disappeared forever. Mawson believed the accident happened because Ninnis was walking next to his sledge rather than sitting on it: 'The whole weight of a man's body bearing on his foot is a formidable load and no doubt was sufficient to smash the arch of the roof.'

Smith, William (c. early 19th century) English sealer. Nothing is known of the merchant-sealer before his journey in his brig, the *Williams*, from Buenos Aires, Argentina, to Valparaíso, Chile, in 1819. As he attempted to sail around Cape HORN on 19 February, strong winds blew Smith far to the southwest where, between snow-flurries, he saw land at 61°10S. When he reached his destination, he reported the discovery to the senior British naval officer on the Pacific coast, Captain William Shirreff, who was sceptical of the claim.

On his return trip around the Cape in June, Smith endeavoured to reach the same high latitude but was unable to penetrate the ice. On his next trip he again sailed southwards and on 15 October spotted land about 14.5 km (9 miles) distant—possibly DECEPTION ISLAND. Taking careful soundings as he veered north, he skirted the northern coasts of a string of islands and the next day landed on the northernmost one, taking possession on behalf of BRITAIN—this was KING GEORGE ISLAND in the SOUTH SHETLANDS.

When news reached London, the Admiralty chartered the *Williams* which, commanded by Edward BRANSFIELD with Smith as pilot, returned to verify the sighting and secure possession. They sailed from Valparaíso on 20 December 1819 and sighted land on 18 January 1820. Turning eastwards through the Bransfield Strait on 30 January, they saw land to the southwest which they named 'Trinity Land', the northern tip of the ANTARCTIC PENINSULA—this was once believed to be the first sighting of the continent, but it now seems more likely that Thaddeus BELLINGSHAUSEN saw it three days earlier.

The *Williams* was returned to Smith and he refitted it for a sealing venture the following summer. He arrived in London with his holds full of skins in September 1821 to find his business partners in financial ruin. The brig and its cargo were seized and he was bankrupted. He died in poverty in the 1840s.

Snares, The The closest SUBANTARCTIC ISLANDS to NEW ZEALAND, the Snares are covered with heavy TUSSOCK grass and a wind-beaten forest of tree daisies. The islands are relatively inaccessible due to their rocky, exposed shores. Only two introduced plant species and a rare 'scurvy grass'

(*Lepidium oleraceum*) grow on the island. Once abundant on the New Zealand mainland, where it was brewed into tea, *L. oleraceum* was drunk by James COOK's crew to prevent scurvy.

Two ships discovered the islands on the same day: 23 November 1791. Captain George Vancouver of *Discovery* named the islands the Snares, and Lieutenant Broughton of *Chatham* called them the Knights Islands. In 1810 Captain Keith, of the ship *Adventure*, abandoned four men on the Snares. One was then murdered by his crewmates, who were rescued in 1817.

Sno-cats, The COMMONWEALTH TRANSANTARCTIC EXPEDITION used four Sno-cats on their crossing of the Antarctic continent. Specially modified for the journey, these snow VEHICLES, which have four independent tracks, were manufactured by the Tucker Corporation of Oregon and originally designed for North American conditions. The massive Sno-cats driven by Vivian FUCHS's party made slow progress over difficult terrain.

Snow The main form of PRECIPITATION in Antarctica, where researchers have found more than eight different types of snow crystals. Water evaporates from the SOUTHERN OCEAN and is carried over the continent. Ice crystals form from water vapour on minute particles of atmospheric dust. When they are heavy enough, these crystals fall as snowflakes—at this point, their density will be relatively low at around 100 kg per cubic m (168 lb per cu yd). When the snow begins to decay and new snow falls on its surface, it begins its metamorphosis into ICE, thus changing the snow crystal's size, structure and texture. If snow is moved by WIND, the crystals become rounded and close-packed, forming 'windpacked' snow, which has a density of around 400 kg per cubic m (674 lb per cubic yd).

Snow blindness Most of the sunlight that hits the white surface of Antarctica is reflected back off the ground. This intense light can quickly damage human eyes if they are not protected by sunglasses or GOGGLES. Eyeballs are literally burnt, causing temporary blindness and, often, extreme discomfort. People liken snow blindness to having sand rubbed into their eyes. In severe cases, eyesight may be permanently damaged.

Snow Hill Island Discovered by James Clark ROSS in 1843, Snow Hill Island lies off the eastern tip of the ANTARCTIC PENINSULA, adjacent to SEYMOUR ISLAND and JAMES ROSS ISLAND. Ross named the 395 m (1296 ft) long island 'Snow Hill', as he thought it was connected to the Peninsula. Otto NORDENSKJÖLD set up a winter base on the island in 1902, and when his ship *Antarctic* was crushed in PACK ICE, four scientists left on the island spent two years sheltering in a small hut that is now designated an HISTORIC SITE.

Snow petrel (*Pagodroma nivea*) Snow petrels are unmistakable—pure white and rather small (32 cm/13 in long) with dark eyes, long wings, short black bills and gunshot-grey feet—like pieces of ice that have come to life. In repose, the tails

appear square, but they open up like fans. There are two subspecies, which differ mainly in size. *P. nivea nivea* is the smaller, and *P. nivea major* nests on more open ground around TERRE ADÉLIE.

Their flight pattern is swift and erratic with frequent, rapid wing-beats that create fluttering effects. They feed on SQUID, FISH and KRILL, which they catch by surface-dipping while on the wing or by surface-seizing. They often hover before diving briefly into the water. Snow petrels are skilled at flying low and in zigzag fashion into the teeth of an Antarctic gale.

Like all FULMARS, they can eject stomach oil as a defence mechanism, which is especially effective against SKUAS. Deposits of this yellowish oil have built up around their nests; the oldest deposits have been carbon-dated as 34,000 years old.

Snow petrels nest on the Antarctic continent and many coastal islands. Their single eggs are laid from late November and the chicks fledge in March or April.

The birds do not venture very far north: they are very much inhabitants of the PACK ICE and flocks will readily roost on ICE FLOES and ICEBERGS, often in the company of CAPE PETRELS and ANTARCTIC PETRELS. They follow ships in the pack ice, swooping down to feed on any food in the disturbed water, and they escort and feed among WHALES.

Snowy sheathbill (*Chionis alba*) These all-white shorebirds are also known as American or greater sheathbills. Rather strange-looking birds, they are podgy and look like albino pigeons, with wingspans of about 80 cm (32 in). The wattles around their faces, often pink, add a little colour, as do their squat blue legs and horny, yellowish-green bills. They have small, sharp carpal spurs on their wings, with which they fight for territory.

Sheathbills feed chiefly through scavenging, eating anything from SEAL and BIRD faeces to the partly digested food dropped by adult PENGUINS when feeding their young. Solitary birds, they build their nests in dark spaces using a variety of materials, including dead chicks, bones, feathers, LICHEN and limpet shells. Both parents incubate their eggs (they usually lay between two and four) for close to a month.

Snowy sheathbills are found on the ANTARCTIC PENINSULA and the nearby islands, including SOUTH GEORGIA and the SOUTH ORKNEYS; in winter they move north to the FALKLAND ISLANDS and the southern coasts of South America.

Soils Because of the limited water supply and subzero temperatures, soils form in Antarctica excruciatingly slowly. Soil formation in temperate regions can be measured on a scale of hundreds of years; the same processes may take millions of years in Antarctica. Most soils on the continent contain PERMAFROST and are permanently frozen just below the surface. The weathering of these soils is due to chemical breakdown of the soil structure and damage by ice crystals.

Distinct soil zones have been identified. Ahumic soils, which coat the ice-free areas of the DRY VALLEYS, are nearly barren, have a high salt content and contain little humus or living matter.

Some of these soils have been dated at around 5 million years old, probably when GLACIERS retreated from these valleys. In coastal areas, soils are still relatively dry, but contain more organic matter, derived particularly from MOSSES and LICHENS. In places, this organic material is stacked into peat banks. As the dead moss cannot decay in the permafrost, deposits can reach up to 1.5 m (5 ft) in thickness and be thousands of years old.

In the west coast of the ANTARCTIC PENINSULA and the MARITIME ISLANDS, soil formation is far more rapid, and the soil is well developed and moisture levels are high. Ornithogenic soil is derived from the excreta of bird colonies, and along with soils near seal colonies, has high concentrations of carbon, nitrogen and living organisms. At lower latitudes, the subantarctic islands have a wide range of soil types. On SOUTH GEORGIA, for example, organic soils, meadow tundra soils, brown soils, mineral soils and peat deposits support a wide range of vegetation.

Solar radiation Antarctica has a net radiation loss—only for a brief period in midsummer does it receive more radiation than it loses to the ATMOSPHERE. The net heat loss is balanced, to some extent, by the flow of warm, moist air onto the land from the ocean; without it, PLANTS and ANIMALS would find it nearly impossible to survive.

In winter no sunlight reaches Antarctica at all (although moonlight, starlight and AURORA AUSTRALIS are seen). Sunlight reaches Antarctica 24 hours a day in summer, but much of the incoming

Below: *Snowy sheathbills* (Chionis alba) *resemble pigeons, and tend to scavenge food from the shores and coastal areas they inhabit.*

solar radiation is reflected by cloud cover and absorbed by gases before it reaches the surface. Of the light that does strike ground, up to 85 percent is reflected back to the atmosphere by the bright ice cover, especially over the POLAR PLATEAU where the blanket of air is thin. In addition, the SUN strikes high latitudes at a very low angle and with low intensity. In midsummer at the SOUTH POLE, for example, the sun is only 23.5 degrees above the horizon and a shaft of sunlight is spread over a much larger area than would be the case near the Equator.

Solar wind Made up of 'plasma'—ionized gas emitted from the sun—the solar wind travels towards Earth, hitting the magnetosphere with a velocity of about 3 million km (1.86 million miles) per hour and compressing its outer surface over much of the globe. The solar wind cannot cross the magnetic field lines at the magnetosphere, but it can follow the lines and enter at the 'magnetic cusps' centred near both poles. Charged plasma penetrates deep into the polar ATMOSPHERE at this 'magnetic cusp', generating the spectacular AURORA AUSTRALIS or 'southern lights' and other OPTICAL PHENOMENA.

When major 'eruptions' occur on the surface of the sun, the plasma is accelerated towards the Earth by huge voltages generated in the magnetosphere, and the solar wind increases from 10,000 megawatts to 15 million megawatts. The storms disrupt the magnetic field right to the surface of Antarctica, affecting RADIO and NAVIGATION systems.

Sooty albatross (*Phoebetria fusca*) Similar to the LIGHT-MANTLED SOOTY ALBATROSS, but with wholly brown plumage, their bills are black with a characteristic yellow stripe. At close range, a semicircle of white feathers is readily visible behind the eyes. The feet are pale grey. Their wings are long and slender with wing-spans of 2 m (6½ ft), their tails are wedge-shaped, and in flight they are extremely graceful.

Sooty albatrosses breed every second year on SUBANTARCTIC ISLANDS between 36°S and 49°S in the southern Indian and Atlantic Oceans. At sea, sooty albatrosses are solitary, but are known to follow ships. They eat mainly SQUID, FISH and CRUSTACEANS.

Sooty shearwater (*Puffinus griseus*) These beautiful, medium-sized, brown, slender SHEARWATERS have long, narrow wings and wing-spans of 105 cm (3½ ft). In flight, they are recognizable by the silvery-white areas beneath their wings. They fly swiftly, low to the water, with short glides intermingled with high arcing on stiff wings, rapid wing-beats and fast directional changes. Sooty shearwaters feed on SQUID, CRUSTACEANS and small FISH such as anchovies. They are fine swimmers and dive for their food, using their wings to row underwater. Although these birds rarely follow ships, they are regular visitors to FISHING trawlers and can gather in large feeding flocks.

They breed during summer in large colonies on some NEW ZEALAND headlands and many nearby islands, including the SNARES, and on islands off the southern coasts of AUSTRALIA and South America. The birds dig into the sandy coastal soil, making burrows in which they lay their single eggs. One parent remains on guard at all times while the other forages at sea then returns at night. The young chicks are a traditional harvest of the Maori of New Zealand—who call them 'titi' or 'muttonbirds': about 250,000 are taken each year. Those birds not breeding disperse throughout the SOUTHERN OCEAN and are found as far south as the PACK ICE. Outside the breeding season, sooty shearwaters are transequatorial migrants in both the Atlantic and Pacific Oceans.

South Africa South Africa's claim to the subantarctic PRINCE EDWARD ISLANDS is recognized internationally. A participant in the INTERNATIONAL GEOPHYSICAL YEAR, South Africa was one of the original 12 CONSULTATIVE PARTIES to the 1959 ANTARCTIC TREATY.

The apartheid policies of the then South African government were a source of strain on the ANTARCTIC TREATY SYSTEM (ATS) in the 1980s and South Africa's participation in the ATS was criticized in the UNITED NATIONS. Although resolutions calling for its expulsion from the ATS were adopted in 1985, South Africa continued to participate.

A member of the Commission for the CONVENTION ON THE CONSERVATION OF ANTARCTIC MARINE LIVING RESOURCES, South Africa was adversely affected by illegal, unregulated and unreported FISHING around its SUBANTARCTIC ISLANDS in the mid-1990s, when the local stocks of PATAGONIAN TOOTHFISH were rendered commercially extinct. South Africa ratified the MADRID PROTOCOL in 1995. It maintains three RESEARCH STATIONS in the Antarctic.

South African National Antarctic Expeditions (SANAE) The agency responsible for SOUTH AFRICA's exploration and research programmes in Antarctica, it operates the SANAE BASE at Vesleskarvet in DRONNING MAUD LAND.

South Geomagnetic Pole A theoretical point calculated mathematically to approximate the source of the Earth's magnetic field. The South Geomagnetic Pole and its counterpart, the North Geomagnetic Pole, represent the end points of the axis of an imaginary bar magnet located in the centre of the Earth's core—the magnetic field of the magnet best describes the actual magnetic field of the Earth.

The South Geomagnetic Pole, the southern point of intersection, is located at 78°30'S, 111°E, near VOSTOK STATION. It is distinct from the SOUTH MAGNETIC POLE, which relates to the actual magnetic field of the Earth. The International Geomagnetic Reference Field is a mathematically modelled field used in research, and against which changes in the actual magnetic field can be measured.

South Georgia GLACIERS cover about 60 percent of this 3755 sq km (1450 sq mile) island in the South Atlantic. It is the subantarctic's most mountainous island, reaching to 2934 m (9626 ft) at Mount Paget in the Allardyce Range. Ernest SHACKLETON, along with Frank WORSLEY and Thomas CREAN, made the first major crossing of this range during the RESCUE of the IMPERIAL TRANSANTARCTIC EXPEDITION members. Shackleton died during a visit to the island in 1922, and is buried there.

The English merchant Anthony de la ROCHE was probably the first to sight the island, in 1675. James COOK made the first landing on 17 January 1775, claiming SOVEREIGNTY for Britain. It was a

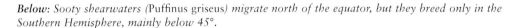

Below: Sooty shearwaters (Puffinus griseus) *migrate north of the equator, but they breed only in the Southern Hemisphere, mainly below 45°.*

Above: The South Orkney Islands group was an important base for sealers in the 19th century. Many of the buildings that were constructed during this period remain.

major SEALING area from the late 18th century, but its populations of SOUTHERN ELEPHANT SEAL and FUR SEAL have now recovered. The first Antarctic WHALING station, Grytviken, was established by a Norwegian company on the sheltered northeastern coast in 1904 and shore-based ventures operated until 1965.

The GERMAN INTERNATIONAL POLAR YEAR EXPEDITION visited the island in 1882–83, setting up a station on the island's southeast coast. In 1949–50 the Falkland Islands Dependencies Survey (now BRITISH ANTARCTIC SURVEY) established a new base at King Edward Point, on the northeast coast. During the 1982 WAR between BRITAIN and ARGENTINA, the Argentinian naval vessels, *Bahia Paraiso* (which was later wrecked near ANVERS ISLAND) and *Guerrico*, landed 200 troops, captured the island and took British soldiers and scientists prisoner. The island was quickly recaptured by Britain.

Wildlife is abundant on South Georgia. There are more than 5 million pairs of nesting MACARONI PENGUINS and several large KING PENGUIN rookeries. As it is rat-free, burrowing birds thrive, along with the SOUTH GEORGIA PIPIT, which is the only songbird in Antarctica. A small population of reindeer, introduced by whalers in 1911, are penned into two areas on the island by GLACIERS.

South Georgia cormorant (*Phalacrocorax georgianus*) Whalers used South Georgia cormorants for food, dubbing them the 'chickens' of the subantarctic. Breeding colonies are found on various islands; the largest, a colony of around 4000 pairs, is on SOUTH GEORGIA. The birds, which are about 72 cm (28 in) long and have wing-spans of 1.2 m (4 ft), have black crests on their foreheads; the head feathers have a bluish-black tinge, and the eye-rings are blue.

South Georgia diving petrel (*Pelecanoides georgicus*) Small and low-flying, these birds have speedy and direct flight patterns and, like PENGUINS, use their wings underwater when diving and looking for food—predominantly KRILL. They are 20 cm (8 in) long, with wing-spans of 32 cm (13 in). The back is glossy black, with shoulders, chin and throat of sooty white. When at sea, they are hard to distinguish from COMMON DIVING PETRELS, which makes it difficult to estimate their at-sea distribution. They breed on SOUTH GEORGIA and Îles KERGUÉLEN, as well as on several other islands.

South Georgia pipit (*Anthus antarcticus*) These small, mottled, brown birds are the only passerines (songbirds) in Antarctica. In summer they eat spiders and INSECTS; in winter when the ground is covered in snow they feed on tide-line debris. Pipits are strong fliers. When they are on the ground they get around by walking and will sometimes break into a run when pursuing insects—but they never hop.

South Magnetic Pole The point, as at the North Magnetic Pole, where the direction of the Earth's magnetic field—which is generated by electric currents flowing both inside the molten core of the Earth and externally—is vertically upwards, and a compass points straight down. The magnetic dip, the angle between the horizontal plane and the magnetic field lines, is 90 degrees at the magnetic poles. The position of the South Magnetic Pole is not fixed, and it currently moves about 5 km (3 miles) per year in a north to northwesterly direction. When it was first reached by Douglas MAWSON, Edgeworth DAVID and Alistair Mackay in 1909, the pole was on land, at about 71°15'S, 148°64'E; in 2000 it was off the coast of TERRE ADÉLIE at 64°51'S, 138°43'E.

South Orkney Islands East of the SOUTH SHETLAND ISLANDS in the SCOTIA ARC, the South Orkneys group consists of four major islands—Coronation, Signy, Powell and Laurie—along with the Inaccessible Islands 29 km (18 miles) to the west. GLACIERS cover 85 percent of the islands, which experience typical MARITIME ISLAND weather conditions: mainly overcast with strong winds and low temperatures.

The islands were discovered independently by two sealers, the American Nathaniel PALMER and the Briton George Powell, on 6 December 1821. The SEALING industry flourished here. FUR SEAL populations were decimated—in 1936, for example, a visitor reported finding only one seal—and have recovered slowly: in 1995, they numbered about 22,000. The first factory WHALING ship arrived in the South Orkneys in 1912, and between 1920 and 1930, 3500 whales were harvested.

ORCADAS STATION on Laurie Island is the oldest continuously run research facility in Antarctica. It was built in 1903 by the SCOTTISH NATIONAL ANTARCTIC EXPEDITION, and has been operated since then by ARGENTINA's meteorological office. BRITAIN established another meteorological station, Base H, on SIGNY ISLAND in 1947–48.

South Polar Times The potential boredom of the dark Antarctic winter of 1902–03 was relieved by the production of the *South Polar Times*, the first publication to be produced on the continent. The monthly was edited by Ernest SHACKLETON, with stories, articles, poems and drawings supplied by various members of the expedition.

South Pole *See following pages.*

South Sandwich Islands Eleven MARITIME ISLANDS, sprinkled across the easternmost end of SCOTIA ARC, included in the TERRITORIAL CLAIMS of both BRITAIN and ARGENTINA. A cold OCEAN CURRENT sweeps north from the WEDDELL SEA making the islands colder than the more southerly SOUTH SHETLAND and SOUTH ORKNEY ISLANDS. Seven of the islands have substantial ice cover, and four, including ZAVODOVSKI, are free of permanent ice cover.

Volcanic activity has been recorded on most of the islands. In 1823 sealer Benjamin MORRELL reported seeing 'nine burning volcanoes' in the South Sandwich Islands: '... three of these islands had vomited out so much of their entrails, that their surfaces were nearly even with the water,' he said. In 1956 a group of Argentinians evacuated their hut on Thule Island when three jets of glowing material shot 300 m (980 ft) into the air over 48 hours.

An underwater VOLCANO lies 25 m (80 ft) below the water surface at the end of the South Sandwich chain, in 1962 a South African research ship trapped in PACK ICE was freed when a submarine eruption sent shock waves through the ice.

Above: A cold ocean current sweeps north from the Weddell Sea, making the South Sandwich Islands colder than the more southerly South Shetlands and South Orkneys.

South Shetland Islands About 1000 km (621 miles) south of TIERRA DEL FUEGO, this major island group lies in SCOTIA ARC and extends from the end of the ANTARCTIC PENINSULA in a 540 km (336 mile) long chain. Consisting of four main groups, they are (from the Peninsula end): Smith and Low Islands; Livingston, Snow, Robert, Greenwich and DECEPTION ISLANDS; King George and Nelson Islands; and ELEPHANT and Clarence Islands.

About 80 percent covered by permanent ICE, the islands were discovered in 1819 by William SMITH. The first of many sealers to work the area was one of Captain Smith's crew who returned a year later. On Livingston Island, Captain Robert

Fildes found that, '... in many places it was impossible to haul a boat up without first killing your way, and it was useless to try to walk through them if you had not a club in your hand to clear your way ...' The FUR SEAL population was almost decimated by the end of 1821, and then again after another bout of SEALING in the 1880s. The whaling boom began in 1907 on King George Island, some beaches are still carpeted in bones.

The treacherous waters around the islands claimed many sealing ships, and in 1819 Antarctica's worst loss of life occurred with the sinking of the *San Telmo* off the coast of Livingston Island. This Spanish navy vessel was caught in severe weather while crossing the DRAKE

*Below: Southern elephant seals (*Mirounga leonina*) are found throughout the subantarctic, on Antarctic islands and on the Antarctic Peninsula.*

PASSAGE, and sunk while being towed by another ship, taking 650 crew with her. James WEDDELL found evidence that survivors of a SHIPWRECK had lived for a long period on Livingston Island, but there is no conclusive evidence that they were from the *San Telmo*.

In 1946–47 BRITAIN took steps to assert its SOVEREIGNTY against rival claims by ARGENTINA and CHILE, setting up a permanent station, Base G, on King George Island. The island now has eight national winter stations, linked by a network of roads. Chile has encouraged families to live at FREI STATION as part of attempts to incorporate its Antarctic territory into the rest of the country. The first tourist vessels visited the South Shetland Islands in 1958 and most CRUISES now stop over at these islands before heading south down the Peninsula.

Southern bottlenose whale (*Hyperoodon planifrons*) One of the TOOTHED WHALES, they range south as far as the PACK ICE, where they feed mainly on SQUID. Distinguished by their high foreheads and pronounced beak-like jaws, bottlenose whales grow to around 7 m (23 ft) long and weigh 3 to 4 tonnes (3–4 tons). Little is known about their numbers, migratory pattern and reproductive cycle.

Southern Cross Expedition (1898–1900) Private British-funded expedition, the first to OVERWINTER on land within the ANTARCTIC CIRCLE. The expedition leader Carsten BORCHGREVINK attempted to obtain funding for a scientific expedition in AUSTRALIA, but eventually obtained British sponsorship.

Borchgrevink purchased the *Pollux*, and refitted and renamed it *Southern Cross*. He recruited physicist Louis BERNACCHI, William COLBECK as magnetic observer, and Norwegian zoologist Nikolai Hanson, who died in Antarctica. The goals, which excluded any attempt to reach the SOUTH POLE, were realistic and the expedition was carefully planned, taking DOGS—the first on the continent—to haul SLEDGES, two Lapp dog-handlers, and food SUPPLIES intended to prevent scurvy.

The expedition sailed from London on 23 August 1898 and on 17 February 1899 reached Cape ADARE, where they built their base. This base included a prefabricated HUT for the 10 men who wintered over. Extensive surveying work was undertaken inland and around the coast. Two important lessons learned were the dangers of FIRE (Colbeck's bunk caught alight) and of coal fumes inside the hut. It was a long, hard winter, with tensions between Borchgrevink and the scientists. After leaving Cape Adare, the expedition explored the ROSS SEA area and a party sledged across the ROSS ICE SHELF, creating a 'farthest south' record to 78°50'S.

Although the expedition's aim of reaching the SOUTH MAGNETIC POLE was not achieved, the Pole's approximate location was fixed. Its pioneering geographical work, the year-long meteorological record obtained, and the successful overwintering were important advances in knowledge of Antarctica. However, it was some years before the expedition's achievements were recognized.

Southern elephant seal (*Mirounga leonina*) Named for the trunk-like proboscis of the bulls, southern elephant seals are the largest seals, and one of the largest mammals, on Earth (after elephants and some species of WHALES). Males can weigh up to 3.6 tonnes (3.6 tons) and grow 5 m (16 ft) long; adult females weigh up to 900 kg (1985 lb) and are up to 3 m (10 ft) long.

The seals have a circumpolar distribution and are found in great numbers around the coastal islands and the ANTARCTIC PENINSULA during the feeding season. Deep divers, they have been recorded at depths of 1700 m (5576 ft) in search of SQUID and FISH. They can stay submerged for up to two hours, during which time they survive without oxygen and their heartbeat slows to as little as one beat per minute. They are believed to spend 80 to 90 percent of their lives underwater. During the breeding season, elephant seals survive on their store of blubber.

In August, the males return to their breeding grounds on the SUBANTARCTIC ISLANDS, where they prefer sandy beaches. They are followed about a month later by the females, many of whom now give birth to the single pups they conceived during the previous mating season. The pups, weighing 34–41 kg (75–90 lb), grow rapidly and can quadruple in size by the time they are weaned, about 22 days after birth. During the weeks the female is nursing, she is also fasting and can lose up to 300 kg (662 lb)—about one-third of her body weight. By the time the pup is weaned, females are on heat again and have been grouped together by the fiercest bulls into harems that can number up to 100 females. Vicious fighting between the enormous males continues for the entire breeding season, accompanied by bellicose roaring. At the end of the breeding season

(February-March), elephant seals return to the ocean for about a month to feed, then come ashore again and find meltwater ponds where they moult—an extraordinary event during which they shed both hair and skin.

From the early 19th century, elephant seals were much hunted for their thick blubber, and populations were seriously depleted. However, commercial SEALING is now outlawed and the present population is estimated at around 750,000.

Southern fulmar (*Fulmarus glacialoides*) See ANTARCTIC FULMAR.

Southern giant petrel (*Macronectes giganteus*) Along with the NORTHERN GIANT PETREL, these are the largest of the true PETRELS. Their massive bills with long and conspicuous horny nasal tubes, their bulky appearance and large, wedge-shaped tails are characteristic of all giant petrels. The southern giants are 87 cm (34 in) long and have wing-spans of around 1.95 m (6½ ft). They can be distinguished from northern giants by their reddish-brown bill tips. Some southern birds are also almost completely white, a colour phase that does not occur in northerns.

Because they can walk quite well, they forage on land as well as in the ocean. Well named as 'vultures' of the sea, they readily follow ships and fishing vessels for any sort of carrion; elsewhere they feed by surface-seizing FISH, SQUID and CRUSTACEANS, and will kill birds as large as KING PENGUINS and scavenge on dead animals such as SEALS.

Southern giant petrels breed in large, loose colonies on SUBANTARCTIC and Antarctic islands. At the nest, they are surprisingly timorous, readily deserting it if disturbed.

*Below: Southern giant petrels (*Macronectes giganteus*) are popularly known as the 'vultures' of the sea because of their tendency to follow fishing vessels for carrion.*

South Pole

The geographic apex of Antarctica, the southern axis of the Earth's rotation. The South Pole is located in the midst of the POLAR PLATEAU. Less than 30 cm (12 in) of SNOW accumulates here each year. It piles into drifts and condenses slowly down onto the ICE SHEET, which moves about 9 m (30 ft) per year; at this rate the ICE now at the Pole will reach the coast in about 120,000 years. The WINDS average just 19 km (12 miles) per hour, but TEMPERATURES average a brutal –50°C (–58°F). In winter, it can plummet to below –110°C (–166°F). Although the elevation of the Pole above sea level is 2835 m (9300 feet), the polar LATITUDE makes the air pressure equivalent to 3230 m (10,600 feet). On arriving at the South Pole most people need time to acclimatize, and some experience altitude sickness.

Reaching the South Pole was one of the great goals of exploration. The first party to step onto this isolated spot was led by the Norwegian Roald AMUNDSEN who first reached the Pole on 14 December 1911, followed on 18 January 1912 by Robert SCOTT and his British party.

The AMUNDSEN-SCOTT SOUTH POLE STATION, known as 'The Dome', was built in the early 1970s by the USA, to replace the station erected in the INTERNATIONAL GEOPHYSICAL YEAR. It is located on six of the seven TERRITORIAL CLAIMS and all lines of LONGITUDE, and occupies all TIME zones.

Near The Dome, the Ceremonial Pole, a chromium globe perched on a striped pole, is surrounded by the flags of the 12 original ANTARCTIC TREATY signatories. Because the ice is constantly moving, the 4 m (13 ft) long South Pole 'marker' is moved each January. The new position is calculated using a Global Positioning System (GPS),

which uses SATELLITES to fix its position. A long chain of markers shows the positions of former 'Poles'.

Because of transport difficulties, the USA has *de facto* control of access to the area. A small number of private expeditions make overland journeys, but usually need logistics support. The USA adopts a 'no-support' policy to TOURISM because of the inherent risks involved, and the cost of rescue. For this reason, all tourist flights must be completely self-sufficient. The first tourist flight to land at the Pole was on 11 January 1988, when ADVENTURE NETWORK INTERNATIONAL flew in two DC6 Twin Otters. Adventure tourism remains highly controversial: in 1997 three skydivers plummeted to their deaths over the South Pole when their parachutes failed to open.

Above: Scott, Wilson and Evans beside Amundsen's tent at the South Pole.

Opposite: Because of the slow but continuous movement of the Ice Sheet, the pole marking the geographic apex of Antarctica must be moved periodically.

Below: The entrance to the original Amundsen-Scott South Pole Station, which was buried by snow and abandoned in the mid-1970s.

Southern lights Another name for AURORA AUSTRALIS.

Southern Ocean This ocean encircles Antarctica and contains the point that is furthest from any land—2575 km (1600 miles). The ocean's northern boundary is the ANTARCTIC CONVERGENCE, which isolates Antarctica from the influence of warmer waters to the north. At this boundary the ocean's mainly eastward-flowing waters (known as the CIRCUMPOLAR CURRENT) meet the Atlantic, Pacific and Indian Oceans. Until recently, the Southern Ocean was considered an extension of these three oceans, but in 2000 the Southern Ocean was officially named by the International Hydrographic Organization, and was given political boundaries that coincide with the ANTARCTIC TREATY's limit of latitude, 60°S.

Composed of three major masses of water, two of which are northward-flowing—ANTARCTIC SURFACE WATER and ANTARCTIC BOTTOM WATER—and the south-flowing CIRCUMPOLAR DEEP WATER, the Southern Ocean produces over half of the world's deep ocean WATER. Heat exchange between the ocean, SEA ICE and ATMOSPHERE drives water circulation and WEATHER patterns in the Southern Hemisphere. The ocean is also a major 'sink' of the CARBON DIOXIDE that is discharged into the atmosphere and is carried into deep waters by OCEAN CURRENTS. Oceanographic researchers are beginning to understand the interaction between the Southern Ocean and the Antarctic continent, and how it affects global environmental change.

The dramatic WINDS, WAVES and ocean currents make the Southern Ocean of intense scientific interest. It is also one of the most feared stretches of water in the world. The full fury of a Southern Ocean storm was described by Reginald Ford, a member of Robert SCOTT's 1901–04 NATIONAL ANTARCTIC EXPEDITION: 'The gale has increased in fury and we are finding it hard to keep our ground ... Our anemometer has recorded the wind speed at 90 miles per hour. The gale bears down with its whirlwinds of blinding snow, which smothers our masts and yards and rigging, and drifts in great heaps on our decks.'

The Southern Ocean water is extremely cold and clear. The surface temperatures range from about 4°C (39°F) in the northerly fringes to –1.8°C (29°F) near the Antarctic continent. PHYTOPLANKTON thrive in this environment and form the basis of a rich and diverse FOOD CHAIN supporting FISH, SEALS, WHALES and BIRDS. However, many marine species are under threat from increased ultraviolet radiation from the OZONE HOLE. This has damaged the DNA of some fish species and is thought to have reduced phytoplankton productivity by 15 percent.

Southern Ocean Coalition See ANTARCTIC AND SOUTHERN OCEAN COALITION.

Southern right whale (Balaena glacialis) Also known as Eubalaena australis, southern right whales are BALEENS. One of the major targets of commercial WHALING expeditions—the species was the 'right' whale to hunt because it is a slow swimmer and floats when dead—their numbers were drastically reduced and they are now rarely seen inside Antarctic waters. Right whales do not resemble other Antarctic baleen whale species: they have massive heads and robust bodies, no dorsals fins or throat grooves, and on the head of each individual is a unique pattern of 'callosities' (naturally occurring growths that look as if a series of pale sponges have been stuck on the head). Right whales can grow up to 17 m (56 ft) long and weigh about 65 tonnes (65 tons).

Also unusual is their feeding method. Rather than taking a massive gulp of water and then pushing this water through their baleen plates, right whales skim the water's surface with their jaws open, using their body movements through the sea to force the water through the baleen fil-

Below: Southern right whales (Balaena glacialis) *rarely enter Antarctic waters, but frequent the subantarctic regions of Australia and New Zealand.*

ters. They breed between August and October, in northerly locations such as South Australia, the AUCKLAND ISLANDS, SOUTH AFRICA and Patagonia. Gestation lasts for 12 to 13 months.

Southern right whale dolphin (*Lissodelphis peroni*) This slender, striking DOLPHIN is distinctive for more than its sharp black-and-white colouring: it also lacks a dorsal fin. The name is derived from the right whale, which is also bereft of this fin. Although they are occasionally found just south of the ANTARCTIC CONVERGENCE, this species is primarily found in subantarctic and temperate waters. Often seen in large pods (1000 animals in one estimate), its natural history is still largely unknown.

Southern royal albatross (*Diomedea epomophora*) Like the WANDERING ALBATROSS, royals are large— 1.1 m (3¾ ft) long, with wing-spans of 3–3.5 m (10–12 ft)—and roam great distances throughout the SOUTHERN OCEAN. They feed mostly at night on SQUID, FISH and CRUSTACEANS and are endangered by longlines used to catch tuna, other large fish and sharks. Although similar-looking to wandering albatrosses, with white bodies and some black markings, their backs appear more bowed in flight. A black line along the cutting edge of the upper mandible is a key feature of this species. Males tend to be whiter and slightly larger than females.

The southern royal albatross is endemic to NEW ZEALAND and breeds on CAMPBELL and AUCKLAND ISLANDS. The birds have very long breeding seasons: a successful pair will raise only one chick every two years, and both parents will be involved in the incubating, brooding and raising. Royal albatrosses do not start breeding until they are about eight or nine years old. They usually pair for life and can live to be over 60 years old. The world population is about 26,000 birds.

Southern skua (*Catharacta lonnbergi*) Large, dark brown, rapacious BIRDS, they have a prominent white band at the base of the flight feathers; their bill and webbed feet are black, and they are 63 cm (25 in) long with wing-spans of 1.3 m (4¼ ft). Like all SKUAS, they are opportunistic feeders, taking carrion of any species; they also prey on FISH, small birds and chicks and will feast on eggs when given the opportunity. They have a feeding range up to 95 km (60 miles) from their nesting site and spend most of their days observing PENGUIN nests or flying around colonies looking for exposed eggs or chicks.

Southern skuas lay two eggs in ground nests, which they strongly defend from allcomers. They have a distribution that encompasses many of the SUBANTARCTIC ISLANDS, the FALKLANDS, GOUGH and TRISTAN DA CUNHA islands and the Argentinian coast.

Souvenirs Raising the finance to support proposed expeditions, to fund short-falls and repairs preoccupied many expedition leaders, even while

Right: *A postcard commemorating Ernest Shackleton's 1907–09 'Dash for the South Pole'.*

Above: *The southern royal albatross (*Diomedea epomophora*) breeds on subantarctic islands, where only one chick is raised every two years.*

they were in Antarctica. One method, widely exploited by Richard BYRD among others, was to licence souvenirs, such as 'official publications', engraved silverware and ashtrays, painted ceramics and tins, in order to raise funds. Obviously, the more publicity an expedition received, the better the sales would be. Souvenirs are now highly prized, fetching good prices in shops and sale rooms.

Sovereignty A state has acquired sovereignty over an area when its ownership is undisputed and recognized by other states. In Antarctica, seven states have made TERRITORIAL CLAIMS, and a number of others have reserved the right to make claims in the future. All these claims are disputed by other states. A wide variety of legal grounds are used by the claimant states to support sovereignty status

in Antarctica. These are the same legal grounds that are used to make claims in other parts of the world.

There are five traditional modes of acquisition to *res nullis* (belonging to no one) or *terra nullis* (land belonging to no one), which are universally recognized in international law. Most common historically is that of effective occupation of the claimed territory, such as happened with the 1885 Treaty of Berlin and the partition of Africa. Discovery of land alone is not sufficient to gain sovereignty over it, although it does grant first rights for a period of time: what is important is the intention and the will to act as sovereign, coupled with effective control. A major problem with this method in Antarctica is that there is no consensus on how requirements of effective occupation are to be determined in an environment that

THE DASH FOR THE SOUTH POLE.

is unsuitable for permanent human occupation. Other grounds used to support Antarctic claims include: historic rights, contiguity or proximity, geological affinity, the sector principle, and symbolic acts. CHILE and ARGENTINA claim historic rights derived from the TREATY OF TORDESILLAS and inherited from SPAIN when they gained independence in 1810. The two South American countries also use geological affinity to link the ANTARCTIC PENINSULA and ELLSWORTH LAND to the Andes Mountains.

Contiguity was used in the 19th century for claiming offshore islands, but is difficult when the distance from a claimant country to the coast of Antarctica is measured in hundreds if not thousands of kilometres, as in the case of European claimants.

The sector principle, which is used by all the claimant states with the exception of NORWAY, was applied in Antarctica by drawing meridian lines down from the western and eastern ends of an explored stretch of coastline towards the SOUTH POLE. Symbolic acts—flying flags and depositing claim markers, setting up post offices and selling STAMPS, or flying in important government officials to visit bases—have been common.

The impact of the ANTARCTIC TREATY has been to freeze the *status quo* on sovereignty issues without resolving any of the DISPUTED CLAIMS. In this, the Antarctic Treaty has proved successful by providing a *modus vivendi*, allowing its members to work together for peace and science in Antarctica. In recent decades the COMMON HERITAGE PRINCIPLE has advanced the idea that Antarctica should be considered as a *terra communis* (land that belongs to everyone). This is similar to the internationalist approach adopted by environmental groups calling for a WORLD PARK in Antarctica.

Specially Managed Areas (SMAs) Annex V of the MADRID PROTOCOL allows areas, including marine areas, to be designated as Antarctic Specially Managed Areas (SMAs). The designation is used to 'assist in the planning and coordination of activities, avoid possible conflicts, improve cooperation between Parties or minimize environmental impacts.' SMAs can contain SPECIALLY PROTECTED AREAS, and may include HISTORIC SITES and monuments. Ten years after the Protocol was signed, no SMAs existed, although one has been proposed for ADMIRALTY BAY on King George Island.

Specially Protected Areas (SPAs) In 1960 the SCIENTIFIC COMMITTEE ON ANTARCTIC RESEARCH drew up general rules of conduct for scientific expeditions in Antarctica to minimize the human impact on species and the environment. These rules led to the AGREED MEASURES FOR THE CONSERVATION OF ANTARCTIC FLORA AND FAUNA, which included establishing Specially Protected Areas (SPAs)— areas considered to be 'unique natural ecological systems' or of 'outstanding' value—in which no active scientific research should be undertaken. In 1985 the committee published an atlas of *Conservation Areas in the Antarctic* setting out SPAs and SITES OF SPECIAL SCIENTIFIC INTEREST

(protected areas in which research can be carried out). SPAs are constantly being reviewed and have been extended to include, for example, areas that are valuable for the study of the electromagnetic field, which need to be protected from artificial radio frequency interference.

Entry to and/or any activities in an SPA requires rigorous planning and an ENVIRONMENTAL IMPACT ASSESSMENT. However, this approach has not always been adopted. King George Island, for example, was designated an SPA by the 1966 ANTARCTIC TREATY meeting because of the biological value of its LAKES. But, in the late 1960s, both CHILE and the USSR not only established bases on the island but disregarded general environmental guidelines (one of the lakes was used as a garbage dump); six other nations subsequently constructed bases on the island, mainly as token stations built so that the nations could become CONSULTATIVE PARTIES inside the Antarctic Treaty decision-making circle. Eventually, in 1975, the island's status as an SPA was revoked because its biological value had been so diminished. In the 1990s, there was a proposal to establish a SPECIALLY MANAGED AREA at ADMIRALTY BAY on King George Island, but by 2001 it had not been adopted.

Speckled teal (*Anas flavirostris*) Relatively recent arrivals in the subantarctic zone, these dabbling ducks reached SOUTH GEORGIA from South America in the 1960s. They are one of only two waterbirds on South Georgia. Similar in appearance to YELLOW-BILLED PINTAILS, with mottled brown bodies and yellow-sided beaks, the teals are slightly smaller and far more sociable. They feed on aquatic INSECTS and ALGAE, and nest in GRASSES close to water.

Spectacled porpoise (*Phocoena dioptrica*) The only porpoise to venture south of the ANTARCTIC CONVERGENCE (a few have been recorded near SOUTH GEORGIA), spectacled porpoises have a blunt snout and triangular dorsal fin. Reaching lengths of up to 2.2 m (7 ft), they are similar in colour to most oceanic cetaceans with a dark back and a white belly. There is a fine white line around the eyes, giving the porpoise the appearance of wearing spectacles. Recent records show that this species is found near SUBANTARCTIC ISLANDS in the Atlantic, Pacific and Indian Oceans, but very little is known of its habits and biology.

Sperm whale (*Physeter macrocephalus*) The whale of Herman Melville's *Moby Dick*, although white sperm whales are extremely rare—most are dark grey. The largest TOOTHED WHALE, females grow to a maximum length of 12.5 m (41 ft) and the males to over 18 m (60 ft) and can weigh 70 tonnes (70 tons). Their heads are huge, comprising up to 30 percent of the total body length. Calves are around 4 m (13 ft) at birth. The largest cetaceans to possess teeth, they have up to 25 pairs of conical teeth in their lower jaws and small vestigial teeth in their upper jaws. Sperm whales do not use their teeth in feeding; it is believed they are used as weapons by the males in fights over females. Only mature bulls migrate as far south as the Antarctic.

Sperm whales' blowholes are shaped like sigmas ('Σ', the 18th letter of the Greek alphabet) and located asymmetrically on the left side of the head. Their distinctive chisel-shaped profile contains a wax-filled 'case' which is set on top of one of the most asymmetrical skulls of any vertebrate. The spermaceti wax contained in this case is believed to aid the transmission of sounds and help buoyancy.

Sperm whales do not sink when killed. This made them ideal targets for commercial whalers, who sold the clear, liquid spermaceti wax for processing into candles, fuel and medicines. Inside the whales is another precious substance: ambergris, a solid waxy secretion found in the intestines. Ambergris was used for perfumes, medicines and aphrodisiacs. Although the WHALING of sperm whales is now illegal, animals are still taken by Iceland and Japan. Nevertheless, numbers are slowly increasing.

These giants are found throughout the world's oceans in waters deeper than 200 m (656 ft); they rarely appear in shallow seas. The stomach capacity of bulls is close to 2000 litres (440 gallons). They feed on up to a tonne a day of SQUID and FISH, and the deep, multiple scars often seen on their skin are believed to be inflicted by giant squid. Dives to depths of 2500 m (8200 ft) have been recorded.

While the adult males spend much of the year in cooler Antarctic waters, females and young males stay in warmer waters where they are joined by the bulls in the winter breeding season. Sperm whales are often seen in groups of between 20 and 40, although migrating groups of up to 4000 have been reported. Older large males usually move around alone. Sperm whales have a four-year reproductive cycle, with gestation taking 14½ months, and suckling lasting about two years. Females become sexually mature at about nine years, males at 19. The average life span is 60 years; some animals live into their late seventies.

Sponges (Porifera) The largest group of filter-feeding invertebrates in the Antarctic BENTHOS. Although usually slow growing, some species can reach an enormous size and may live for several centuries. Sponge survival is influenced by ice formation in shallow water and predation by ECHINODERMS at greater depths.

Sponges are simple INVERTEBRATE animals without a nervous system, mouth or digestive sac. Essentially a giant water-sieving system (the Latin 'porifera' means 'pore-bearing'), they have a porous construction with a maze of water channels that is unique among animals. They feed mainly on organic detritus (such as DIATOMS) and PLANKTON and, in some cases, absorb nutrient molecules directly from the water.

Antarctic sponges come in a variety of shapes, sizes and species. They may resemble large fans, volcanoes, bushes, staghorn corals, hedgehogs or cacti. They have evolved a variety of methods to combat predation by SEASTARS, NUDIBRANCHS and other benthic predators. The slimy white sponge (*Myacale acerata*) reproduces and grows quickly: one individual had a measured growth rate of 67 percent in one year then maintained the same size

Above: Sponges like this Dendrilla antarctica *comprise the largest group of filter-feeding invertebrates in the benthos.*

SOUTHERN BOTTLENOSE WHALES, PENGUINS, SEALS, ANTARCTIC TOOTHFISH, BIRDS and FISH—thus, squid are an important component of the BENTHIC and PELAGIC marine ECOSYSTEMS.

Thirty to 40 squid species are found in Antarctic waters, ranging in size from the small *Brachioteuthis* at 15 cm (6 in) to the 10 m (33 ft) long *Mesonychoteuthis hamiltoni*, which uses large hooks as well as suckers to capture prey, and the mysterious giant squid, *Architeuthis dux*, which reaches 18 m (60 ft). Squid have 10 sucker-studded arms and large eyes that allow them to see well in low light. They feed on small fish, CRUSTACEANS such as KRILL, and other cephalopods. Occuring at all depths, they make vertical migrations to the surface, usually at night. They have well-developed nervous systems, and move fast, often in shoals, and are able to change direction very rapidly.

Stamps See PHILATELY.

Starfish See SEASTARS.

Stations See RESEARCH STATIONS.

Stonington Island The site of two abandoned research stations, Stonington Island is located off the west coast of the ANTARCTIC PENINSULA south of ADELAIDE ISLAND. East Base was built by the 1939–41 UNITED STATES ANTARCTIC SERVICE EXPEDITION, and was also used by the RONNE ANTARCTIC RESEARCH EXPEDITION, when the first WOMEN to overwinter in Antarctica, Edith Ronne and Jeannie Darlington, stayed at the base with their husbands. Base E, established by BRITAIN in 1945–46, is about 200 m (656 ft) away.

for the next nine years. White volcano-like sponges, *Rosella nuda* and *Scolymastra joubini*, grow up to 2 m (6½ ft) tall and 1.5 m (5 ft) wide.

Most Antarctic sponges are 'glass sponges', with silicon dioxide skeletons, in contrast to the flexible, absorbent protein skeletons of many tropical sponges. Spiky sponges have long sharp glass needles for protection. Other sponge species, such as the cactus sponge, have chemical defences to discourage predators. Scientists have recently discovered that some of these chemical compounds inhibit the digestive enzymes of sponge-eating seastars.

Springtails (Collembola) The only free-living INSECTS that survive the harsh environment of the Antarctic continent, springtails get their common name from their jumping ability. All eight species of Antarctic springtails are also found on the SUB-ANTARCTIC ISLANDS.

Primitive, iridescent black, wingless insects, springtails are about 1 to 2 mm (0.04–0.08 in) long. They occur, often in dense clusters, on NUNATAKS and in coastal regions of the Antarctic continent, mainly in areas of MOSS and LICHEN where some moisture can be found, and they make up 90 percent of all the insects and MITES around PENGUIN colonies. Springtails avoid freezing by supercooling: keeping their body fluids liquid at temperatures as low as −35°C (−31°F) with the aid of ANTIFREEZE which is synthesized within their body. If ice crystals do form in their body fluids it is fatal. Because food in their gut increases the risk of ice forming, springtails must find a balance between starving and freezing. During cold periods they go without food. Over winter they may become completely frozen in ice, unable to breathe, but are believed to be able to survive for up to a month without oxygen. Some springtails are comatose for over 300 days a year.

The most common species, the Antarctic springtail (*Cryptopygus antarcticus*) is the dominant terrestrial arthropod south of the ANTARCTIC CONVERGENCE. It grows to 1 mm (0.04 in) long,

feeds on microfungi and ALGAE, and is preyed upon by predatory MITES. A single female lays between three and 20 eggs at a time, but hundreds of springtail eggs are often found together, suggesting that springtails aggregate for egg-laying.

Squid Scientists estimate there are about 100 million tonnes of squid in Antarctic seas, of which SPERM WHALES may consume as much as 50 million tonnes (50 million tons) each year. Around 18,000 squid beaks (which are indigestible) have been found in the stomach of a single sperm whale. Squid are also major food sources for

Below: Springtails are the only free-living insect to survive in Antarctica.

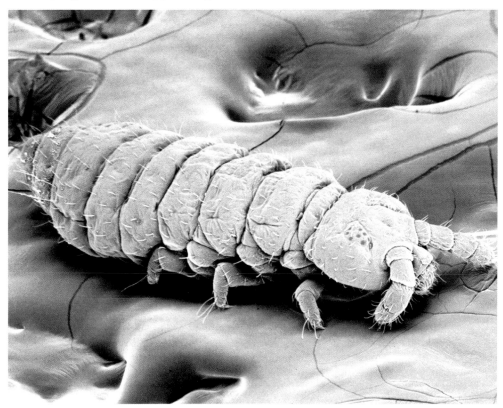

Storm petrels

Storm petrels (Oceanitidae) The smallest of the PETRELS. Of the 10 Southern Hemisphere species, only three are counted as Antarctic petrels: WILSON'S STORM PETREL, BLACK-BELLIED STORM PETREL and GREY-BACKED STORM PETREL. As the storm petrels fly over water they patter on its surface—the word 'petrel' is derived from St Peter walking on water. They are easily distinguished by dancing flight patterns and square-shaped tails, and are noted for their small, slender bodies and very long legs.

Storms See BLIZZARDS, WINDS and SOUTHERN OCEAN.

Streams These occur mainly in OASES in summer, when glacial meltwater flows into LAKES as short-lived, or 'ephemeral', streams. Unlike streams in warmer regions, the flow of Antarctic streams is not influenced by PRECIPITATION. The daily input of meltwater is controlled by the angle at which the SUN hits a GLACIER face. This produces a 24-hour cycle of melt and freeze, during which the flow may stop altogether.

Species that inhabit streams must be equipped to survive being frozen and thawed, often several times a day over summer, as well as a long winter freeze. For this reason no large grazing animals, such as CRUSTACEANS, FISH and INSECTS, live here. However, there are large populations of CYANBACTERIA, which stain the stream beds dark brown; this pigmentation allows them to absorb summer radiation more effectively. Studies in dry stream beds have found that within a few hours of meltwater returning after winter, cyanbacteria 'wake up' and begin to photosynthesize.

Subantarctic fur seal See KERGUELEN FUR SEAL.

Subantarctic islands Scattered across the vast SOUTHERN OCEAN near the ANTARCTIC CONVERGENCE, where large water masses mix to produce wet, cloudy, squally WEATHER, these isolated islands have CLIMATES moderated by the effects of the ocean and mean annual TEMPERATURES ranging from around freezing to about 10°C (50°F). In the path of the powerful ROARING FORTIES, the prevailing westerly WINDS frequently reach gale force. When the wind belts against steep cliffs of the AUCKLAND ISLANDS, for example, it is powerful enough to blow water that has cascaded part way to the ground up again, creating a 'reverse waterfall'. Most subantarctic islands have no permanent snow or ice and PRECIPITATION mainly falls as rain or sleet.

On some islands glacial action has carved and scoured deep fiords and lakes; on others, extinct volcanoes have left dramatic geological features, such as steep collapsed craters and lava flows that radiate to the sea.

Geologically, most of the islands are very young and have a variety of origins. TRISTAN DA CUNHA, GOUGH and PRINCE EDWARD ISLANDS were formed from the buckling of oceanic ridges. Those in NEW ZEALAND terri-tory—the SNARES, BOUNTY, ANTIPODES, AUCKLAND and CAMPBELL ISLANDS—are associated with New Zealand's CONTINENTAL SHELF. MACQUARIE ISLAND is unique in

Above: Typically, subantarctic islands were created by volcanic activity and are edged by forbidding cliffs that rise steeply from the sea.

Below: Auckland Islands' 'Victoria Tree' was named after the ship that replenished emergency supplies left on a number of New Zealand's subantarctic islands for survivors of shipwrecks.

Above: The few hardy trees that grow in the subantarctic do so in 'flag-form', that is, leaning away from the prevailing wind and the harsh sea salt it carries.

that it is the only area in the world where rock from deep within the Earth's mantle has been exposed by PLATE TECTONICS. Volcanic 'hot spots,' such as Îles KERGUÉLEN, HEARD and MCDONALD ISLANDS rest on oceanic plateaux. The only breeding grounds for many species within the Southern Ocean, the subantarctic region is a thriving oasis. MAMMAL and BIRD populations are often greater here than on the Antarctic continent, and modern conservation practices are fostering the recovery of many species that had been ravaged by hunting and by INTRODUCED SPECIES. No known exotic species have lived on Heard, McDonald and Adams Island in the Auckland Islands. On all other subantarctic islands, introduced animals have ravaged native vegetation and wildlife. Rats devour birds' eggs, and rabbits, pigs and goats—brought to the islands as food for sailors and sealers—destroy fragile plant species. House cats were introduced to some of the early meteorological stations to control rodents, or found their way to shore from SHIPWRECKS.

The subantarctic islands are known for the abundant populations of nesting seabirds, including the world's largest population of WANDERING ALBATROSSES on the Auckland Islands, and an estimated 3 million pairs of SOOTY SHEARWATERS on the Snares Islands. The vegetation on most of the islands is lush and green, with TUSSOCK, GRASSES and FLOWERING PLANTS as well as LICHENS, cushion-like MOSS, FUNGI and ribbons of giant KELP.

Subglacial lakes There are up to 70 subglacial lakes, including Lake VOSTOK hidden below thick layers of ice in the Antarctic ICE SHEET. Their extent is being investigated using SATELLITE surveying and RADAR mapping. It is believed that the lakes may have been formed before the Ice Sheet. If this is the case, they have been cut off from the ATMOSPHERE for hundreds of thousands of years, and the sediment on the lake floors would provide a unique archive of the global climate before the last glaciation.

Submarines German submarines used the SOUTHERN OCEAN as a hiding ground during World War II. The U-boat *Pinguin* captured a Norwegian WHALING fleet off the coast of DRONNING MAUD LAND. It was sunk by the British *Cornwall* in 1941, but two other German submarines continued to patrol the Southern Ocean, making excursions north to disrupt enemy shipping.

Early submarine observations of the sea-floor sediment beneath the Southern Ocean were used to support the theory of CONTINENTAL DRIFT.

Summer The austral summer is the season of continual DAYLIGHT and concentrated biological activity. Summer lasts about four months—from November to February. The SUN begins to rise in August, and the PACK ICE breaks up in November. BIRDS begin their courtship rituals and WHALES return to Antarctic feeding grounds. Summer is also the season of most human activity. The population of Antarctica swells to thousands as support staff arrive at the bases to relieve those who have overwintered. They are closely followed by scientists, who come south to participate in research programmes.

Sun In Antarctica, the appearance of the sun is a seasonal phenomenon. In SUMMER it makes no sense to talk about night or morning because the sun is shining as brightly at midnight as it is at noon, circling overhead without dipping below the horizon. With the beginning of the WINTER season, the sun sets and Antarctica is sheltered by the Earth from any glimpse of sunlight until the summer sunrise.

This seasonal pattern can be distorted by the 'Novaya Zemlya' effect, which is caused by refraction of the sun's rays once it has dipped below the horizon. If the rays are parallel with the Earth's curvature, the sun may miraculously 'reappear', depending on atmospheric conditions. In *The Crossing of Antarctica*, his account of the COMMONWEALTH TRANSANTARCTIC EXPEDITION, Vivian FUCHS wrote: 'On the 23 April the sun initially left us and although some hours of twilight continued for many days we decided to postpone some of the seismic work until the return of the moon a fortnight later. We were therefore astonished when we went outside three days afterwards and found the whole base bathed in sunlight! This was due to refraction, but our newborn sun remained over the horizon for only twenty minutes, and the next day the same phenomenon again occurred.'

Sun dogs A name for PARHELIA, types of HALOS.

Sun pillar A type of HALO. Sun pillars take the form of vertical streaks of light, extending above or below the SUN.

Sunrise In mid-August, the sun begins to relieve the WINTER darkness in Antarctica. Gradually, day by day, it lights the continent, until between late September and the end of February it is circling in the sky for 24 hours a day.

Sunset In late March, overwinterers in Antarctica are treated to one of the most spectacular sunsets in the world. Russell Owen, in *The Antarctic Adventure*, marvelled at the event: '... the sun is surrounded by false suns and halos, when gold drips from a rough-edged golden sphere to the snow plain below and ripples like waves on a lake, when the sky is coloured with red and green and blue segments, with the great soft disk of the sun in the center ...'.

Supplies Although modern TRANSPORT allows parties to be sent from a base into the field within hours, they are still subjected to the same environmental conditions as the early explorers faced. All groups must still carry out meticulous preparations to ensure that they have the appropriate supplies and EQUIPMENT. This is even more crucial

Below: A sun pillar stretches upwards from the sun at McMurdo Sound.

Above: Maintaining supplies of food and equipment at Antarctic bases is a difficult logistical operation for the countries involved.

for overland expeditions and field parties that will not be close to a permanent base—they must be completely self-sufficient. Bases are stocked with supplies at the beginning of each season, and over summer re-supply SHIPS and AIRCRAFT bring fresh food and eagerly awaited mail.

Surface water See ANTARCTIC SURFACE WATER.

Surgery Evacuating a sick patient from isolated Antarctic stations may be impossible, and doctors may have to perform surgery on the ice. Large bases have fully equipped hospitals, and station staff are trained in anaesthetics and sterile operating theatre procedures so they can assist doctors.

In some situations, surgery has been performed using improvised equipment and inexperienced assistants. On an expedition across the ROSS ICE SHELF during the 1901–04 NATIONAL ANTARCTIC EXPEDITION, Charles ROYDS performed the first known SURGERY on the continent when he operated on the hand of one of the party, using carbolic toothpaste as an antiseptic.

In 1961, USSR physician Dr L I Rogozov operated on himself for acute appendicitis, while two co-workers held mirrors and surgical equipment. In the same year an Australian doctor, with no neurosurgical experience, performed brain sur-

gery with instruments improvised from a catalogue, and radiogram instructions from a neuro-surgeon in Australia.

Serving as doctor to the OVERWINTERING staff at the AMUNDSEN-SCOTT SOUTH POLE BASE in 1999, Jerri Nielsen made headlines when she discovered a lump in her breast that turned out to be an aggressive, fast-growing cancer. After a daring airdrop of drugs and EQUIPMENT by the US Air Force (see EMERGENCIES), she performed a biopsy and e-mailed photographs of slide samples of the tumour to doctors in the USA. After confirmation she had cancer, she began chemotherapy with the help of colleagues at the base.

Sweden The 1901–04 SWEDISH SOUTH POLE EXPEDITION, led by Otto NORDENSKJÖLD, explored the east coast of the ANTARCTIC PENINSULA, and was the first to voluntarily OVERWINTER in Antarctica. In 1949–52 Sweden took part in the NORWEGIAN-BRITISH-SWEDISH ANTARCTIC EXPEDITION, but during the INTERNATIONAL GEOPHYSICAL YEAR it chose to concentrate its efforts on the Arctic rather than Antarctica.

In 1984, motivated by a desire to express its 'strong support for the basic principles of the Antarctic Treaty', such as the demilitarization of the treaty area, Sweden acceded to both the

ANTARCTIC TREATY and the CONVENTION ON THE CONSERVATION OF ANTARCTIC MARINE LIVING RESOURCES. In 1988 it became a CONSULTATIVE PARTY to the treaty. Sweden maintains two RESEARCH STATIONS in Antarctica, both in DRONNING MAUD LAND.

Swedish Polar Research Secretariat The government organization responsible for managing SWEDEN's research expeditions to both poles. It cooperates with the Swedish Royal Academy's Polar Research Committee in research planning for Antarctic expeditions, and has two Antarctic RESEARCH STATIONS.

Swedish South Polar Expedition (1901–04) Swedish scientific expedition. Financed by businessmen and the Swedish government and led by Otto NORDENSKJÖLD, the *Antarctic* set out under the command of Captain Carl LARSEN from Gothenburg on 16 October 1901. The aim of the expedition was to carry out scientific investigation and exploration on the eastern ANTARCTIC PENINSULA coast.

In February 1902, after a prefabricated HUT for the OVERWINTERING party was erected on SNOW HILL ISLAND near SEYMOUR ISLAND, the ship turned north to the FALKLAND ISLANDS for the winter. Scientific investigations were carried out on boat and SLEDGE journeys, important FOSSIL discoveries were made at Seymour Island and the Prince Gustav Channel to the south of JAMES ROSS ISLAND was explored.

The *Antarctic* left the Falkland Islands on 5 November to pick up the winter party. After being repeatedly trapped in PACK ICE, the ship became ice-bound on 12 February 1903, was crushed and sank (along with much of their SUPPLIES) about 40 km (25 miles) from PAULET ISLAND, which the men reached on foot on 28 February. They were destined to overwinter there, surviving mainly on PENGUIN meat.

Before the *Antarctic* sank, a three-man party from the ship had set off overland in an attempt to reach Nordenskjöld's party at Snow Hill Island. CASTAWAYS trapped in a tent inside a stone hut at HOPE BAY, it wasn't until October 1903 that they encountered the Snow Hill party on Vega Island. At the same time, six of the stranded *Antarctic* men, led by Carl Larsen, set off for Hope Bay. On 7 November, just hours after the Argentinian rescue ship *Uruguay* had found Nordenskjöld and others at Seymour Island, Larsen's party from the *Antarctic* walked into the base. 'No joy can describe the boundless joy of this first moment,' Nordenskjöld wrote. 'I learned at once that our dear old ship was no longer in existence, but for the instant I could feel nothing but joy when I saw amongst us these men.'

Syowa Station Established in 1957, on East Ongul Island off the coast of DRONNING MAUD LAND, as part of JAPAN's contribution to the INTERNATIONAL GEOPHYSICAL YEAR. Syowa has operated continuously since then, apart from the winter of 1958. The station has a summer population of around 40, and can accommodate up to 29 OVERWINTERING.

t

Tardigrades Minute INVERTEBRATE animals, about 0.25 mm (0.009 in) in size, tardigrades are able to survive severe desiccation through a process known as CRYPTOBIOSIS for up to 60 years. Only a few of the 23 species known to live in Antarctica are endemic. Tardigrades have been found surviving in VICTORIA LAND at 77°S.

Flat-bodied, slow-moving animals which typically have four pairs of short legs with claws and a sucking mouth, tardigrades live mainly in moist habitats, such as MOSSES, ALGAE and dead plant material. Densities can reach 14 million animals per sq m (1.2 sq yds). They are mainly HERBIVORES, feeding on living moss cells and plant detritus, but some species also eat ROTIFERS and NEMATODES. Sometimes called 'water bears', like rotifers they also inhabit fresh water.

Tasman, Abel Janszoon (*c.* 1603–1659) Dutch navigator, born in Lutjegast, near Groningen. Tasman was sent by Anthony van Diemen, governor-general of Batavia (present-day Djakarta), on several voyages in quest of the 'Unknown Southern Land', or *TERRA AUSTRALIS INCOGNITA.* He sailed from Batavia on 14 August 1642 in the *Heemskerk* and *Zeehaen* and became the first European to discover Van Diemen's Land (Tasmania), NEW ZEALAND, Tonga and Fiji.

Taylor Glacier This 90 km (56 mile) long GLACIER dominates the Upper Taylor Valley in the DRY VALLEYS. Although in recent geological history it flowed into the ROSS SEA, it has now receded and terminates at Lake Bonney on the valley floor. Unlike others in the area, the glacier is polythermal—that is, it is partly wet-based (a section of its base is wet rather than frozen, as is usual with polar glaciers)—and it has surface movements of up to 14 m (46 ft) per year.

The spectacular Blood Falls, a red-ice frozen waterfall, tumbles from the terminal (the ice cliff at the end) of the glacier. This discoloration is caused by iron oxide deposits and salt precipitating from salty water formed at the base of the glacier and bursting through CREVASSES to the surface.

Robert SCOTT became the first person to cross the line joining the SOUTH MAGNETIC POLE and the geographical SOUTH POLE when he traversed the head of the glacier. He described the Taylor Valley as 'a valley of the dead: even the great glacier which once pushed through it has withered away.' The area is named after Griffith TAYLOR, who was the first scientist to study this area as part of Scott's 1910–13 BRITISH ANTARCTIC EXPEDITION.

Technology The difference between the achievements of early explorers and those of present-day Antarctic expeditions and programmes can be directly related to technological developments. Such inventions as the Nansen Primus stove and ICEBREAKERS, the introduction of mechanized surface TRANSPORT and RADIO communication systems, the use of AIRCRAFT and increased knowledge about, and consequential improvements in, living in polar conditions have all made it easier for humans to adapt to the Antarctic environment.

Technology provides life-support systems: heat, water, food and TRANSPORT. It also plays a key role in the development of specialized scientific equipment, to allow researchers to extend their view from the continent increasingly higher into space, and further down beneath the ICE SHEET. Specialized polar technology allows equipment to be adapted to extreme conditions, such as freezing temperatures and high winds, so that expeditions can venture further into the field. It also helps stations to design waste disposal and power generation facilities that minimize the human impact on the Antarctic environment.

Temperature Antarctica is the coldest CONTINENT on Earth; the world's record low temperature of –91°C (–131.8°F) was recorded in 1997 at VOSTOK STATION. Low temperatures result largely from the enormousness of the Antarctic ICE SHEET and the lack of SOLAR RADIATION absorbed at the surface. Temperatures also decrease with elevation and are lower towards the interior. The mean annual temperature of the POLAR PLATEAU, which is several thousand metres higher than the coastal areas, is –57°C (–70.6°F).

Cold temperatures are often associated with an inversion layer: the air temperature increases

Below: The terminal face of Taylor Glacier dominates the Upper Taylor Valley in the Dry Valleys region.

(rather than decreases) with height above the ground surface. This effect is most exaggerated over the Polar Plateau because the sky is clear and often cloud-free, allowing solar radiation to 'bounce' off the surface. Cooled by the cold SNOW surface, the lowest layer of air can be 30°C (86°F) colder than the air temperature at 1000 m (3280 ft); however, when this layer is disturbed by strong WINDS it mixes with the air above. For this reason, surface temperatures in Antarctica are very erratic and can rise and fall quickly over a few hours. In coastal areas, the SOUTHERN OCEAN has a warming effect and mean annual temperatures are around –17°C (1.4°F). Winter temperatures rarely drop below –40°C (–40°F) and summer temperatures may reach 15°C (59°F). The ANTARCTIC PENINSULA is referred to as the 'banana belt' of Antarctica because it is relatively mild, especially in summer.

Tents A 'polar pyramid', or 'Scott tent', based on a traditional Eskimo skin tent, is still widely used in Antarctic field camps. It can sleep up to three people, although this is quite cramped. The entrance is an oval hole that can be closed shut with a drawstring. The modern version is insulated with a double skin that traps air between the layers. Although a waterproof groundsheet is sewn in, field parties still need to sleep on thick mats, and use double-down sleeping bags with waterproof shells. Along with sleeping tents, most field camps have a large 'mess tent', in which they can cook and sit in comfort. Small mountain tents may also be used for sleeping accommodation.

Terns (Laridae) These gregarious and graceful birds are sometimes called 'sea swallows'. Most terns have grey backs, white underbellies, black caps and forked tails. Although most of the 22 tern species inhabit warm climes, they are found from the Arctic to the Antarctic. Two species breed in the Antarctic and subantarctic regions— the ANTARCTIC TERN and the KERGUELEN TERN— and another, the ARCTIC TERN, makes an annual journey from the northern end of the world. True migratory birds, many tern species move from warmer areas to cooler waters with more abundant summer food resources. They feed mainly on FISH, but will also eat CRUSTACEANS and SQUID, catching their prey close to the surface of the water, usually during the day.

For most terns, breeding takes place in large, noisy, and often densely packed, breeding grounds. Pairing usually lasts for life, which can be over 30 years, and pairs will continue to return to successful breeding sites. Courtship begins with intricate and visually breathtaking displays of gliding and high, fast dives. After this, the courting male feeds the female before continuing ground displays of pirouettes and strutting. Terns lay one to three eggs, depending on location, and both parents share incubation, using voice recognition to locate and feed their chicks. Famously noisy, terns use a variety of calls, clicks and mews to communicate. Predated by SKUAS, terns band together to protect nesting sites.

Terra Australis Incognita The notion of Terra Australis Incognita, or an 'unknown southern land', originated with the early Greeks. In the fourth century BC Aristotle postulated the theory that the Earth is spherical, and the idea that a southern polar region existed arose from the feeling that the two hemispheres must be counterbalanced. The name 'Antarktos' means opposite the Bear, the northern constellation that contains the pole star Polaris. By the 17th century, as the navigation and shipbuilding techniques of the European nations developed and the New World was 'discovered', voyages of exploration extended geographical knowledge. But when Edmond HALLEY sailed from Britain in the Paramore in 1698 with the aim of determining the position of Terra Australis Incognita, an enormous area of the Southern Hemisphere was still unknown.

In the mid-18th century, the French mathematican Pierre Louis Moreau de Maupertuis introduced the idea that, because of its isolation, if a southern continent did exist, the land and things on it would be dramatically different from what was known elsewhere. Within two decades James COOK had circumnavigated the Antarctic continent.

Terra Nova British whaling ship, built in Dundee, Scotland, in 1884, weighing 858 tonnes (858 tons). A relief vessel for the 1901–04 NATIONAL ANTARCTIC EXPEDITION, it was towed halfway across the world at a speed which 'must have surprised the barnacles on her stout wooden sides,' according to Robert SCOTT.

Scott acquired the Terra Nova for the 1910–13 BRITISH ANTARCTIC EXPEDITION, and appointed Lieutenant Edward EVANS as captain. At the port of Lyttelton, NEW ZEALAND, it spent a month in repairs, leaving in December 1910, six weeks behind schedule. It was further delayed by unseasonably heavy PACK ICE. The Terra Nova sailed into MCMURDO SOUND in three successive Januarys: in 1911 to allow the explorers to disembark; in 1912 to resupply them; and in 1913 to bring them home.

On 12 September 1943, the ship was crushed by ice and sank off the Greenland coast.

Terra Nova Bay The bay formed by the Drygalski Ice Tongue and the Campbell and Priestly Glaciers on the VICTORIA LAND coast north of MCMURDO SOUND, and named after the ship of the 1910–13 BRITISH ANTARCTIC EXPEDITION. Today, Terra Nova Bay is the site of ITALY's only research base in Antarctica, a summer-only station that can accommodate up to 50 people.

Terre Adélie An area of EAST ANTARCTICA between GEORGE V LAND and WILKES LAND (136°E to 142°E), first sighted by Jules-Sébastien DUMONT D'URVILLE on 19 January 1840. D'Urville delayed naming the land (which bears his wife's name) until the expedition had been able to explore the coast further and confirm it was land rather than a large mass of ice. On 24 March 1924 FRANCE issued a decree exerting a TERRITORIAL CLAIM over Terre Adélie. Terre Adélie has the Antarctic continent's thickest ice, which extends to a depth of 4750 m (15,580 ft).

Terres Australes et Antarctiques Françaises Established by FRANCE on 6 August 1955 to administer its Antarctic TERRITORIAL CLAIMS, which previously had been under the jurisdiction of the French governor of Madagascar.

Territorial claims Although the continent has never been permanently settled, conventional territorial claims to Antarctica were made during the first half of the 20th century. The first defined claim was made by BRITAIN in 1908 to the region

Below: Small mountain and pyramid tents are the tents most commonly used in Antarctica today.

Above: The Terra Nova, *a veteran of two British Antarctic expeditions and three successive sailings into McMurdo Sound.*

from 20°W to 80°W and below 60°S. Despite this, there are overlapping claims by ARGENTINA and CHILE. The British followed with a 1923 claim on behalf of NEW ZEALAND (then with British Dominion status) to the ROSS DEPENDENCY in the area from 150°W to 160°E and below 60°S. In 1933 and 1936 similar claims were made for AUSTRALIA to the AUSTRALIAN ANTARCTIC TERRITORY in the area 45°E to 160°E and below 60°S; this excludes TERRE ADÉLIE between 136°E to 142°E, which was claimed by FRANCE in 1924. The Australian claim covers about 42 percent of the Antarctic continent.

In 1939 NORWAY claimed DRONNING MAUD LAND between 20°W to 45°E, but did not specify a northern limit to the claim.

Chile claimed the area between 53°W and 90°W in 1940. An earlier undefined Chilean claim dates back to 1906.

In 1943–47 Argentina defined its claim to the area 25°W to 74°W, below 60°S. Its claim to the SOUTH ORKNEYS had been made in 1925, and to the FALKLAND ISLANDS region in 1937—both of which had been claimed by Britain several decades earlier.

The only remaining land unclaimed is in MARIE BYRD LAND, between 90°W and 150°W, and the hinterland of the undefined Norwegian claim in Dronning Maud Land.

Because the grounds for establishing SOVEREIGNTY over a territorial claim in Antarctica are unclear, all the claims are disputed. In addition, the territories of Argentina, Britain and Chile all overlap each other in the ANTARCTIC PENINSULA

and so are DISPUTED CLAIMS. Since the signing of the ANTARCTIC TREATY in 1959, all claims have been effectively 'frozen'.

See also LAWS, INTERNATIONAL, REGARDING ANTARCTICA.

Territories Seven states had made TERRITORIAL CLAIMS to areas below 60°S before 1959, when the ANTARCTIC TREATY was signed. A number of other countries, including the USSR and the USA, either declined to claim SOVEREIGNTY in Antarctica or DISPUTED CLAIMS already made. The treaty accepted the *status quo* on claims: they are neither recognized nor disputed.

Most of the states claiming territory in Antarctica have organized the claimed areas into administrative units that are part of their governments. They are ANTÁRTIDA ARGENTINA, the AUSTRALIAN ANTARCTIC TERRITORY, the BRITISH ANTARCTIC TERRITORY, the TERRITORIO CHILENO ANTÁRTICO, TERRES AUSTRALES ET ANTARCTIQUES FRANÇAISES. NEW ZEALAND and NORWAY manage their claims as dependencies: ROSS DEPENDENCY and DRONNING MAUD LAND respectively.

Territorio Chileno Antártico Although CHILE announced in 1906 that a TERRITORIAL CLAIM was to be defined in the Antarctic, it was not until 1940 that it formally claimed Territorio Chileno Antártico, a wedge-shaped sector running between 53°W and 90°W, down to the SOUTH POLE, with no northern boundary specified—an area more than one and a half times larger than mainland Chile. Like ARGENTINA, Chile makes ref-

erence to the 1494 TREATY OF TORDESILLAS in claiming that it inherited historic rights to SOVEREIGNTY over Antarctic territory from SPAIN.

Chilean territorial claims in Antarctica are disputed by ARGENTINA and BRITAIN, and Chile and Argentina also had a long-running dispute over the Beagle Channel Islands in the DRAKE PASSAGE that was not resolved until the 1980s. The DONOSO-LA ROSA DECLARATION of March 1948 affirmed the existence of a South American Antarctic to which only Argentina and Chile possessed rights of sovereignty, and this was reaffirmed with the ACT OF PUERTO MONTT in February 1978. One of the original CONSULTATIVE PARTIES to the ANTARCTIC TREATY, Chile has always opposed efforts at internationalizing Antarctica and it continues to maintain the legitimacy of its claim. In January 1977, when President Augusto Pinochet visited Antarctica, he declared Chile's claim was a continuation of its mainland territory.

Terror, Mount An extinct volcano on ROSS ISLAND, Terror was discovered along with the nearby Mount EREBUS in 1841 by James Clark ROSS, who named the two mountains after his ships.

Thaler Committee In 1984, pressure on the government of FRANCE about the environmental impact of building an AIRSTRIP in the hard rock near the DUMONT D'URVILLE BASE led to the appointment of a *comité des sages*, which became known as the Thaler Committee.

The Thaler Committee had eight members: two foreign scientists and six French members. Its role was to examine the ENVIRONMENTAL IMPACT ASSESSMENT published on the airstrip project and to make recommendations. Its conclusions were kept secret for several months, but were finally released in September 1984. The Thaler report recommended that the airstrip project be abandoned, that consideration be given to less harmful options, and that a new impact study be done.

Despite the second environmental impact assessment acknowledging potential damage to the BIRD colonies that were in the way of the proposed airstrip, France resumed construction, which involved blasting the tops off five small islands and linking them with the left-over gravel. In 1994 a tidal wave caused by falling ice damaged the airstrip, which was abandoned.

Thin-billed prion (*Pachyptila belcheri*) Small birds, closely related to the ANTARCTIC PRION, they have distinctive facial patterns that are discernible at close range. The palest PRION, they are pale grey. Only 26 cm (10 in) long, with wing-spans of 61 cm (24 in), their blue bills are long and slender and their feet pale blue with cream webbing. In flight, they are fast and stay close to the water.

Also known as slender-billed or narrow-billed prions, these birds breed on Îles KERGUÉLEN, the FALKLAND ISLANDS and Îles CROZET. They range widely in Antarctic and subantarctic waters and are common around the southern tip of South America in winter. They feed at night by surface-seizing CRUSTACEANS, and also take some small SQUID and FISH.

Thomson, Charles Wyville (1830–82) Scottish marine biologist and oceanographer, born in Bonsyde, Scotland. He conducted scientific work off the northwest coast of the British Isles in 1868 before becoming professor of natural history at Cork, Belfast and Edinburgh universities. He persuaded the ROYAL SOCIETY to mount an expedition to discover 'everything about the sea', and in 1872 he was appointed chief scientific officer on the *CHALLENGER* EXPEDITION, the round-the-world voyage that is regarded as having founded the science of oceanography. The *Challenger* spent four months in and around the SOUTHERN OCEAN and in late February 1874 reached 72°22'E, 66°33'S—the first steamship to cross the ANTARCTIC CIRCLE.

On the expedition's return, Thomson described his deep-sea researches in *The Depths of the Ocean* and published *The Voyage of the Challenger* in 1877. He was engaged in editing the expedition's scientific findings when he died; the work was completed by John MURRAY.

300 Club An elite group of streakers at the SOUTH POLE who wait until the winter TEMPERATURE has dropped to below –73°C (–100°F), take a 93°C (200°F) sauna then run over the ice wearing nothing but shoes. The temperature drop is 300°F (or 150°C).

Thurston Island A large ice-covered island lying between the BELLINGSHAUSEN and AMUNDSEN SEAS at the northwest end of ELLSWORTH LAND. It was discovered by Richard BYRD on a 1940 flight, and

charted as a peninsula. It was found to be an island in 1960.

Tides Tidal changes in water level around the Antarctic coast are difficult to predict. They change ICE SHEET elevations by 1–2 m (3–7 ft), and during spring the FILCHNER and LARSEN ICE SHELVES by more than 5 m (16 ft). ICE SHELVES tend to flex in response to these water-level changes, and it has been suggested that this may have even deflected the moon.

According to Christopher Doake of the BRITISH ANTARCTIC SURVEY, 'It is generally agreed that tidal friction slows the rotation of the earth and decelerates the moon in its orbit. Flexing of ice shelves may dissipate enough tidal energy to have a significant effect on the moon's retreat from the earth.' Rapid ice movement around the coast makes taking direct tidal measurements difficult. A series of gauges installed at coastal research stations relay tidal data by SATELLITE to provide information about sea-level variations.

Tierra del Fuego The archipelago that lies at the southern tip of South America. In 1520 Ferdinand MAGELLAN discovered the strait, which bears his name, separating Tierra del Fuego from the South American mainland. This discovery rekindled speculation about the existence of a large undiscovered southern continent and many mapmakers believed Tierra del Fuego was the northern tip of *TERRA AUSTRALIS INCOGNITA*.

The name Tierra del Fuego—Spanish for 'Land of Fire'—comes from fires lit by the archipelago's indigenous people that were seen from European ships.

Tierra del Fuego is buffeted by Antarctic storms, particularly on its mountainous southern coast. The archipelago is made up of one large island, Isla Grande, and many smaller islands, only a few of which are inhabited. Possession of Tierra del Fuego is shared between ARGENTINA and CHILE.

Time Because all the global time zones converge on Antarctica, there is no real time zone for the continent itself. Most Antarctic bases run on their home time, or the time at their logistics bases. For American bases, for example, it is easiest to run on NEW ZEALAND time, as this is where their resupply flights depart from. SHIPS tend to stay on the same time as their port of departure. Except for coordinating flights and radio schedules with base stations, the time of day is irrelevant to summer field parties, as they have the luxury of 24 hours of sunlight to work in. Often groups adjust their timetables so that they work at night and sleep into the morning.

Titan Dome An ice dome near the SOUTH POLE at approximately 88°30'S, 165°E. With an elevation of approximately 3100 m (10,168 ft) above sea level, Titan Dome is one of the three highest areas on the POLAR PLATEAU.

Below: Beech trees (Nothofagus) are common to all the lands that once formed part of Gondwana, including Tierra del Fuego.

Above: Tourists look out over the Lemaire Channel, which runs between the Antarctic Peninsula and Booth Island and is nicknamed 'Kodak gap' for its popular photogenic qualities.

Toothed whales The best known species of Odontocete or toothed whales found south of the ANTARCTIC CONVERGENCE are the SPERM and KILLER WHALES. Of the others, the largest are BAIRD'S BEAKED WHALE (12 m/39 ft and 13–15 tonnes/13–15 tons), ARNOUX'S BEAKED WHALE (10 m/33 ft long and 7–8 tonnes/7–8 tons) and the SOUTHERN BOTTLENOSE WHALE (7 m/23 ft and 3–4 tonnes/3–4 tons). Smaller species include the HOURGLASS DOLPHIN, SOUTHERN RIGHT WHALE DOLPHIN, LONG-FINNED PILOT WHALE and SPECTACLED PORPOISE.

Unlike other mammals, toothed whales have only one nostril, although they have two internal nasal passages. Some, including sperm and beaked whales, have the ability to undertake deep and prolonged dives. SQUID are an important food source and sperm whales catch them at great depths. Killer whales (also known as orcas) are voracious hunters, feeding on FISH, SEALS, PENGUINS, BIRDS and other whales.

Of sperm whales, only the adult males venture into Antarctic waters. Killer whales are abundant in the SOUTHERN OCEAN, in open water and in PACK ICE, and southern bottlenose and Arnoux's beaked whales are found near pack ice and around SUBANTARCTIC ISLANDS. Pilot whales are widely distributed in the subantarctic latitudes of the Southern Ocean, and hourglass dolphins are seen around the southern coast and offshore islands of South America. Little is known about the distribution of the other toothed whales, or about their breeding cycles.

Toothfish fishing In the mid-1980s longlines were introduced to catch PATAGONIAN TOOTHFISH. Called 'white gold' for their high value, in 2001 toothfish were fetching as much as US$10 per kg (just over 2 lb).

The toothfish fishery has suffered from the problem of illegal, unregulated and unreported fishing (known as IUU fishing) and, as a result of over-exploitation, some stocks of Patagonian toothfish are already commercially extinct. Illegal fishing usually takes place using vessels registered under a flag of convenience, which makes enforcement of regulations difficult. The 'pirate' fishing vessels do not take precautions to reduce incidental BY-CATCH mortality of BIRDS on the hooks of the longlines and, consequently, several bird species are now endangered.

Organizations such as GREENPEACE have called for a moratorium on all fishing for toothfish in the areas managed by the CONVENTION ON THE CON- SERVATION OF ANTARCTIC MARINE LIVING RESOURCES until the fishery is brought under control.

Tourism Tourism in Antarctica began in the late 1950s and is now a significant commercial activity. There are three main forms: CRUISES on SHIPS or YACHTS around the coast of Antarctica, overflights of the continent on AIRCRAFT, and landing on the continent itself for extended land expeditions. Ship-based tourism is the most common.

'Expedition cruising' was pioneered by Lars-Eric Lindblad, who incorporated educational elements into the cruises and in 1969 launched the *Lindblad Explorer*, a liner specifically built to carry tourists to Antarctica.

The first tourist flight to Antarctica was conducted from CHILE in 1956. In 1977 both Qantas and Air New Zealand began regular sightseeing flights over the continent, but after the 1979 Air New Zealand DC10 crash at Mount EREBUS that killed 257 people, overflights ceased until 1994. A scenic flight takes up to 12 hours, of which four hours are spent over the continent. Videos and talks by Antarctic experts fill the rest of the time. Flights take place in the summer only.

Land expeditions operate out of PATRIOT HILLS in the ELLSWORTH MOUNTAINS and from DRONNING

Above: Uplifted about 45 million years ago, much of the Transantarctic Mountain range lies beneath the ice cover; only the peaks are visible.

MAUD LAND and may include MOUNTAINEERING, camping, SKIING, and even visits to the SOUTH POLE. Groups are small due to the costs and risks involved with 'adventure tourism'. In 1998 three people were killed while skydiving at the South Pole.

In the 1999–2000 season the INTERNATIONAL ASSOCIATION OF ANTARCTIC TOUR OPERATORS estimated that 14,762 people travelled to Antarctica with private expeditions, mostly on cruises. Although the number of visitors is expected to continue growing, Antarctica will probably remain an expensive and therefore relatively limited tourist destination.

As tourism has grown so have concerns about the potential impact on the environment, the safety risk to tourists, and possible disruption to SCIENTIFIC RESEARCH activities. The bulk of tourist visits are concentrated in a few locations, and the desire for photographs can put pressure on the fragile environment, crushing minute PLANT life or disrupting BIRD breeding. Not all bases will accept visitors; the argument is that a disaster involving

a large number of tourists would divert transport and RESCUE crews needed to support research. The opposite argument is that tourism enhances public awareness of Antarctica, the value of its unique environment and the importance of Antarctic science.

Tractors Douglas MAWSON's 'air-tractor', a Vickers REP monoplane which crashed in 1911 during a demonstration flight in Adelaide before it had even arrived in Antarctica, was the first AIRCRAFT in Antarctica—albeit no flight was attempted with it there. It was also the first tractor: the strange machine managed several short land journeys at Cape DENISON before being abandoned.

By the late 1950s tractors were operating successfully in Antarctica. On the COMMONWEALTH TRANSANTARCTIC EXPEDITION, Edmund HILLARY drove modified Massey Ferguson farm tractors, fitted with tracks for use in the snow, to the SOUTH POLE, and Vivian FUCHS used the more lumbering, purpose-built SNO-CATS for the first transcontinental crossing. Now, bulldozers up to 30 tonnes (30

tons), with wide tracks and specialized fuel systems, are used for long overland journeys—each can pull a train of sledges carrying fuel, generators, living caravans and laboratories slowly over the ICE SHEET for thousands of kilometres.

Transantarctic Mountains The longest mountain chain in Antarctica, the Transantarctic Mountains run north-south, along the eastern side of the ROSS ICE SHELF and stretch to COATS LAND on the WEDDELL SEA. They divide the giant Antarctic ICE SHEET into eastern and western sheets. Extending for more than 3200 km (2000 miles), the mountains rise to 4530 m (14,860 ft), include some of the world's largest GLACIERS and conceal the DRY VALLEYS region near the ROSS SEA. Branches splaying out from the main chain of mountains include the SHACKLETON, PENSACOLA, HORLICK, QUEEN ALEXANDRA, BRITANNIA, Prince Albert and Admiralty ranges.

During the early Tertiary period, about 45 million years ago, the Transantarctic Mountains were uplifted. About 15 million years ago, they were buried by the growing Ice Sheet, and now only the peaks show above the ice cover. Made up of horizontal layers of sedimentary rocks, the mountains contain some of the world's largest coal deposits, the seams of which hide fossilized remains of species that inhabited Antarctica in warmer climates. The USA, BRITAIN and NEW ZEALAND have operated RESEARCH STATIONS along the base of the mountains.

Both Ernest SHACKLETON and Robert SCOTT toiled through the Transantarctic Mountains on SLEDGING journeys, and many of their successors, including Vivian FUCHS and Edmund HILLARY, negotiated routes through the icy passes. 'These mountains are not beautiful in the ordinary acceptance of the term, but are magnificent in their stern and rugged grandeur,' said Shackleton.

Trans-Globe Expedition (1980–81) Private three-man British global expedition led by Ranulph Fiennes—the first surface journey around the world via the two poles. The expedition took seven years of planning and three years of travel, covered a total of 160,000 km (99,000 miles), and achieved only the second crossing of the Antarctic continent. The Antarctic part of the expedition took place in 1980–81. The group consisted of Fiennes, Charles Burton and Oliver Shepard.

They left SANAE BASE in DRONNING MAUD LAND on 29 October 1980 and arrived at SCOTT BASE on the ROSS SEA 75 days later, on 11 January 1981, after covering 3600 km (2232 miles), with air support provided by ADVENTURE NETWORK INTERNATIONAL. En route, the team made valuable scientific observations, including taking an ice CORE SAMPLE at every degree of LATITUDE.

Transport The first journeys to Antarctica were undertaken in small wooden sailing SHIPS and sturdy WHALING boats—a far cry from today's summer-time flights in purpose-built AIRCRAFT. Most countries operating in Antarctica use a mixture of land- and sea-based transport, along with a network of AIRCRAFT, including HELICOPTERS. The logistics of ferrying staff, scientists and equipment to Antarctica requires a great deal of planning and skill. Many supplies are transported by sea on strengthened vessels that sail through the ice in channels cut by ICEBREAKERS. Large cargo aircraft also carry huge amounts of EQUIPMENT and personnel to the continent every season.

Travel within Antarctica involves negotiating the most difficult terrain in the world in the most inhospitable conditions. Traditionally, SLEDGES were man-hauled or pulled by DOGS—and, less successfully, other animals, such as PONIES. Explorers trialled various motorized VEHICLES: the most reliable have proved to be customized TRACTORS, first used extensively in the 1950s. Now light aircraft and helicopters are used to travel to field sites, and an array of vehicles have been adapted to Antarctic conditions for 'over-snow' transport.

Treaties The first known treaty to mention Antarctica was the 1493 TREATY OF TORDESILLAS in which the rival states of SPAIN and PORTUGAL agreed to a longitudinal boundary between 46°W

Above: Whether Hägglund, helicopter or skidoo, transport in Antarctica must be reliable and able to cope with the world's most inhospitable conditions.

and 49°W, 'from the Arctic to the Antarctic Poles'. The ANTARCTIC TREATY, which was signed by 12 countries and has since been acceded to by 32 other states, has operated continuously since it entered into force on 23 June 1961. The treaty is an innovative approach to overcoming DISPUTED CLAIMS about territory that had previously prevented agreement on how to manage Antarctica. In addition, a number of international CONVENTIONS govern the management of RESOURCES in Antarctica and the SOUTHERN OCEAN.

Treaty of Tordesillas The 1493 Treaty of Tordesillas was the result of an arbitration attempt by Pope Alexander VI concerning a dispute over territories discovered by PORTUGAL and SPAIN. The two countries agreed to a longitudinal boundary between 46°W and 49°W—'from the Arctic to the Antarctic Poles'—with Spain gaining the territories to the west of the boundary and Portugal those to the east of the boundary. Reference is made to the Treaty of Tordesillas by both ARGENTINA and CHILE in claiming inherited historic rights to SOVEREIGNTY over TERRITORIAL CLAIMS in Antarctica from Spain.

Tristan da Cunha Located midway between South Africa and Brazil, Tristan da Cunha rises to a 2060 m (6760 ft) high volcanic cone, and is the world's most remote inhabited island. Smaller islands in the group are scattered to the south, and include Inaccessible and Nightingale Islands. The only permanent human settlement is at Edinburgh, on Tristan, which has a population of around 300. The Portuguese admiral Tristão d'Acunha discovered the islands in 1506, and in 1775 Guilleme Bolts landed and claimed sovereignty for Austria. Sealers arrived in 1790, and by 1802 the mammals were obliterated.

Tundra A vast treeless zone lying between the ice cap and the timberline of North America and

Eurasia, which has a permanently frozen subsoil. Many of the SUBANTARCTIC ISLANDS also possess tundra-like vegetation with a limited range of short FLOWERING PLANTS and ferns, very few shrubby species and no trees.

Tussock The characteristic and dominant vegetation on SUBANTARCTIC ISLANDS. Typically, the islands have a coastal fringe of tall tussock grass (*Poa* spp.), that covers most of the available soil near the shoreline and extends to an elevation of around 250 m (820 ft). Tussock grows close together in thick clumps and can reach 2 m (6½ ft) in height. Clumps of tussock provide good habitats for BIRDS such as PETRELS, which make their nests by burrowing into the hillsides. The tussock provides shelter from wind and rain and the birds, in turn, provide guano as a source of nutrient fertilizer for the growing tussock.

22° halo Type of OPTICAL PHENOMENON, the most common HALO over Antarctica: if a line were extended from two opposite outer edges of the halo back to an observer's eye, the angle would be 44°. Formed from the interaction of light with plate-shaped crystals contained in water vapour in the upper atmosphere, this halo occurs when the crystals lie flat and in line with the rays of the sun or moon.

Tyree, Mount Antarctica's second highest peak, 4845 m (15,896 ft) above sea level. It lies 13 km (8 miles) northwest of the VINSON MASSIF along the main ridge of the SENTINEL RANGE. It was discovered by US Navy pilots in January 1957, and named after Rear Admiral David Tyree, who was head of US Antarctic operations from 1959 to 1962.

Regarded as Antarctica's most difficult peak, Mount Tyree was first climbed by members of a private expedition led by Nicholas Clinch on 6 January 1967.

Ukraine After the USSR broke up in 1991, RUSSIA assumed its place in the ANTARCTIC TREATY SYSTEM, and the Ukraine acceded to the ANTARCTIC TREATY in its own right. The Ukraine operates VERNADSKY STATION, which was transferred from BRITAIN on condition that the long-term monitoring of ATMOSPHERE and CLIMATE would continue.

Ultraviolet radiation Emitted by the SUN, ultraviolet (UV) light is hot and damaging to many forms of life. Most ultraviolet radiation is absorbed in the stratosphere (see ATMOSPHERE) by OZONE. However, as ozone is depleted over Antarctica and the OZONE HOLE increases, the stratosphere is becoming less effective at filtering harmful UV light. In the SOUTHERN OCEAN, it has disrupted the photosynthesis of PHYTOPLANKTON, which absorbs light near the water surface; ultraviolet radiation in the form of high-energy UV-B penetrates the ocean to a depth of about 20 m (66 ft) and researchers have found that in spring and early summer the productivity of phytoplankton is 6 to 12 percent less in the 'ozone-hole zone' than in the area outside the zone. This has the potential to disrupt the entire marine ecosystem, as a decrease in this species could be relayed through the FOOD WEB to larvae, FISH, BIRDS and MAMMALS. Biologists have also found that UV-B can alter the DNA of marine and land mammals, and the eggs and larvae of ICEFISH, a unique species lacking in haemoglobin. In humans, UV-B can weaken immune systems, and cause sunburn and possibly cancer.

United Nations The United Nations (UN) was established on 24 October 1945 by countries seeking to preserve international peace in the wake of World War II. In 2001 it comprised 189 member states. The UN has been involved in the broad development of international environmental law, which has influenced developments in Antarctica, but at various times there have been calls for the UN to play a greater role in the management of the continent.

In the 1940s efforts by the USA to declare Antarctica an international trusteeship through the UN failed due to opposition by states with TERRITORIAL CLAIMS in Antarctica. Of the claimant states, only NEW ZEALAND ever expressed support for the trusteeship proposal. INDIA attempted to place the topic of Antarctica on the UN Agenda in 1956–58, but was persuaded to withdraw the proposal as negotiations for the ANTARCTIC TREATY were progressing.

In the 1970s, development of the COMMON HERITAGE PRINCIPLE saw renewed interest in the potential RESOURCE wealth of Antarctica. The principle's supporters sought the establishment of a UN management regime over Antarctica that would allow broader participation in decision-making about the continent, and particularly the sharing of its potential MINERAL wealth. Sri Lanka's attempt to have Antarctica included in the UN Convention on the LAW OF THE SEA negotiations failed because it was felt that Antarctica would further complicate already difficult negotiations. Anxiety among the CONSULTATIVE PARTIES about the possibility of UN arrangements for Antarctica was one impetus for negotiating the CONVENTION ON THE REGULATION OF ANTARCTIC MINERAL RESOURCE ACTIVITIES (CRAMRA). The interest of the UN Food and Agriculture Organization (FAO) in the marine resources of the SOUTHERN OCEAN was a spur to the development of the CONVENTION ON THE CONSERVATION OF ANTARCTIC MARINE LIVING RESOURCES.

In 1983 the 'question of Antarctica'—intended to explore a 'proper and representative international regime beyond the Antarctic Treaty'—first appeared on the agenda for debate at the UN General Assembly, and it has appeared at regular intervals since then. Early compromises ended in 1985 when the consultative parties boycotted UN votes on Antarctica.

The consultative parties have been criticized in the UN by countries such as Malaysia for the secrecy of the ANTARCTIC TREATY SYSTEM (ATS), the exclusivity of membership, and for including the then-apartheid regime of SOUTH AFRICA. The consultative parties responded by inviting observers to ATS meetings and publishing key documents. Membership of the ATS expanded in the 1980s as countries such as BRAZIL, CHINA and INDIA joined as consultative parties, thus increasing the proportion of the world's population represented. Interest in Antarctica's resource wealth declined when CRAMRA failed to be ratified and was replaced by the MADRID PROTOCOL in 1991.

United Nations Convention on the Law of the Sea See LAW OF THE SEA.

United States Antarctic Developments Project (1946–47) See OPERATION HIGHJUMP.

United States Antarctic Expedition (1928–30) See BYRD'S FIRST EXPEDITION.

United States Antarctic Expedition (1933–35) See BYRD'S SECOND EXPEDITION.

United States Antarctic Service Expedition (1939–41) Official expedition, led by Richard BYRD, to consolidate USA exploration in Antarctica. The aims were largely strategic and included examining the area bordered at 80°W by GRAHAM LAND and at 150°W by the ROSS DEPENDENCY. Byrd was instructed by President Franklin D Roosevelt to make TERRITORIAL CLAIMS at various locations. Brass plaques declaring the presence of USA personnel were screwed into Mount McKinley and other features.

The expedition consisted of 59 men, 130 DOGS, three AIRCRAFT, two light army tanks and two TRACTORS. Most of the technicians came from the American Army, Navy and Marine Corps; the scientists were civilian. Two ships, the *North Star* and the *Bear*, transported the expedition to the BAY OF WHALES. A base station, the West Base, was established about 8 km (5 miles) from LITTLE AMERICA, Byrd's former base. The East Base was constructed 3200 km (1987 miles) to the east, on STONINGTON ISLAND.

A considerable amount of scientific work was accomplished during 1940. Scientists from disciplines including geology, glaciology, meteorology, seismology and communications collected data. Aerial photography played an important part in surveying and the SHACKLETON GLACIER was discovered. From the West Base, research parties surveyed the Edsel Ford Mountains, the Rockefeller Mountains and the coast. At the East Base a weather station was set up, and SLEDGING parties went to the Eternity Range, the WEDDELL SEA coast and King George VI Sound.

The expedition trialled a giant 'snow cruiser' that was 16.7 metres (55 ft) long and weighed 35 tonnes (35 tons). It contained living quarters and carried a year's supply of food and fuel, and a small aeroplane on its roof. Although it had performed well in tests on sand in the USA, it proved too heavy for snow and moved less than 5 km (3 miles).

United States Antarctic Program In 1971 the NATIONAL SCIENCE FOUNDATION was given overall responsibility for managing the USA's Antarctic activities in addition to the research responsibilities it had assumed in 1959 with the US ANTARCTIC RESEARCH PROGRAM. Each year around 3000 Americans are involved in the US Antarctic Program.

United States Antarctic Research Program Following the success of co-operative research during the INTERNATIONAL GEOPHYSICAL YEAR, the USA decided to continue its Antarctic work through a long-term research programme. The NATIONAL SCIENCE FOUNDATION was allocated responsibility for the project and the US Antarctic Research Program began in 1959.

United States Expedition (1955–57) A government expedition, in two parts. Although the first was known as the 1955–56 OPERATION DEEPFREEZE (Naval), the expedition as a whole is generally referred to as 'Operation Deepfreeze'.

United States Exploring Expedition (1838–42) Government-sponsored voyage of exploration. In 1836, at the urging of shipowners, sealers, whalers and scientists, the USA government granted $30,000, the sloop *Vincennes* and five inadequate support SHIPS to an expedition to chart the SOUTHERN OCEAN. Charles WILKES was appointed commander (the fifth choice), and the expedition included 83 officers, nine scientists, including naturalists and artists, and 342 sailors; of the sailors, 62 were sacked, 42 deserted, and 15 lost their lives. The flotilla set out in August 1838 and made its base at Orange Harbour near the tip of TIERRA DEL FUEGO. From here, three separate expeditions set out: to the SOUTH SHETLAND ISLANDS, where surveying work was carried out; southwest in a failed attempt to exceed James COOK's most southern point of 71°10'S; and to carry out sur-

Above: The USA's McMurdo Base is dubbed 'Mactown' by 'locals'.

veying work around Tierra Del Fuego and through the MAGELLAN STRAIT. One ship was lost at sea, a second was sent back to the USA, and the remaining four explored the Pacific during the 1839 winter, reaching Sydney later in the year. Setting out from here on 26 December 1839, the plan was to sail south then west along the coast of the Antarctic continent. Three ships succeeded in reaching the coast of what is now known as WILKES LAND, and over 2000 km (1240 miles) of coastline between 160°E and 98°E were charted. Sailing northward, the expedition explored the Pacific and surveyed parts of the North American coast before heading home. Wilkes's claim to have discovered the continent has always been disputed and, on the expedition's return in 1842 (when it was greeted with indifference), he was subjected to a court-martial on five petty charges but convicted of only one. The five-volume *Narrative of the United States Exploring Expedition* and 20 volumes of scientific records were published.

United States Geological Survey Antarctic mapping, geology, geophysics, glaciology and ecological monitoring are carried out by the United States Geological Survey (USGS) organization as part of the UNITED STATES ANTARCTIC PROGRAM. The USGS has worked in Antarctica since 1947. Initially, projects focused on geophysical and geologic surveys, and topographic mapping began in 1957. The USGS's present-day research includes the geological and glacial histories of Antarctica and monitoring global CLIMATE.

USA Americans were involved in SEALING in the SOUTHERN OCEAN from the late 18th century, and the sealer Nathaniel PALMER claimed to have discovered the Antarctic continent in 1820. The US Congress voted to fund the 1838–42 UNITED STATES EXPLORING EXPEDITION. Ninety years later, Richard BYRD led two private expeditions to Antarctica.

At various times in the first half of the 20th century, the US government considered making a territorial claim in Antarctica, but after World War II it promoted the idea of internationalization of the continent. In 1946–47 the USA mounted OPERATION HIGHJUMP, the largest expedition sent to Antarctica, involving the American navy, over 4700 people, nine aircraft and 12 ships. It was followed by OPERATION WINDMILL in 1947–48.

During the INTERNATIONAL GEOPHYSICAL YEAR, the USA established seven bases, including the strategically located AMUNDSEN-SCOTT SOUTH POLE BASE and MCMURDO STATION. It hosted the 1959 WASHINGTON CONFERENCE, and was one of the original 12 CONSULTATIVE PARTIES to the ANTARCTIC TREATY.

The USA, which has refused recognition of the existing claims in Antarctica, and 'reserved' its own rights so that it can maintain open access to the entire continent, favoured the treaty because it preserved its rights and access in Antarctica. The USA has been the most active inspector of bases under the provisions of Article VIII of the Antarctic Treaty. The USA has generally promot-

ed a more open ANTARCTIC TREATY SYSTEM (ATS), and was one of the first states to introduce representatives from non-governmental organizations into its delegations.

It is a member of the Commission for the CONVENTION ON THE CONSERVATION OF ANTARCTIC MARINE LIVING RESOURCES (CCAMLR), and has usually adopted a pro-conservation stance at CCAMLR meetings. Interested in the possibility of MINING in Antarctica, the USA influenced the MADRID PROTOCOL by pushing for a 50-year moratorium rather than a permanent ban.

The USA currently operates three year-round bases and numerous summer-only camps, and sends about 3000 people to Antarctica each year.

Uruguay In 1776 Uruguay issued FISHING licences in the SOUTHERN OCEAN zone. Two centuries later, in 1980, Uruguay acceded to the ANTARCTIC TREATY; five years later it became a CONSULTATIVE PARTY and signed the CONVENTION FOR THE CONSERVATION OF ANTARCTIC MARINE LIVING RESOURCES. From 1985 Uruguay has maintained one year-round research station, Artigas, on the ANTARCTIC PENINSULA.

Usarp Mountains A mountain range which extends nearly 200 km (124 miles) inland from the coast of OATES LAND, parallel to the Rennick Glacier. It takes its name from the US ANTARCTIC RESEARCH PROGRAM.

USSR See RUSSIA.

V

Vanda Base Vanda Base was established by NEW ZEALAND in 1967 near Lake VANDA in the Wright Valley to facilitate SCIENTIFIC RESEARCH into the DRY VALLEYS region. The base, which operated as a summer-only station until 1994, was popular because of the 'Lake Vanda Swim Club': many visitors came to swim in the lake rather than conduct science. It was demolished because of rising lake levels.

Vanda, Lake A SALINE LAKE in the Wright Valley, in the DRY VALLEYS region of VICTORIA LAND.

Vehicles A wide variety of overland vehicles are used by modern EXPEDITIONS and field parties in Antarctica to travel across SEA ICE and inland terrain. They need to be adapted to combat the cold, which cracks metal, freezes lubricating oil and makes FUEL waxy so that it will not vaporize and ignite. They must also have specialized wheels and tracks to handle soft snow, rock-hard ice and steep, undulating terrain. There are many hazards on unexplored terrain and unstable sea ice, particularly CREVASSES. Navigation can also be difficult in WHITEOUT conditions. Many vehicles are equipped with SATELLITE navigation technology and RADAR.

Ernest SHACKLETON introduced motorized vehicles to Antarctica on his 1907–09 expedition. It was a failure—a MOTOR CAR that could not cope with the soft snow and the cold. Robert SCOTT tried motor SLEDGES and Richard BYRD modified Citröen cars. By the late 1950s, TRACTORS had become the first vehicles to successfully operate in Antarctica. Modified Massey Ferguson farm tractors, fitted with tracks for use in the snow, were used on the COMMONWEALTH TRANSANTARCTIC EXPEDITION, along with four massive SNO-CATS. Other vehicles were soon developed to replace DOG teams. With a track at the back and a ski on the front, a Skidoo carries up to two people, and can easily negotiate soft snow. On hard ice, four-wheel motorbikes, or Quikes, are useful. Swedish-designed HÄGGLUNDS are used along with massive 30 tonne (30 ton) bulldozers, with wide tracks and specialized fuel systems, for long overland journeys.

Ventifacts Highly polished pieces of ROCK, carved by millions of years of wind erosion. Dense outcrops of rocks on ridges and exposed areas become sculpted into pockets, scalloped patterns and thin, intricate wafers. A particular feature of the DRY VALLEYS, where they are shaped by the ferocious KATABATIC WINDS, they range in size from small pebbles, constantly rolled by gusts around rock depressions, to large granite boulders undercut by the wind into graceful curves.

Vernadsky Station Originally known as FARADAY STATION. When control of this station was transferred from BRITAIN to the UKRAINE in February 1996, it was renamed after scientist Vladimir Vernadsky. The oldest operational station on the

Below: The ice covering Lake Vanda is patterned with finely ridged sastrugi.

Above: Because of their specialised features, vehicles manufactured for use in Antarctica are not cheap. A skidoo carries a price tag of more than US$10,000, while Hägglund prices start at around US$150,000.

ANTARCTIC PENINSULA, it is located at Marina Point on Galindez Island. RESEARCH is conducted into meteorology, geomagnetism, human physiology and atmospheric science.

The station accommodates 24 people and is popular with other research parties and tourists because of its bar. The bar at Vernadsky's, inherited from the British along with the station, is made of hardwood carved from a shipment intended for a jetty. When the British authorities found out what had happened to their wood, the carvers were sacked. However, the bar survived. Only two drinks are available: Ukrainian wine and Ukrainian vodka.

Vertebrates ANIMALS with a bony skeleton and a well-developed brain. FISH, BIRDS and mammals such as WHALES and SEALS make up the modern-day vertebrate FAUNA of Antarctica, but FOSSIL records reveal that marine reptiles, DINOSAURS and MARSUPIALS were also once present.

Although Antarctic birds and some mammals breed on land, with the exception of a few sub-antarctic land birds such as the SOUTH GEORGIA PIPIT, almost all Antarctic vertebrates feed on marine organisms and therefore belong to the marine ECOSYSTEM.

Vestfold Hills An OASIS covering 400 sq km (154½ sq miles), on the coast of PRINCESS ELIZABETH LAND. Named in 1935 by Klarius and Caroline Mikkelsen (the latter was the first WOMAN to land on the Antarctic continent) after their home county in Norway, the Vestfold Hills contain a number of freshwater and SALINE LAKES.

Victoria Land The sector of land that borders the western edge of the ROSS SEA. It was first discovered by James Clark ROSS on 12 January 1841,

who named it for Queen Victoria. Ross claimed it for BRITAIN, raising a British ensign on POSSESSION ISLAND, just off the coast. The Victoria Land coast is dominated by glaciers and ice tongues, as well as a number of volcanic mountains. Victoria Land falls within the boundaries of ROSS DEPENDENCY.

Vincennes Bay Located in WILKES LAND, Vincennes Bay was the site of the now-abandoned WILKES STATION, established by the USA during the INTERNATIONAL GEOPHYSICAL YEAR and subsequently handed over to AUSTRALIA.

Vinson Massif The highest mountain in Antarctica at 4897 m (16,066 ft) above sea level, it is part of the SENTINEL RANGE in the northern part of the ELLSWORTH MOUNTAINS and lies inland of the RONNE ICE SHELF. It was first discovered in January 1957 by US Navy pilots. In December of that year, the Vinson Massif, along with other peaks in the Sentinel Range, were surveyed from the ground and their heights established. The Vinson Massif is named for US congressman Carl G Vinson, who was a vigorous campaigner for US government support for Antarctic exploration from the time of BYRD'S SECOND EXPEDITION of 1933–35. Vinson is often summitted by MOUNTAINEERING expeditions. Although the mountain is not considered a technically difficult climb, the extreme CLIMATE and unpredictable WEATHER patterns make any attempt to scale the peak a serious undertaking. An American Alpine Club team, led by Nicolas Clinch, was the first group to climb Vinson in December 1966.

Vocabulary As is typical in isolated regions, a specialized lexicon has developed in and about Antarctica. This is compounded by the fact that communities on the continent are temporary and

transient, that the majority of those who have spent time in the region belong to a small number of professions (predominantly explorer, sailor, whaler, sealer, fisherman, scientist and military) and are overwhelmingly male, and that many words have been transposed across centuries and across languages.

The Antarctic English names for landforms, species, physical conditions and other phenomena peculiar to Antarctica include words introduced by early explorers from a number of European countries: for example, SASTRUGI is Russian, KATABATIC comes from the Greek, KRILL is Norwegian. The whalers and sealers used colourful terms to describe harvesting and processing techniques— for example, 'flense' (to strip off skin or blubber); they spoke of flensers and lemmers (from the Norwegian, meaning to dismember); and they gave species common names such as BALEEN whales because they were harvested for their whalebones, HUMPBACK after their shape, BLUE for their colour, or terms already used in the Arctic (MINKE). Likewise, sailors and scientists have given birds familiar names such as 'nellies', 'wanderers', 'stinkers' and 'GPs' (giant petrels).

Natural physical processes have spawned a sub-lexicon of their own, some of which have entered the scientific literature, others which remain local jargon: FRAZIL ICE, PANCAKE ICE, berg, SEA ICE, névé (last year's snow), anchor ice, growlers and bergy bits (floating ice), GREASE ICE, BRASH ICE, PACK ICE, BLACK ICEBERG, bummock (an ice formation hanging beneath pack ice), POLYNA, blizz static (electric charge that builds up during a blizzard), calving, nilas (thin crust of floating ice), LEAD, snow bridge, SUN DOGS and SUN PILLARS, WHITEOUT and WINDCHILL.

Clothing and equipment first used in the Arctic, such as mukluk (skin boots) and polar pyramids (tents) retain their original names, as do more recent additions such as the Swedish-made tracked vehicles, HÄGGLUNDS (or Haggs). Other contemporary modes of transport include skidoos, helos (American for helicopter), Hercs (Hercules cargo planes) and Zodiacs (an inflatable rubber dinghy with outboard engine). The most desirable food is 'freshies' (fresh fruit and vegetables); in earlier years pemmican (a North American food of ground dried meat mixed with fat and sometimes cereal) and sledge biscuits (or 'sledgies') were combined to make a stew called 'hoosh'.

Antarctic English is also distinguished by national linguistic characteristics—for example, 'donga' (a separate bedroom in a base), 'chompers' (snacks), and 'apple' (small, round prefabricated hut) are Australian; a 'jolly' (sightseeing trip), 'scradge' (food) and a 'wobbly' (panic attack) are indisputably British, and 'hero picture' (photo of self taken at SOUTH POLE sign) is American. And the affectionate reference to the Antarctic continent as 'Big Pav' comes from the NEW ZEALAND meringue known as the pavlova.

Volcanoes Active volcanoes are found in WEST ANTARCTICA, in the western ROSS SEA, along the ANTARCTIC PENINSULA and on offshore MARITIME and SUBANTARCTIC ISLANDS.

Above: Mount Erebus is an active volcano, and it erupts regularly—up to 10 times a day.

Volcanoes form where hot mantle rises between rifts in the continental plate beneath Antarctica. Some of the magma is stored in chambers beneath the crust; when it reaches the surface it erupts from volcanoes as lava or volcanic ash. There are also about 70 extinct volcanoes in Antarctica.

In West Antarctica, only Mount BERLIN in MARIE BYRD LAND is still active. However, a circular depression on the West Antarctic ICE SHEET suggests there may be a subglacial volcano beneath the sheet supplying meltwater to the base of the ICE STREAMS. Most volcanic activity in the western Ross Sea is associated with the TRANSANTARCTIC MOUNTAINS. The Pleiades, a group of volcanic cones in northern VICTORIA LAND, probably erupted less than 1000 years ago and it is estimated that Mount MELBOURNE, near TERRA NOVA BAY, erupted less than two centuries ago and still has steam seeping from the summit.

The Peninsula and SCOTIA ARC zone is a boundary between two continental plates, where the Pacific Ocean sea floor is being subducted, or sunk back, into the Earth's interior. The area is currently active and volcanoes tap magma from deep within the Earth. There was an eruption on DECEPTION ISLAND in December 1967, which caused the evacuation of the Argentinian, British and Chilean bases.

The southernmost active volcano in the world, Mount EREBUS, is also Antarctica's largest and most famous. Very large eruptions of Erebus have occurred in the past, and ash from the volcano has been found in ICE as far as 300 km (186 miles) away. The volcano erupts about 10 times daily, but it is rare for lava to be thrown beyond the crater. However, in 1984, gas trapped under a solid lava crust was released and a series of cannon-like explosions showered lava 'bombs', some of which were as big as cars, 3 km (almost 2 miles) beyond the crater rim. The eruptions lasted for three months, and scientists had to abandon a RESEARCH STATION near the crater rim.

On the slopes of Antarctica's volcanoes some unique organisms have adapted not only to the light-dark cycle but also to extreme heat and cold. Mount Erebus supports the only Antarctic ALGAE that need high temperatures to survive. These thermophilic algae will not grow at temperatures below 25°C (77°F) and thrive at 45°C (113°F). At Mount Melbourne, a leafy MOSS that exists nowhere else in Antarctica grows in abundance. Scientists believe that many other species may have lived near Antarctic volcanoes but have been obliterated by eruptions.

In October 2000, the *Journal of Geophysical Research—Atmospheres* published a report on CORE SAMPLES extracted from the interior of EAST ANTARCTICA, at 84°S, 43°E, which have provided a 4100-year record of volcanic activity: 54 volcanic events were identified and a comprehensive list of eruptions in the Southern Hemisphere over this time was constructed. Samples drilled during the Cape Roberts Project (see DRILLING) show signs of major volcanic eruptions that shook Antarctica about 25 million years ago. Although scientists are still unsure of their exact source, it is believed that up to four major eruptions, which were large enough to impact global climate, occurred.

Vostok Base Established in December 1957 by the USSR near the location of the SOUTH GEOMAGNETIC POLE in the interior of Antarctica, Vostok is named after one of the ships used by Thaddeus BELLINGSHAUSEN. Sited at an altitude of 3448 m (11,316 ft), the coldest temperature on record, −89.2°C (−103°F), was recorded here on 31 July 1983, although there is an unconfirmed lower recording of −91°C (−106°F) from 1997. Below the base, scientists have located one of the SUBGLACIAL LAKES that lie under the Antarctic ICE SHEET.

Vostok, Lake One of the SUBGLACIAL LAKES that lie below thick layers of ICE in the Antarctic ICE SHEET. Some of these lakes may have been formed before the Ice Sheet and thus have been cut off from the ATMOSPHERE for hundreds of thousands of years. Located near the SOUTH GEOMAGNETIC POLE, Lake Vostok is one of the most intensively studied of these lakes.

The size of Lake Ontario, Vostok lies under more than 4 km (2½ miles) of ice. It is so large that it creates a flat surface on the Ice Sheet that has been mapped from space. Microbial life may have developed in this incredibly cold, dark environment in isolation from the rest of the biosphere. Scientists believe that conditions in Lake Vostok may be comparable to those on Europa, one of Jupiter's frozen moons. If so, these organisms will have evolved in a very specialized way.

An international DRILLING team came within 120 m (394 ft) of the lake's surface in 1998. Further exploration has been halted so that methods of sending robotic probes into the lake without contaminating this pristine environment can be developed.

Below: The world's lowest temperatures have been recorded at the Russian research station Vostok.

W

Wales, William (*c.* 1734–98) British astronomer. Born in Yorkshire, England, the mathematician Wales was appointed by the ROYAL SOCIETY to observe the transit of Venus at Hudson Bay, and sailed as astronomer on the *Resolution* with James COOK in 1772–75. Along with William Baly, he was instructed to test methods of determining longitude.

Wales's first experience of ICEBERGS was in January 1773; he monitored the continuous break-up and erosion of bergs and measured the saltiness and temperatures of the ocean, noting that the CIRCUMPOLAR DEEP WATER was warmer than the SURFACE WATER. He recorded a close encounter with an iceberg on 18 February: 'About noon came under the lee of the ... island of ice, and were by a kind of indraught ... sucked so near that we had scarce any [means] of escaping being drove against it, which must have been inevitable destruction; and it was equally as unknown how we got off without, and we had scarce got to a cable's length from it before several pieces almost as large as the ship broke off from that very part where we were.' On 19 March he wrote of the AURORA AUSTRALIS that 'these lights were so bright that we could discern our shadows on the deck.'

On the expedition's return, Wales and Baly published their *Original Astronomical Observations, made in the course of a Voyage towards the South Pole, and Round the World*, and Wales became master of the Mathematical School at Christ's Hospital.

Wandering albatross (*Diomedea exulans*) One of the greatest oceanic travellers, wandering albatrosses often follow ships and fishing trawlers using longlines to catch tuna, and other big fish and sharks. This habit has had a disastrous effect on wandering albatross populations: about one in 10 are killed on longlines every year. Fatalities can upset the sex ratio of males to females—the latter travel great distances to seek out the squid bait used by the longliners. Severely depleting these fish stocks or shifts in longline fishing away from the birds' feeding grounds may offer the albatross a reprieve. However, any recovery will be slow because of the protracted pre-breeding stage and two-year breeding cycle of these magnificent birds.

Wandering albatrosses look similar to the ROYAL ALBATROSS. Mature adults usually have a completely white body, head and neck with upper wings of varying degrees of black. The darker plumage of the younger birds lightens as they age. Wandering albatrosses only moult every second year, in contrast, most genera in the family Diomedeidae moult annually.

These great birds—with wing-spans of 2.5–3.5 m (7–10 ft)—may fly several thousand kilometres on a single foraging trip. One post-breeding female from SOUTH GEORGIA was recorded covering a distance of 25,075 km (15,546½ miles) in 36½ days. They feed mainly by surface-seizing SQUID, but catch some FISH, jellyfish and CRUSTACEANS in their large, hooked bills. Natural prey is taken from the surface of the sea at night, but they also scavenge meat offal from passing vessels during the day.

Breeding takes place on SUBANTARCTIC ISLANDS such as South Georgia, Îles KERGUÉLEN, PRINCE EDWARD, AUCKLAND and MACQUARIE ISLANDS, and the birds range throughout the SOUTHERN OCEAN and as far north as the Tropic of Capricorn. They breed once every two years, the time it takes to raise the single chick. The egg weighs about 500 g (just over 1 lb) and both parents share in raising and feeding the chick. Once it has fledged, the young albatross remains at sea for five to 10 years before returning to the island of its birth to begin its own breeding cycle.

Wars Antarctica has been largely protected from warfare by its inhospitable climate and its distance from the rest of the world. The few conflicts fought around Antarctica have involved naval battles.

In World War I, a German raiding squadron, under the command of Admiral von Spee, defeated a British squadron at Coronel, CHILE, on 1 November 1914. Five weeks later, after the Germans encountered the British squadron re-fuelling at Port Stanley in the FALKLAND ISLANDS, the German ships were pursued and destroyed.

Just before World War II began, the GERMAN ANTARCTIC EXPEDITION, under the command of Captain Alfred Ritscher, surveyed the coast of DRONNING MAUD LAND in preparation for a German TERRITORIAL CLAIM. German submarines used the SOUTHERN OCEAN as a hiding ground during World War II, and in January 1941 the German U-boat *Pinguin* captured three Norwegian whaling factory ships, and a supply of fuel. BRITAIN responded with OPERATION TABARIN.

After decades of squabbling over territory—which included Argentinian warships entering the harbour at DECEPTION ISLAND and establishing a permanent base there—Britain and ARGENTINA finally went to war. On 2 April 1982 Argentina invaded the Falkland Islands, including SOUTH GEORGIA, over which the British had claimed SOVEREIGNTY in 1908 and administered ever since. The small garrison in the Falklands quickly surrendered, but Britain responded by sending forces to recapture the islands and establishing an exclusion zone to blockade them. On 25 April the British recaptured South Georgia, then gained control of the sea around the islands and won air battles. The British invaded the Falklands, and over 10,000 Argentinian forces surrendered on 12 June. Argentinian war casualties numbered 655, and Britain lost 236.

Despite this conflict, the area defined as 'Antarctica' by the ANTARCTIC TREATY—the area below 60°S—had remained a zone of peace, highlighted by the fact that during the war Argentinian and British diplomats continued meeting to discuss the issue of marine living resources and a minerals regime.

Washburn, Lake A glacial LAKE in the DRY VALLEYS region which once inundated the entire Taylor Valley. Today paleo-lake sediments left by lake stands (perched deltas) of Washburn approximately 12,000 to 24,000 years ago are the subject of extensive research.

Washington Conference The meeting at which the ANTARCTIC TREATY was negotiated. Negotiations began on 15 October 1959 after approximately 60 preparatory meetings in the preceding 15 months, and the 12 states involved signed the treaty on 1 December 1959.

Waste disposal Until the 1980s, waste disposal practices at Antarctic bases had little or no regard for the environment. As most waste decays very slowly in Antarctica's arid climate, a legacy of debris and toxins will remain in some areas for decades to come. Bases such as MCMURDO STATION once burned or 'ice-staged' garbage—waste was

Below: A waste dump at McMurdo Station in 1976/77. Nations are now more conscious of the human impact on Antarctica's fragile environment. The site has since been cleaned up and McMurdo Station has an extensive waste recycling system.

either dragged under the SEA ICE and into the ocean, or piled up on the ice and the ICE SHEET blown up, dropping the garbage into the sea. Consequently, the sea floor off the coast of McMurdo has an extensive collection of refuse, including toxic PCBs. For decades fuel drums, chemical waste and garbage from ARGENTINA's MARAMBIO STATION were thrown down a slope into the sea, leaking toxins into the soil, streams and ocean. Excess building materials, mainly iron and wood, from the construction of bases were also considered too expensive to remove, and debris from abandoned bases, such as HALLETT STATION, litters some parts of the continent.

In 1987 GREENPEACE sent biologist Maj De Poorter to report on the state of the environment at scientific bases located on the ANTARCTIC PENINSULA. She reported that 'Many of the bases took absolutely no care whatsoever in protecting the environment. There was stuff everywhere: rubbish, garbage bins, old tyres, pierced oil drums that had spilt their foamy contents on the ground ...'. Dr Ron Lewis Smith, a member of the SCIENTIFIC COMMITTEE ON ANTARCTIC RESEARCH, visited AUSTRALIA's abandoned WILKES STATION in 1986 and reported discarded machinery, buildings, flares, explosives and boxes of caustic soda spilling their contents over the ice. Subsequently, most stations began 'repatriating' and recycling their rubbish. In 1994, for example, a portable wood-chipping mill was delivered to McMurdo Station to pulp waste timber and return it to the USA. Most bases now have environmental officers, and staff and visitors undergo environmental awareness and safety training.

Water Most water in Antarctica is locked up in SNOW and ICE. Some meltwater is found near the coast, but this generally seeps into underlying snow and freezes. In OASES there is enough surface run-off over summer to form STREAMS. The only water on the continent to remain unfrozen all year round is in SALINE LAKES.

Waves These originate when WIND blows across the surface of the ocean to form 'wind waves'. Initially chaotic and disorganized, they eventually group into 'swell', which can travel over large distances. The SOUTHERN OCEAN is characterized by consistently powerful winds. It also has a large surface area over which swell can form and grow, unobstructed by land. Massive waves roll across the sea, breaking up SEA ICE and pounding against the shoreline of the continent. The highest wave recorded—37 m (121 ft) high—was in the Southern Ocean.

Over the centuries, sailors have left many terrifying accounts of encountering Southern Ocean waves. Derek Lundy, a skipper in an around-the-world yacht race, described them as '... a never-ending series of five-or-six story buildings, with sloping sides of various angles, moving toward [you] at about forty miles per hour. Some of the time, the top one or two storeys of the buildings will collapse ...'.

Weather The harshest and most challenging weather conditions on Earth are found in Antarctica and the surrounding SOUTHERN OCEAN. Changes are dramatic. WINDS can shift from calm to full gale force almost instantaneously or TEMPERATURES can plummet. Offshore cyclones occur with little warning and persist for days at a time. The weather systems that circle Antarctica drive storms across the Southern Ocean and beyond. For explorers, scientists and adventurers, living and working on the coldest, driest, windiest continent poses formidable challenges.

The weather over most of Antarctica is affected more by the surface layer of air than by large-scale atmospheric circulation. The 'boundary layer' is air that is directly affected by the ground surface; in Antarctica this is generally cooled by SNOW and ICE cover, and moves with the shape of the land to produce GRAVITY WINDS, such as the KATABATIC, which flow downhill from the POLAR PLATEAU and interact with warmer coastal air.

A narrow storm belt, known as the ROARING FORTIES, circles Antarctica. Cyclonic systems, or depressions, continually form over the ocean, around the CIRCUMPOLAR TROUGH, spiralling polewards and eastwards. They last for about a week and break up near the coast. This ring of low pressure produces dense CLOUDS, fog and extremely severe BLIZZARDS. Occasionally, depressions will reach as far south as the POLAR PLATEAU, bringing warmer, more humid air and snow. However, generally the interior of Antarctica has calmer, more stable and clearer weather conditions than coastal areas.

Any outside activities in Antarctica are constrained by weather conditions, which have been divided into three categories. Condition 3 is the mildest, with good visibility and relatively mild temperatures. Condition 1, however, shuts work down at field camps, and even at permanent bases it is unwise to venture outside for any reason: the wind gusts at more than 102 km (63 miles) per

Below: James Weddell: the account of king penguins included in his 1825 book on Antarctica is recognised as remarkably astute.

hour and blowing snow creates WHITEOUT conditions, making the chances of becoming disorientated, even when walking to the next building, very high.

Weddell, James (1787–1834) British naval officer, sealer and explorer. Born in Ostend, in the Netherlands, Weddell spent some years at sea before joining the British Navy and serving in the Napoleonic war against France. He became a captain with the ENDERBY BROTHERS, and made three SEALING voyages to the SOUTHERN OCEAN: in 1819–21, 1821–22 and 1822–24. The first of these was for seal prospecting, and the second combined sealing with surveying work around SOUTH GEORGIA, the SOUTH SHETLAND and the SOUTH ORKNEY ISLANDS.

On his third voyage, in 1822–24, in the brig *Jane* accompanied by the cutter *Beaufoy* under Captain Matthew Brisbane, he beat James COOK's 'farthest-south' record: sailing in the sea he named after King George IV—now the WEDDELL SEA—he reached 71°15', a record in open water that was unsurpassed for 90 years. Encouraged by the Enderbys to explore the Southern Ocean, he kept accurate records of sea surface temperatures, measured magnetic declination, and recorded seismic activity in the South Shetlands. He observed 'a pinnacle of an ICEBERG so thickly incorporated with black earth as to present the appearance of a rock'; these unusual forms are called 'BLACK ICEBERGS'. *Leptonychotes weddelli*, the WEDDELL SEAL, was discovered during the expedition.

Weddell published *A Voyage towards the South Pole*, which is recognized as one of the classics of Antarctic exploration, in 1825. It included an account of KING PENGUINS: almost nine decades later, the naturalist R C Murphy wrote, 'The details of his [penguin] study have long been overlooked, or perhaps disbelieved, by ornithologists, but they actually comprise the best account of the bird's life history that has yet been published. Nothing in my own observations would lead me to change a line of Weddell's almost forgotten history.'

Weddell Drift Occurs in the north WEDDELL SEA and is the only area around Antarctica where SEA ICE flows from west to east. This flow forms the northern edge of the Weddell Sea Gyral, a clockwise eddy first observed by crew aboard ships wintering over in sea ice in the area.

See also WEST WIND DRIFT.

Weddell Sea Extends east from the ANTARCTIC PENINSULA, and has an area of about 2,800,000 sq km (1,081,080 sq miles). Along with the ROSS SEA, it is one of two principal embayments in the otherwise circular coastline of Antarctica and is bound to the south by the vast RONNE and FILCHNER ICE SHELVES.

In the winter, PACK ICE extends from the Weddell Sea 3000 km (1860 miles) into the SOUTHERN OCEAN: further than anywhere else around the continent. The slow clockwise drift of the pack ice is known as the Weddell Sea Gyral. In 1912, the *Deutschland*, commanded by Wilhelm

Above: A Weddell seal (Leptonychotes weddelli) dines on Antarctic cod. Weddell seals keep their ice holes open year-round by scraping their teeth on the ice edge.
Right: Weddell seal pups are born on the ice and quadruple their birth weight in around six weeks.

FILCHNER, was trapped in ice over winter in the Weddell Sea. Expedition members left the ship on 22 June, and returned eight days later to find it had drifted about 61 km (38 miles) southwest.

The Weddell Sea was named after its discoverer, James WEDDELL. It was further investigated by William BRUCE between 1902 and 1904. During the INTERNATIONAL GEOPHYSICAL YEAR, bases were established along the sea's south and southeast coasts. It is a focus of continued scientific interest, particularly in relation to the enormous mass of ANTARCTIC BOTTOM WATER generated in the sea.

Weddell seal (*Leptonychotes weddelli*) At around 3 m (10 ft) in length and 400–500 kg (882–1102 lb), Weddell seals are one of the large Antarctic phocid seals. Their faces are small and appealing, with slightly upturned mouths, short whiskers and dark brown eyes. Towards humans, Weddells are one of the least aggressive seals, and even underwater they are quite affable.

Weddells are the southernmost naturally occurring mammals on Earth. They are found under and around FAST ICE, staying close to breathing holes, although immature males spend some time on more northerly PACK ICE. During the winter when the ice freezes over, the seals maintain breathing holes using their canine teeth and incisors to scour away the ice as it refreezes. They

are extremely proficient divers and when searching for their main food sources—FISH and CRUSTACEANS, including KRILL—can dive to depths of 720 m (2362 ft) and stay underwater for up to 80 minutes.

It is thought that females are sexually mature at the age of three and may breed for up to nine years, although some have been known to breed as early as two years of age. Copulation occurs in the water beneath the ice, where the males engage in violent underwater territorial battles. A wide range of calls, reminiscent of sound effects in old science-fiction movies, accompanies the under-

water mating and territorial rituals. The single pups weigh around 25–30 kg (55–66 lb) at birth and nurse until they are around six weeks, by which stage they have usually quadrupled their birth weight.

West Antarctica West Antarctica faces the Pacific sector of the SOUTHERN OCEAN, and joins EAST ANTARCTICA at the TRANSANTARCTIC MOUNTAINS. Below its ICE SHEET are a number of separate land masses that would become a group of islands if the ice melted. The bedrock of West Antarctica is less than 600 million years old, and is much younger than that of EAST ANTARCTICA. It is buckled and folded into enormous mountain ranges that, in some places, are as much as 2500 m (8200 ft) below sea level. Among the MOUNTAINS above the surface is the highest on the Antarctic continent, the VINSON MASSIF, which lies 4897 m (16,066 ft) above sea level in the SENTINEL RANGE.

Unlike the EAST ANTARCTIC ICE SHEET, the ice cover over West Antarctica is unstable and has grown and shrunk many times since its formation. Fast-moving ICE STREAMS drain ICE to the West Antarctic coast, where it breaks off ICE SHELVES into the ocean.

West Antarctic Ice Sheet Although constituting only about 12 percent of the Antarctic ICE SHEET, the West Antarctic Ice Sheet is marine-based and therefore more dynamic and responsive to CLIMATE change than the EAST ANTARCTIC ICE SHEET—it has expanded and contracted many times since its formation 20 million years ago, and the weight of the ice has pushed its base below sea level. ICE STREAMS drain through the western Ice Sheet, transporting large quantities of ice from the interior to the coast. They appear to be wet-based, and flow considerably faster than the surrounding ice. Most of the ice drained from WEST ANTARCTICA flows into either the ROSS ICE SHELF in the ROSS SEA or the FILCHNER ICE SHELF in the WEDDELL SEA.

The West Antarctic Ice Sheet Program (WAIS) is a multinational initiative designed to find out what triggers marine ice-sheet collapse and to predict the future of the western Ice Sheet. Scientists are using a radar satellite (RADARSAT) to map the Ice Sheet and monitor its behaviour. Ice CORE SAMPLES give some clues about the Ice Sheet's past; however, scientific research has not yet established the history of the Ice Sheet's fluctuations or its future. Part of the uncertainty comes from lack of knowledge of how the climate and sea level influence the West Antarctic Ice Sheet.

West Ice Shelf Off PRINCESS ELIZABETH LAND between Barrier Bay and Posadowsky Bay, the West Ice Shelf is 290 km (180 miles) wide. It was discovered by the 1901–03 GERMAN SOUTH POLAR EXPEDITION and named because it was first viewed from a westerly direction.

West Wind Drift Also referred to as the Circumpolar Current. One of six major ocean circulation systems in the world, the West Wind Drift is a surface OCEAN CURRENT that flows 24,000 km (14,880 miles) around the Antarctic continent from west to east.

It is driven by fierce westerly WINDS that blow over the SOUTHERN OCEAN, driving the waters east. Its wide sweep around Antarctica is only stalled when it passes through a constriction in DRAKE PASSAGE (1000 km/620 miles wide) between South America and the tip of the ANTARCTIC PENINSULA. At this point it intensifies and becomes the strongest and deepest ocean current in the world, moving at least 25,000,000 cubic metres (975,000,000 cu ft) per second—almost twice as much water as the Gulf Stream or 400 times the amount of the Mississippi River.

Inside the West Wind Drift, the EAST WIND DRIFT is a shallower, slower-moving current that flows off the Antarctic land mass; the two drifts are separated by the ANTARCTIC DIVERGENCE.

Whale (Cetacea) The largest mammals on Earth, whales are also highly intelligent. Before commercial WHALING decimated their populations, Antarctic whales comprised the largest stocks—in sheer weight—of mammals ever to have existed on Earth. The harvesting of whales provided the impetus for many early expeditions to Antarctic waters: notably, Scottish expeditions in 1892–93, two led by Carl LARSEN between 1892 and 1894, and the 1894–95 Norwegian expeditions that included Henryk BULL. Since the 1970s, the commercial exploitation of whales has become one of the most controversial conservation issues, and is today controlled by the INTERNATIONAL WHALING COMMISSION.

Whales belong to the mammalian order of Cetacea, and are further divided into BALEEN WHALES (Mysticete) and TOOTHED WHALES (Odontocete). The six species of mysticete baleen whales that occur in Antarctic waters—BLUE, FIN, SEI, HUMPBACK, MINKE and SOUTHERN RIGHT WHALE—belong to the family Balaenopteridae. The Antarctic species of Odontocete or toothed whales include members of four families: Physeteridae (SPERM WHALES), Ziphiidae (beaked whales), Delphinidae (DOLPHINS) and Phocoenidae (SPECTACLED PORPOISE).

Whalebird See PRION.

Whaling When the SEALING industry began exhausting seal populations in the southern oceans, the sealing firms turned to whaling. Some of the earliest exploratory expeditions to Antarctic waters were sponsored by firms such as Enderbys, which in the 1820s and 1830s encouraged their captains—including John BISCOE, John BALLENY and possibly Peter KEMP—to explore for new lands in the course of their sealing and whaling trips.

Until the late 19th century, WHALES in Antarctic waters were protected by climate and distance. But the introduction of harpoon guns, steam-powered whalers and factory ships changed this. The first whaling occurred around the ANTARCTIC PENINSULA, where several ice-free anchorages for factory ships were available. In 1923, the Norwegian whaling captain Carl LARSEN took a factory ship and five whale catchers into the ROSS SEA and 221 whales were slaughtered. Other companies followed, and between 1927 and 1931, over 14,100 whales were killed and processed in the ROSS SEA area alone.

ARGENTINA was the first country involved in commercial whaling in the SOUTHERN OCEAN and it continued until the late 1920s. The increasing interest in whaling acted as a spur to political interests. NORWAY and BRITAIN both made TERRITORIAL CLAIMS to protect their whaling interests. These two countries took most of the catch for the first six decades of the 20th century, and hunted in these waters until the 1960s. JAPAN and, for a time, the USSR became predominant from about 1969.

The INTERNATIONAL WHALING COMMISSION (IWC) was established in 1946, initially to govern the conduct of whaling, and from 1975 to limit catches to sustainable levels. But in 1982 the IWC voted to stop all commercial whaling from 1985–86. This does not affect 'aboriginal subsistence' whaling, which is confined to certain Northern Hemisphere nationals and permitted in limited numbers.

Despite the total ban, commercial whaling has not ceased, although it has been significantly curtailed, and Japan has continued to catch a limited number of MINKE WHALES in particular, under the stated aim of conducting scientific research. IWC members are subject to ongoing, intensive lobbying from several countries intent on continuing to hunt whales.

Below: Former whaling stations throughout the subantarctic islands are littered with bleached whale bones.

Above: Relic of a distant age: an old whaling station on Macquarie Island.

White Island In MCMURDO SOUND, White Island is the location of a population of WEDDELL SEALS, the most southern mammal habitat in the world, which gains access to the sea through cracks in the McMurdo Ice Shelf.

White-capped mollymawk (*Thalassarche cauta*) Large, pale-headed MOLLYMAWKS with mostly white underwings and pale green bills, they have a wing-span of 2.2–2.5 m (7–8½ ft) and are 90–99 cm (35–39 in) long. They are seen throughout the subantarctic waters of the Pacific and Indian Oceans and breed only in the AUCKLAND and ANTIPODES ISLANDS and on islands around Tasmania, south of mainland AUSTRALIA. Breeding colonies are found on cliffs and cliff ledges overlooking the sea: their conical nests are made of mud and grass and are reused each year.

In flight, white-capped mollymawks have a distinctive black spot where the front margin of the underwings joins the body. They readily follow FISHING trawlers and ships for offal thrown overboard. The world population is declining, with up to 10,000 birds killed on longlines every year.

White-chinned petrel (*Procellaria aequinoctialis*) The largest burrowing PETREL, these blackish-brown birds are 55 cm (22 in) long with wingspans of 1.4 m (4½ ft) and have powerful straw-coloured bills and white chin patches. White-chinned petrels are summer breeders, laying single eggs in burrows on many SUBANTARCTIC ISLANDS, including SOUTH GEORGIA.

Also known as 'cape-hens' and 'shoemakers' (because of their tapping call), the birds often follow FISHING trawlers in pursuit of edible scraps, over which they fight noisily. Their main diet consists of SQUID, CRUSTACEANS and FISH. They have a smooth, soaring, gliding style and when they do flap their wings, the beats are precise and slow. At-sea distribution is circumpolar and wide-ranging, especially in winter when they venture north into subtropical waters.

White-headed petrel (*Pterodroma lessonii*) These fairly large, white-headed GADFLY PETRELS of the open ocean have dark underwings and conspicuous black eye-patches. They grow about 43 cm (17 in) long, and their wings, which they hold stiffly in flight, span 1.1 m (3½ ft). Their bill is short, stout and black, and their feet fleshy pink with dark extremities. Solitary birds at sea, they do not follow ships.

These petrels breed in burrows during summer on MACQUARIE, AUCKLAND, ANTIPODES and POSSESSION ISLANDS and Îles KERGUÉLEN, and in the Pacific region are key indicators for subantarctic waters. They range further south in the colder waters of the SCOTIA SEA.

Although their numbers are thought to be in the hundreds of thousands, populations have been affected by the introduction of cats, especially on Macquarie Island; a cat eradication programme has been implemented there.

Whiteout Navigational hazard in which travellers often lose their perception of distance and sense of direction. When a calm, overcast sky meets the snow surface, it is hard to distinguish between sky and snow; whiteout conditions have been linked to a condition called 'empty field myopia', when the eye focuses at only a short distance. In the flat light, there are no shadows and it is difficult to see bumps and hollows, or even to know if the terrain is uphill or downhill. This is a serious problem on sledging journeys. Ernest SHACKLETON described the lack of shadows when mist or clouds diffused light, giving the impression that the surface was level; in reality SASTRUGI, or rippled snow surfaces, would suddenly stop sledges, or travellers would descend rapidly down an unforeseen slope. Whiteout was implicated as contributing to the DC10 AIR CRASH on Mount EREBUS in 1979.

Wild, Frank (1874–1930) British explorer, born in Yorkshire, England. He joined the merchant navy at 16, and the British Navy 10 years later. He volunteered for the 1901–04 NATIONAL ANTARCTIC EXPEDITION, on which he met Ernest SHACKLETON.

Shackleton invited Wild to join his 1907–09 BRITISH ANTARCTIC EXPEDITION. The two, together with Jameson Adams and Eric Marshall, reached 88°23'S, less than 180 km (112 miles) from the SOUTH POLE, the most southerly point reached thus far—'May none but my worst enemies spend their Xmas in such a dreary God forsaken spot as this,' Wild wrote. They walked a total of 2736 km (1700 miles). On the 1911–14 AUSTRALASIAN ANTARCTIC EXPEDITION, Wild led a group of eight men who OVERWINTERED on the SHACKLETON ICE SHELF, 2410 km (1500 miles) from the main base at Cape DENISON, and mapped 500 km (310 miles) of new coastline.

Wild returned to the Antarctic as Shackleton's second-in-command on the 1914–17 IMPERIAL TRANSANTARCTIC EXPEDITION. When their ship, the *ENDURANCE*, was crushed by ice, he steered the *James Caird*, one of her lifeboats, to ELEPHANT ISLAND. There, he was left in charge of 22 men while Shackleton's party left to find help. He made a hut from two boats and scraps of old tents and kept morale high over the 105 days the men awaited rescue. Wild was 'a magnificent leader here, scrupulously fair in everything, and popular and respected by everyone,' the expedition's surgeon A H Macklin wrote.

After serving in World War I, in 1917 he was asked to join Shackleton once more on the *QUEST* EXPEDITION of 1921–22 to circumnavigate the Antarctic continent. After Shackleton's sudden death in 1922, Wild took command and completed the voyage. He spent his later years cotton farming in Africa and died in the Transvaal of pneumonia aged 56.

Wiencke Island East of ANVERS ISLAND, on the west coast of the ANTARCTIC PENINSULA, this is one of the most popular Antarctic tourist destinations. SHIPS anchor at Port Lockroy, an 800 m (2600 ft) long harbour, which was frequented by whaling fleets and contains a number of relics of that era. An old British meteorological station, Base A, was restored by the ANTARCTIC HERITAGE TRUST in 1996 and includes a museum known as Bransfield House. There is a GENTOO PENGUIN rookery on the island.

Wilhelm II Land Formerly known as Kaiser Wilhelm II Land, Wilhelm II Land lies between

PRINCESS ELIZABETH LAND and QUEEN MARY LAND in EAST ANTARCTICA. On 21 February 1902 the GERMAN SOUTH POLAR EXPEDITION spotted a cliff-covered coastline, and the expedition's leader, Erich von DRYGALSKI, named the new land after the German emperor and claimed it for Germany. The Treaty of Versailles settlement after World War I annulled German claims to the area and Wilhelm II Land became part of the AUSTRALIAN ANTARCTIC TERRITORY.

Wilkes, Charles (1798–1877) American naval officer, scientist and explorer. Born in New York City, he joined the US Navy as a 20-year-old, and in 1830 was placed in charge of the navy's charts and instruments.

When Congress voted to fund the UNITED STATES EXPLORING EXPEDITION of 1838–42, Wilkes (the fifth choice) was put in command. Known for his decisiveness, Wilkes was also quick-tempered and stubborn, and the expedition was not a happy one. Although it was a sizeable expedition, it was under-funded and ill-equipped. Wilkes reported 'I was turned loose to manage everything … [despite] the disgrace which had attended the getting up of the Expedition and its failure and folly, as well as the honest expectations of the whole country … I had made up my mind the Expedition should not fail in my hands and believed I could carry it out to a successful issue.'

In the course of the four-year voyage, not quite three months in total were spent in Antarctic waters. In February 1839, Wilkes sailed with two ships from Cape Horn to carry out surveying work between the ANTARCTIC PENINSULA and the SOUTH ORKNEY ISLANDS. From here he went on to

Below: The controversial US naval explorer Charles Wilkes, whose achievements were recognized by Britain's Royal Geographical Society.

the SOUTH SHETLANDS and west into the WEDDELL SEA, from where he was forced back to winter in the Pacific.

A second voyage south set out from Sydney, AUSTRALIA, in December 1839. On 10 January 1840 they encountered ICEBERGS for the first time and at 64°S came to a vast wall of ice at the edge of a plateau, which Wilkes was convinced was part of an Antarctic continent, and sailed along what is now known as WILKES LAND, at one point encountering Jean-Sébastien DUMONT D'URVILLE'S expedition.

On the expedition's return in 1842, Wilkes was subjected to a court-martial on five petty charges but convicted of only one. His subsequent career was as controversial as his Antarctic adventure: he was promoted within the US Navy, was court-martialled for a second time in 1864 and this time found guilty, but nevertheless was promoted to rear-admiral. The five-volume *The Narrative of the United States Exploring Expedition* was published in 1845, and 20 volumes of scientific records followed. Although his achievements were not recognized in his homeland until decades after his death, Wilkes was awarded a gold medal by Britain's ROYAL GEOGRAPHICAL SOCIETY.

Wilkes Land A large section of EAST ANTARCTICA, between QUEEN MARY LAND and TERRE ADÉLIE. It is named after Charles WILKES. Much of Wilkes Land is covered by some of the thickest ice in Antarctica, and includes the DOME CIRCE or Dome Charlie region, the thickest part of which lies in the adjacent Terre Adélie.

Wilkes Station The USA built Wilkes Station on Clark Peninsula in WILKES LAND in 1957 during the INTERNATIONAL GEOPHYSICAL YEAR, and conducted research into glaciology. AUSTRALIA took over operations two years later, and replaced Wilkes with CASEY STATION. The remains of the old station can still be seen.

Wilkins, George Hubert (1888–1958) Australian scientist, photographer and explorer. Born in South Australia, the son of a farmer, Wilkins studied at the Adelaide School of Mines. In 1912 he was Turkey's official photographer during the Balkan War, and between 1913 and 1917 was photographer on several Arctic expeditions. He then enlisted in the Australian Flying Corps.

A member of the ill-equipped 1920–22 BRITISH IMPERIAL EXPEDITION, Wilkins withdrew after a frustrating few months in Antarctica. The following year he joined the QUEST EXPEDITION as ornithologist, and from 1923 to 1925 collected specimens for the British Museum.

Wilkins mounted an expedition to Antarctica in 1928–30 sponsored by newspaper publisher William Randolph Hearst; the aim was to fly over the Antarctic continent. From DECEPTION ISLAND he made the first Antarctic flight, in November 1928, in a Lockheed Vega monoplane piloted by Carl Ben Eielson. The 10-hour flight down the coast of the ANTARCTIC PENINSULA demonstrated that AIRCRAFT were viable means of TRANSPORT in Antarctica. He made a number of geographical

discoveries but, because his aerial views were rather distorted, the Peninsula was charted as an archipelago; it was not until John RYMILL tried to penetrate some of Wilkins's 'channels' and 'straits' that these gaps were found to be GLACIERS. He also carried out aerial surveying around ALEXANDER ISLAND and CHARCOT ISLAND.

Following the expedition, Wilkins returned to the Arctic to work in commercial aviation and in 1931 bought a surplus US Navy submarine for $1 with the intention of sailing it under the North Pole. Refitting the submarine cost $250,000, and it was renamed *Nautilus*. Wilkins reached 81°59'N but could not penetrate the ice, so the plan was abandoned and the submarine scuttled.

In 1937 he undertook a fruitless search for a Russian party missing near the North Pole. After his death, his ashes were taken to the North Pole by submarine and scattered on the ice.

Willis Islands A group of small islands adjacent to Bird Island at the northern tip of SOUTH GEORGIA, with a well-established population of FUR SEALS.

Wilson, Edward Adrian (1872–1912) British scientist, artist and explorer, born in Cheltenham, England. He was educated at Cambridge University, graduating in natural history and medicine. His college rooms were described by a contemporary as 'more like a museum than a dwelling place'.

Wilson joined the 1901–04 NATIONAL ANTARCTIC EXPEDITION, serving as second surgeon, artist and vertebrate zoologist. Together with Robert SCOTT and Ernest SHACKLETON, he reached 82°16'S in 1902, closer to the SOUTH POLE than any other party had reached.

Wilson was appointed head of scientific staff and second-in-command on Scott's 1910–13 BRITISH ANTARCTIC EXPEDITION. In the winter of 1911 he trekked 105 km (65 miles) with Henry BOWERS and Aspley CHERRY-GARRARD to Cape CROZIER to collect EMPEROR PENGUIN eggs, with the object of explaining the origin of all birds. This was the first journey ever attempted during the depths of the Antarctic winter and was described by Cherry-Garrard as the 'worst journey in the world'. Wilson reached the SOUTH POLE as part of Scott's party on 17 January 1912. He died from starvation on the return journey, on about 29 March 1912.

Wilson's storm petrel (*Oceanites oceanicus*) With wing-spans of only 40 cm (16 in) and weighing around 40 g (1½ oz), these are the smallest and lightest seabirds in the world. They are also thought to be the most abundant, and are found in every ocean at various times of the year. They are sooty brown with a white 'U'-shaped patch near the tail, and black feet with yellow webbing. Their diet mainly consists of planktonic CRUSTACEANS, which they catch while skimming over the water and pattering their feet on the surface; this behaviour prompted early sailors to dub them 'Mother Carey's chickens'—'Mother Carey' was another name for the Virgin Mary, and the sailors believed the birds came to collect the souls of dead sailors.

Breeding occurs on the SUBANTARCTIC ISLANDS and on the Antarctic continent, including the ANTARCTIC PENINSULA. Their single eggs are laid in rock crevices in December.

Windchill Low TEMPERATURES and strong WINDS combine in Antarctica to create dangerous wind-chill conditions. The effect of COLD depends on the rate at which the human body loses heat to the air surrounding it. When the air is moving rapidly past the body, the heat is quickly swept away. American scientist Paul SIPLE devised an index of windchill, which relates to the loss of heat and the effects on the body of wind velocity and temperature. In a 10 km per hour (6 mph) wind at –10°C (14°F), for example, the body loses heat as quickly as it would at –70°C (–94°F) in still air.

Winds Antarctica is continually buffeted by strong winds. Around the coast, POLAR EASTERLIES predominate, but this pattern may be severely disrupted by the much stronger KATABATIC WINDS that stream off the POLAR PLATEAU. The highest recorded wind velocity was 327 kph (203 mph), measured at DUMONT D'URVILLE BASE in July 1972.

The 1911–14 AUSTRALASIAN ANTARCTIC EXPEDITION led by Douglas MAWSON experienced some of the worst wind conditions on record. At Cape DENISON they unwittingly built a camp on one of the windiest corridors of land in the world. They continually experienced BLIZZARDS and 'Herculean gusts' of wind. Frank HURLEY, the expedition's photographer, wrote, 'On one occasion when the wind attained a velocity of 120 miles [193 km] per hour, I was lifted bodily, carried some 15 yards [14 m] with my camera and tripod, which together weighed 80 lbs [36 kg], and dumped on the rocks. I was reduced to crawling on all fours ...'.

They also had to contend with 'whirlies'. Mawson commented that a man skinning the head of a seal might be in perfect calm, while his companion at the seal's other end was on the edge of a furious whirlie in gusts of up to 290 km (180 miles) per hour. The expedition's data showed that in the first year the wind averaged over 80 km (50 miles) per hour. This was considered so extreme that meteorologists did not accept the findings until Mawson's instruments were tested in London.

In his book *Antarctica: The Last Horizon*, John Bechervaise, who has led three Australian expeditions to Antarctica, recalls losing two aircraft in the worst winds he had experienced: the planes were '... dismembered of wings and tailplanes and torn to pieces, although they were firmly anchored to the plateau ice. ... The storm increased until gusts exceeded 300 kph [186 mph] ... the main plane even was detached from an aircraft and carried away bodily, like a blown leaf.'

When the relentless winds stop, Antarctica is known for its silence. According to Captain Reginald Ford, it '... transcends the silence of the most silent sea. There is not a cry of a bird ... nothing but a deathlike and fathomless quiet. It is almost overwhelming.' As Carsten BORCHGREVINK wrote, 'The silence roared in our ears, it was centuries of heaped up solitude.'

Wingless midge (*Belgica antarctica*) The largest INVERTEBRATE in Antarctica, the wingless midge is a terrestrial arthropod, an INSECT found on the ANTARCTIC PENINSULA. It grows approximately 12 mm (½ in) long and may sometimes be seen crawling over MOSS on warm days. As its name suggests, it has lost the ability to fly through wing reduction—which frequently occurs in insects from cold and windy areas.

Unlike those of MITES and SPRINGTAILS, larvae of the wingless midge are able to survive some freezing; larvae can survive temperatures as low as –15°C (5°F), but exposure to temperatures of –20°C (–4°F) is fatal.

Winter During the months of polar darkness—from March to August—Antarctica is isolated from the rest of the world. Marine mammals retreat to warmer waters, and a mantle of SEA ICE encloses the continent, barricading out any ships. The human population drops from thousands to skeleton crews ('overwinterers') who spend the winter working long shifts, collecting scientific data and servicing laboratory instruments. The pitch-black night is broken by dazzling displays of the AURORA AUSTRALIS.

Winter in Antarctica is bitterly COLD—so cold that in many places when boiling water is thrown into the air it turns instantly to SNOW. Refrigerators have to be heated to prevent produce from freezing. At the SOUTH POLE during winter, AIRCRAFT cannot land because the cold freezes hydraulic fluid in landing gear.

Wireless See RADIO.

Wisting, Oscar (1871–1936) Norwegian explorer. He worked on whaling ships, then joined the Norwegian Navy. Recommended to Roald AMUNDSEN, he sailed on the *FRAM* on what turned out to be the 1910–12 NORWEGIAN ANTARCTIC EXPEDITION.

During the expedition's 1911 Antarctic winter, Wisting managed supplies, constructed rawhide lashings for sledges, and sewed tents and clothing. He was a member of Amundsen's party, which reached the SOUTH POLE for the first time, on 14 December 1911.

He joined Amundsen's 1918 expedition on the *Maud* to navigate the north-east passage, and became commander in 1921 when Amundsen left the expedition. In 1926 Wisting, along with Amundsen and Lincoln ELLSWORTH, made the first trans-Arctic flight over the North Pole in the airship *Norge*, and he led one of the search parties when Amundsen disappeared in the Arctic in 1928. Two years later, when the *Fram* was exhibited in a museum in Oslo, Wisting was granted permission to sleep aboard, and was found dead the next morning in his old cabin.

Women It is likely that an unknown woman accompanying her husband on a WHALING ship may have been the first of her sex to have sailed into Antarctic waters. Certainly, Norwegian whalers carried women, including Caroline Mikkelsen, the wife of a whaling ship captain, who set foot on the continent at VESTFOLD HILLS

on 20 February 1935. The first women to OVERWINTER in Antarctica were Edith Ronne and Jennie Darlington, who spent 1947 at STONINGTON ISLAND with their husbands.

The USSR included women scientists in their INTERNATIONAL GEOPHYSICAL YEAR expeditions, but it was not until the late 1960s or later that women were permitted to join many national programmes—indeed, the first time American women were permitted to winter-over at an official base—MCMURDO STATION—was in 1974. Now, almost half the McMurdo staff are women.

Even the 'men-only' club of Antarctic explorers has been broken into. The 'NINETY DEGREES SOUTH' EXPEDITION of 1986–87 was led by Monica Kristensen. The American Women's Expedition reached the SOUTH POLE in January 1993, and the Norwegian skiier Liv Amesen made a solo trek to the Pole in 1994. It is interesting that none of these achievements received media attention comparable to that afforded men's exploits of the same time.

World Heritage Convention In 1972 the World Conference of the United Nations Educational, Scientific and Cultural Organization (UNESCO) adopted the Convention Concerning the Protection of the World Cultural and Natural Heritage (World Heritage Convention). It is dedicated to the preservation of the common heritage of the world. As at May 2001, the agreement had been signed by 164 states. There are 690 sites on the current World Heritage List, including HEARD, MCDONALD and MACQUARIE ISLANDS, and NEW ZEALAND's SUBANTARCTIC ISLANDS.

The states that are party to the convention meet every two years. They elect a 21-member World Heritage Committee, which meets annually to make decisions about sites to be listed under the convention. A secretariat for the convention is provided by UNESCO. Three bodies advise the committee: the INTERNATIONAL UNION FOR THE CONSERVATION OF NATURE AND NATURAL RESOURCES, the International Council on Monuments and Sites, and the International Centre for the Study of the Preservation and Restoration of Cultural Property.

At one time, environmentalists who were lobbying for Antarctica to be declared a WORLD PARK considered the World Heritage Convention to be a potential mechanism for establishing better protection status for Antarctica and allowing more states to be involved in that protection. However, this approach may have been superseded by the MADRID PROTOCOL.

World Park The idea of establishing Antarctica as a world park was first mooted in 1972, and a formal proposal was submitted by NEW ZEALAND at the 1975 ANTARCTIC TREATY CONSULTATIVE MEETING. It was not well received, attracting conditional support only from CHILE. Conservationists then advocated extending the moratorium on MINING by arguing that Antarctica be made a 'world preserve'.

The INTERNATIONAL UNION FOR THE CONSERVATION OF NATURE AND NATURAL RESOURCES was critical of the proposal, questioning the acceptability of the ANTARCTIC TREATY SYSTEM as custodian, and the difficulty of ensuring that a lasting commitment could be made in the face of expected commercial pressures for EXPLOITATION of the continent's resources.

In the 1980s the ANTARCTIC AND SOUTHERN OCEAN COALITION campaigned for Antarctica to be declared a world park. GREENPEACE established the World Park Base in Antarctica in 1987 in support of a world park. It was removed in 1992–93, after the MADRID PROTOCOL, which banned mining for 50 years, had been agreed. Since then, the world park campaign has faded into the background, and the emphasis has been on campaigns against illegal FISHING and WHALING, and on attempts to ensure the implementation of the Madrid Protocol.

Worsley, Frank (1872–1943) Navigator and explorer, born in Akaroa, NEW ZEALAND. He went to sea at the age of 16 and served as a reserve officer in the British Navy. Worsley was appointed captain of the ENDURANCE on the IMPERIAL TRANS-ANTARCTIC EXPEDITION, led by Ernest SHACKLETON. After the *Endurance* became stuck in PACK ICE in the WEDDELL SEA and eventually sank, Worsley's brilliant navigational skills played an important part in the rescue of the expedition members. He steered flimsy lifeboats to ELEPHANT ISLAND, then on to SOUTH GEORGIA. This latter journey, across one of the worst seas in the world, has been described as one of the greatest feats in maritime history. Worsley, who was only able to take four navigational readings on the voyage, managed to successfully navigate across 1300 km (800 miles) of ocean.

At South Georgia, Worsley accompanied Shackleton and Thomas CREAN on foot across the uncharted, mountainous interior of the island to the whaling settlement of Stromness.

Worsley commanded two ships in World War I, and with Shackleton organized supplies for the forces on the north Russian front.

He sailed on the QUEST EXPEDITION, during which Ernest Shackleton died before reaching Antarctica. Afterwards, Worsley undertook several other expeditions, among them one to the Arctic.

Writers 'Everyone has an Antarctica,' says a character in Thomas Pynchon's 1963 novel *V*, and many writers have indeed attempted to explore the paradox that is Antarctica. Physically—and metaphorically—located at the end of the Earth, the continent weighs heavy on the human psyche. Polar geographer Hugh Mill wrote in 1909 that '… the desire to wipe out *terra incognita* appeals more deeply to certain instincts of human nature than [pursuing] either science or trade.'

The experience of Antarctica has led many returned explorers and adventurers to put pen to paper and publish their journals, memoirs, travel books and adventure stories. The idea of Antarctica has inspired poets and writers of fiction. Samuel Taylor Coleridge's 1798 long poem *The Rime of the Ancient Mariner* and Edgar Allan Poe's 1837 novella *The Narrative of A Gordon Pym* are among the best-known works and writers such as T S Eliot, Doris Lessing, Beryl

Bainbridge and Jules Verne have also explored the Antarctic theme in their poetry and fiction. Today, many national Antarctic organizations sponsor 'Writers in Antarctica' programmes.

y

Yachts The development of satellite NAVIGATION technology, advanced radio communications and weather forecasting techniques has allowed many more yachts to travel into Antarctic waters. They include private vessels, on expeditions or scientific missions, and tourist charters. To reach these southerly LATITUDES, sailors must first negotiate waters swept by the ROARING FORTIES and the FURIOUS FIFTIES: 'Beyond 40°S there is no law … Beyond 50°S there is no God,' goes an old mariner's saying. Despite the use of modern technology, the environment is still extremely unpredictable, much of the coastline remains uncharted, and 'new' rocks have an uncanny habit of appearing and damaging ships' hulls.

The first solo yachting voyage to Antarctica was made in 1972–74 by David LEWIS, who sailed from Hobart, Tasmania, to the ANTARCTIC PENINSULA, then via the SOUTH ORKNEYS to Cape Town.

Yalour Islands Spanning an area of about 2.5 km (1½ miles), these islands are located off the west coast of the ANTARCTIC PENINSULA, near the

ARGENTINE ISLANDS. They are known for LICHENS, MOSSES, and ANTARCTIC HAIRGRASS (*Deschampsia antarctica*). About 8000 pairs of ADÉLIE PENGUINS nest on the islands.

Yellow-billed pintail (*Anas georgica*) Only two waterbirds are found on SOUTH GEORGIA, both dabbling ducks of very similar appearance. Yellow-billed pintails are slightly larger and have longer necks than SPECKLED TEALS, although both are mottled brown with yellow-sided beaks. Yellow-billed pintails are shy birds, and are not usually found in groups. They feed on aquatic insects and ALGAE, and nest in grasses close to water.

z

Zavodovski Island An ice-free island in the SOUTH SANDWICH ISLANDS, Zavodovski has one of the world's largest penguin colonies with an estimated 1 million breeding pairs of CHINSTRAP PENGUINS.

Zoological research Most early expeditions to Antarctica carried out zoological research, simply describing and classifying animal groups. Many new species were dredged from the depths of the SOUTHERN OCEAN *en route* to the continent. On the GERMAN INTERNATIONAL POLAR YEAR EXPEDI-

Below: Small planktonic organisms are visible in an ice hole.

TION of 1882–83, for instance, Dr K von den Steinen made pioneering investigations into PENGUIN behaviour and population size. Edward WILSON, who accompanied Robert SCOTT on two expeditions, made many detailed studies of the lifecycle of Antarctic BIRDS.

As SEALS were often killed for DOG food on expeditions, zoologists took the opportunity to obtain measurements and compare the characteristics of different species. These studies were land-based, and it was not until the 1970s that zoologists began to consider the behaviour of seals in the water—the ways in which they swim, maintain their body temperature and breathe. At this time, scientists also studied the recovery rates of FUR SEAL populations after their near-extinction in the 19th century.

Most early zoologists were preoccupied with evolution. The idea that the development of an individual reflects the history of its species inspired Wilson to undertake an overland winter journey to Cape CROZIER with Aspley CHERRY-GARRARD and Henry BOWERS to collect EMPEROR PENGUIN embryos in 1911.

Unfortunately, this yielded little scientific information about the evolution of these ancient birds and the eggs the trio brought back were relegated to a storeroom in BRITAIN's Museum of Natural History.

Until the 1970s, studies of Antarctic birds were uncoordinated, and a variety of different methods were employed. In 1978 the SCIENTIFIC COMMITTEE ON ANTARCTIC RESEARCH began to coordinate banding programmes, and SOUTH AFRICA began an international data bank of observations. Aerial surveys also began to be used around this time to estimate bird numbers and monitor population changes. In 1990, WANDERING ALBATROSSES were fitted with transmitters and successfully tracked by SATELLITE. This provided data on their flying speeds and the distance they travel on foraging trips. Other technological advances have included artificial nests, which have built-in weighing equipment. A number of other devices, such as depth recorders and instruments that measure energy consumption, have been successfully fitted to Antarctic birds.

In 1900, on the BELGIAN ANTARCTIC EXPEDITION, Emile Racovitza made the first discovery of INVERTEBRATES on the continent. He described finding a small fly, MITES and SPRINGTAILS, the remarkable survival skills and habits of which remain the subject of ongoing research. Even more fascinating was the discovery, in the DRY VALLEYS, of tiny invertebrates, such as ROTIFERS and TARDIGRADES, which could survive exposure to extremely low temperatures and repeated freezing and thawing.

Zooplankton Tiny animals that drift in the ocean, browsing continuously among the PHYTOPLANKTON and, in turn, providing the basic diet for larger Antarctic organisms, such as WHALES, SEALS, FISH, SQUID and BIRDS.

KRILL and COPEPODS are the main animal species making up the zooplankton, but other CRUSTACEANS, salps, arrow worms, jellyfish and larval fishes are also present. Many of these are carnivorous. The lifecycle of many benthic, or bottom-living, fish includes a planktonic larval stage: a season is spent feeding in the surface waters before descending to the seabed.

Zooplankton distribution is closely linked to that of the phytoplankton on which they feed. The rich inshore waters support huge zooplankton stocks. The highest concentrations are found in the disturbed waters in the EAST WIND DRIFT zone off the northern and western ANTARCTIC PENINSULA and the SCOTIA ARC; birds and seals mass in these rich feeding areas, many of which are near their breeding grounds.

Photographic credits

Natural History New Zealand Ltd and David Bateman Ltd would like to thank all photographers and organisations for use of their photographs. Copyright to the photographs remains with the owners.

Jeanie Ackley/Natural History New Zealand p. 12 (top); p. 13; p. 14 (top left); p. 17; p. 18 (bottom); p. 24 (Treaty sign); p. 32; p. 43; p. 51 (bottom); p. 56 (top); p. 57; p. 60; p. 64; p. 65 (top); p. 69 (bottom); p. 70; p. 82 (top); p. 83 (petrified wood); p. 88 (top); p. 91; p. 95; p. 98 (top); p. 100; p. 103 (bottom); p. 104 (top left); p. 105 (bottom); p. 111; p. 113; p. 116; p. 135 (top); p. 139; p. 140 (bottom); p. 142 (bottom left and right); p. 157 (bottom); p. 165 (bottom); p. 175; p. 176; p. 183; p. 186; p. 191; p. 202
Alexander Turnbull Library, National Library of New Zealand Te Puna Mātauranga o Aotearoa p. 15 F- 8677-1/4; p. 18 Schmidt Collection/G- 1206-1/1; p. 27 Scott Album/F- 11354-1/2; p. 40 Herbert George Ponting/F-11351-1/2; p. 41 Herbert George Ponting, Scott Album/F – 2931-1/4; p. 56 Painted by J Webber., F Bartolozzi R. A. sculp/A-218-009; p. 61 Herbert George Ponting, Scott Album/F- 11372-1/2; p. 63 F- 22343-1/2; p. 71 F- 112426-1/2; p. 74 F- 20142-1/2; p. 77 (top) Scott Album/ F-2932-1/4; p. 77 (bottom) F- 8699-1/4; p. 80 (top) Gaumont Co. Ltd/Eph-E-ANTARCTICA-1913-01; p. 84 F- 8077-1/2; p. 97 John Pascoe Collection/F- 20196-1/2; p. 106 F- 8763-1/4; p. 129 F- 32270-1/2; p. 132 C- 25699-1/2; p. 134 Scott Album, F-11224-1/2; p. 147 Scott Album, F- 30595-1/2; p. 158 J.H. Kinnear Collection/G- 16452-1/2; p. 164 Norman Judd Collection/G- 100388-1/2; p. 164 F- 8703-1/4; p. 165 F- 20146-1/2; p. 177 Morrison Collection/F- 8261-1/2; p. 185 F- 96875-1/2; p. 196 PUBL-0182-132; p. 198 F- 26950-1/2; p. 200 PUBL-0182-214
Don Anderson/Natural History New Zealand p. 52
Antarctica New Zealand Pictorial Collection p. 29; p. 58 (top); p. 72; p. 84 (top); p. 112; p. 114; p. 118 (top); p. 126; p. 156; p. 157 (top); p. 172 (top); p. 174 (bottom); p. 194
Archives New Zealand/Te Whare Tohu Tuhituhinga O Aotearoa, p. 86 National Publicity Studios Photographic Collection, Alexander Turnbull Library 920, A42432)]
William Bolten/Antarctica New Zealand Pictorial Collection p. 45
John Bradshaw/Antarctica New Zealand Pictorial Collection p. 146 (top); p. 155
British Antarctic Survey p. 88 (bottom)
Elizabeth Bulleid/Antarctica New Zealand Pictorial Collection p. 182
Canterbury Museum p. 31 J.J. Kinsey Collection 266/1; p. 76 2569; p. 110 13134; p.121; p. 129 Quartermain Collection 13906; p. 131 J.J. Kinsey Collection 127/1; p. 153 National Maritime Museum 8826; p. 174 (top) 426b
George Chance Back cover (albatross); p. 16 (bottom); p. 177 (top)
Lloyd Davis p. 152
Paul Donovan/Natural History New Zealand pp. 2-3; p. 46; p. 50; p. 69 (top); p. 79 (top); p. 86; p. 96; p. 101 (bottom); p. 162 (bottom); p. 180; p. 181 (top); p. 184
David Geddes/Antarctica New Zealand Pictorial Collection p. 34; p. 39; p. 49 (top); p. 62 (bottom); p. 79 (bottom); p. 94; p. 137 (bottom); p. 140; p. 166; p. 171; p. 194 (top)
Cath Gilmour/Antarctica New Zealand Pictorial Collection p. 203
Global Publishing p. 20
Greenpeace p. 149
Peter Harper/University of Canterbury p. 36; p. 49 (bottom); p. 118 (bottom); p. 148; p. 170
Tim Higham/Antarctica New Zealand Pictorial Collection p. 16 (top); p. 25 (bottom); p. 26; p. 30; p. 154
Nicola Holmes/Antarctica New Zealand Pictorial Collection p. 82 (bottom); p. 199
Ed Jowett/Natural History New Zealand p. 79 (middle); p. 80 (bottom); p. 142 (top); p. 160; p. 161 (top)

Katsu Kaminuma/Antarctica New Zealand Pictorial Collection p. 62 (top); p. 110 (bottom)
Hugh Logan/Antarctica New Zealand Pictorial Collection p. 14 (top right)
William Logie/Antarctica New Zealand Pictorial Collection p. 137 (top)
Richard McBride/Antarctica New Zealand Pictorial Collection p. 42; p. 141; p. 162 (top)
John Macdonald/Antarctica New Zealand Pictorial Collection p. 106
John Macdonald/University of Auckland p. 19; p. 90
Steve Mercer/Natural History New Zealand p. 35; p. 179
Colin Monteath/Antarctica New Zealand Pictorial Collection p. 44; p. 108; p. 144; p. 169
National Aeronautics and Space Administration (NASA) p.138
The Natural History Museum, London p. 122
Natural History New Zealand p. 68
Norsk Polarinstitutt p. 38
John D. Palmer/Antarctica New Zealand Pictorial Collection p. 195
Eric Phillips p. 75; p. 92 (bottom); p. 150; p. 168
Herbert Ponting, courtesy Antarctica New Zealand Pictorial Collection p. 58; p. 65 (bottom)
Max Quinn/Natural History New Zealand p. 1; p. 59; p. 73; p. 74; p. 102; p. 104 (bottom); p. 133; p. 151; p. 161 (bottom); p. 187; p. 193
Chris Riley/Natural History New Zealand p. 99; p. 197 (bottom)
Nigel Roberts/Antarctica New Zealand Pictorial Collection p. 92 (top); p. 117 (top)
N. L. Round-Turner/Antarctica New Zealand Pictorial Collection p. 172 (bottom)
Royal New Zealand Navy/Alexander Turnbull Library, National Library of New Zealand Te Puna Mātauranga o Aotearoa p. 53 New Zealand Free Lance Collection/C- 22748-1/2
Chris Rudge/Antarctica New Zealand Pictorial Collection p. 66; p. 101 (top); p. 124; p. 127; p. 136; p. 189
Lou Sanson/Antarctica New Zealand Pictorial Collection p. 38 (top); p. 119 (top)
Scott Polar Research Institute p. 48 Ponting 305
Rod Seppelt/Antarctica New Zealand Pictorial Collection p. 117 (bottom)
Brent Sinclair/University of Otago p. 179 (bottom)
Mike Single/Natural History New Zealand p. 14 (bottom); p. 51 (top); p. 78; p. 85; p. 87; p. 97; p. 103; p. 105 (top); p. 128; p. 143 (top); p. 145; p. 159; p. 188; p. 192
Dot Smith/Antarctica New Zealand Pictorial Collection p. 173
Tim Stern/Antarctica New Zealand Pictorial Collection p. 37
Rowan Strickland/Natural History New Zealand p. 109
United States National Archives p. 24
Garth Varcoe/Antarctica New Zealand Pictorial Collection p. 81
Mike Watson/Ardea p. 23
Kim Westerskov p. 5; p. 55 CCLAMR; p. 180 (top)
Kim Westerskov/Antarctica New Zealand Pictorial Collection p. 21; p. 47 (top); p. 114; p. 115; p. 119 (bottom); p. 125; p. 167; p. 181 (top); p. 197 (top)
David Wharton/University of Otago p. 107; p. 123; p. 130
Wheeler & Son, courtesy Canterbury Museum p. 164 (left)
Gillian Wratt/Antarctica New Zealand Pictorial Collection p. 12 (bottom); p. 28; p. 47 (bottom); p. 58; p. 98 (bottom); p. 135 (bottom); p. 143 (bottom); p. 146 (bottom)

Selected bibliography

Caroline Alexander, *The Endurance: Shackleton's Legendary Antarctic Expedition,* HarperCollins, New York, 1997.

Roald Amundsen (trans. A. G. Chater), *The South Pole,* vols. 1 & 2, John Murray, London, 1912; reprinted 1976 by C. Hurst & Co, London.

J C Beaglehole, *The Life of Captain James Cook,* A & C Black, London, 1974.

John Bechervaise, *Antarctica, The Last Horizon,* Cassell, Sydney, 1979.

Philip J Beck, *The International Politics of Antarctica,* Croom Helm, London, 1986.

Lothar Beckel, ed., *The Atlas of Global Change,* Macmillan, New York, 1988.

Kenneth J Bertrand, *Americans in Antarctica, 1775–1948,* American Geological Society, Burlington, 1971.

Creina Bond and Roy Siegfried, *Antarctica: No Single Country, No Single Sea.* C. Struik, Cape Town, 1979; reprinted in 1988 by New Holland, Cape Town.

W N Bonner and D W H Walton, eds., *Key Environments: Antarctica,* Pergamon Press, Oxford, 1985.

Carsten Borchgrevink, *First on the Antarctic Continent: Being an Account of the British Antarctic Expedition,* Newnes, London, 1901.

Peter Brent, *Captain Scott and the Antarctic Tragedy,* Weidenfeld & Nicolson, London, 1974.

Barney Brewster/Friends of the Earth, *Antarctica: Wilderness At Risk,* Reed, Wellington, 1982.

British Antarctic Survey, *Antarctica—A Continent for Science,* British Antarctic Survey, Cambridge, 1987.

British Antarctic Survey, *British Scientific Research in Antarctica,* British Antarctic Survey, Cambridge, 1991.

Richard Byrd, *Alone,* G. P. Putnam's, New York, 1938.

Apsley Cherry-Garrard, *The Worst Journey in the World,* Constable, London, 1922; reprinted many times.

L J Conrad, *Bibliography of Antarctic Exploration: Expedition Accounts from 1768 to 1960,* privately published, 1999.

Frederick A Cook, *Through the First Antarctic Night 1898–1899,* William Heinemann, London, 1900.

Louise Crossley, *Explore Antarctica,* Cambridge University Press, Melbourne, 1995.

Frank Debenham, *Antarctica, the Story of a Continent,* Herbert Jenkins, London, 1957.

Department of Conservation, *New Zealand's Sub-Antarctic Islands,* Reed, Auckland, 1999.

Jules-Sébastien-César Dumont d'Urville (trans. Helen Rosenman), *Two Voyages to the South Seas,* vol. 2, Melbourne University Press, Melbourne, 1988.

Lorraine Elliott, *Protecting the Antarctic Environment: Australia and the Minerals Convention,* Australian National University, Canberra, 1993.

Lorraine Elliott, *International Environmental Politics: Protecting the Antarctic,* St. Martin's Press, New York, 1994.

G E Fogg, *A History of Antarctic Science,* Cambridge University Press, Cambridge, 1992.

G E Fogg and David Smith, *The Explorations of Antarctica,* Cassell, London, 1990.

Alastair Fothergill, *Life in the Freezer,* BBC Books, London, 1993.

Robert Fox, *Antarctica and the South Atlantic: Discovery, Development and Dispute,* BBC Books, London, 1985.

Vivian Fuchs and Edmund Hillary, *The Crossing of Antarctica,* Cassell, London, 1958.

Alan Gurney, *Below the Convergence: Voyages Toward Antarctica 1699–1839,* Norton, New York, 1997.

Alan Gurney, *The Race to the White Continent,* Norton, New York, 2000.

James D Hansom and John E. Gordon, *Antarctic Environments and Resources: A Geographical Perspective,* Addison Wesley Longman, New York, 1988.

C Harris and B Stonehouse, eds., *Antarctica and Global Climate Change,* Belhaven Press, London, 1991.

David L Harrowfield, *Icy Heritage: Historic Sites of the Ross Sea Region,* Antarctic Heritage Trust, Christchurch, 1995.

Trevor Hatherton, ed., *Antarctica: the Ross Sea Region,* DSIR Publishing, Wellington, 1990.

Robert Headland, *The Island of South Georgia,* Cambridge University Press, Cambridge, 1984.

Robert Headland, *Chronological List of Antarctic Expeditions and Related Historical Events,* Cambridge University Press, Cambridge, 1989.

Roland Huntford, *Scott and Amundsen,* Hodder & Stoughton, London, 1979.

Roland Huntford, *Shackleton,* Hodder & Stoughton, London, 1985.

Frank Hurley, *Shackleton's Argonauts: the Epic Tale of Shackleton's Voyage to Antarctica in 1915,* Angus and Robertson, Sydney, 1948; reprinted 1979 by William Collins, Auckland.

Bertrand Imbert, *North Pole, South Pole: Journeys to the Ends of the Earth,* Thames & Hudson, London, 1987.

Fred Jacka and Eleanor Jacka, eds., *Mawson's Antarctic Diaries,* Unwin Hyman, London, 1988.

Christopher C Joyner, *Governing the Frozen Commons: The Antarctic Regime and Environmental Protection,* University of South Carolina Press, Columbia, 1998.

George A Knox, *The Biology of the Southern Ocean,* Cambridge University Press, Cambridge, 1994.

Phillip Law, *Antarctic Odyssey,* William Heinemann, Melbourne, 1983.

David Lewis, *Ice Bird,* Norton, New York, 1976.

J F Lovering and J R V Prescott, *Last of Lands … Antarctica,* Melbourne University Press, Melbourne, 1979.

Stephen Martin, *A History of Antarctica,* State Library of New South Wales Press, Sydney, 1996.

Douglas Mawson, *The Home of the Blizzard,* William Heinemann, London, 1915; reprinted 1996 by Wakefield Press, Kent Town.

Paquita Mawson, *Mawson of the Antarctic,* Longmans, London, 1964.

John Maxtone-Graham, *Safe Return Doubtful: the Heroic Age of Polar Exploration,* Scribner, New York, 1988.

John May, *Greenpeace Book of Antarctica,* Dorling Kindersley, London, 1988.

Roger Mear and Robert Swan, *In the Footsteps of Scott,* Jonathan Cape, London, 1987.

Philip L Mitterling, *America in the Antarctic to 1840,* University of Illinois Press, Urbana, 1959.

Colin Monteath, *Antarctica: Beyond the Southern Ocean,* David Bateman, Auckland, 1996.

Sanford Moss, *Natural History of the Antarctic Peninsula,* Columbia University Press, New York, 1988.

David Mountfield, *A History of Polar Exploration,* Hamlyn, London, 1974.

Baden Norris, *Antarctic Reflections,* New Zealand Antarctic Society, Christchurch, undated.

W S B Patterson, *The Physics of Glaciers,* 2nd ed., Pergamon Press, Oxford, 1981.

L B Quartermain, *Historic Huts,* DSIR, Wellington, 1965.

L B Quartermain, *South to the Pole: The Early History of the Ross Sea Sector, Antarctica,* Oxford University Press, Wellington, 1967.

L B Quartermain, *New Zealand and the Antarctic,* Government Printer, Wellington, 1971.

Philip W Quigg, *A Pole Apart: The Emerging Issue of Antarctica,* McGraw Hill, New York, 1983.

Beau Riffenburgh, *The Myth of the Explorer,* Oxford University Press, Oxford, 1994.

Finn Ronne, *Antarctica, My Destiny,* Hastings House, New York, 1979.

Jeff Rubin, *Antarctica,* Lonely Planet, Hawthorn, 1996.

Emilie J Sahurie, *The International Law of Antarctica: New Haven Studies in International Law and World Public Order,* vol. 6, Kluwer Law International, 1992.

Ann Savours, ed., *Scott's Last Voyage: through the Antarctic camera of Herbert Ponting,* Sidgwick & Jackson, London, 1974.

Robert Scott, *The Voyage of the Discovery,* Smith, Elder, London, 1905.

Robert Scott, *Scott's Last Expedition,* Smith, Elder, London, 1913.

Ernest Shackleton, *The Heart of the Antarctic: the Farthest South Expedition 1907–1909,* William Heinemann, London, 1909; reprinted 2000 by Signet, New York.

Ernest Shackleton, *South,* William Heinemann, London, 1919; reprinted 1989 by Century, London.

Deborah Shapley, *The Seventh Continent: Antarctica in a Resource Age,* Resources for the Future, Washington DC, 1985.

Paul Simpson-Housley, *Antarctica—Exploration, Perception and Metaphor,* Routledge, London, 1992.

Tony Soper, *Antarctica—A Guide to the Wildlife,* Globe Pequot Press, Old Saybrook CT, 1994.

John Stewart, *Antarctica: An Encyclopedia,* vols. 1 & 2, McFarland and Co., Jefferson, 1990.

Bernard Stonehouse, *Animals of the Antarctic: The Ecology of the Far South,* Holt, Rinehardt & Winston, New York, 1972.

Bernard Stonehouse, *The Last Continent: Discovering Antarctica,* Shuttlewood Collinson, Burgh Castle, 2000.

Ian Strange, *Field Guide to the Wildlife of the Falklands Islands and South Georgia,* HarperCollins, New York, 1992.

W Sullivan, *Assault on the Unknown: The International Geophysical Year,* McGraw Hill, New York, 1961.

Keith Suter, *Antarctica: Private Property or Public Heritage,* Pluto Press Australia, Leichardt, 1991.

G Tetley, *Antarctica 2001—a Notebook.* Proceedings of the Antarctic Futures Workshop, 28–30 April 1998, Antarctica New Zealand, Christchurch, 2001.

David Thomson, *Scott's Men,* Allen Lane, London, 1977.

John Thomson, *Shackleton's Captain: a Biography of Frank Worsley,* Hazard Press, Christchurch, 1998.

Various authors, *Antarctica: The Extraordinary History of Man's Conquest of the Frozen Continent,* 2nd edition, Reader's Digest, Sydney, 1998.

W F Vincent, ed., *Environmental Management of a Cold Desert Ecosystem: the McMurdo Dry Valleys.* Report of the National Science Foundation, Washington DC, 1995.

D W H Walton, ed., *Antarctic Science,* Cambridge University Press, Cambridge, 1991.

Sara Wheeler, *Antarctica, The Falklands and South Georgia,* Cadogan Books, London, 1997.

F A Worsley, *Endurance: An Epic of Polar Adventure,* Philip Allan & Co, London, 1931.

Websites

Alfred Wegener Institute, www.awi-bremerhaven.de

Amundsen-Scott South Pole Station (virtual tour), astro.uchicago.edu/cara/vtour/pole/

Antarctic Sun (online newspaper), www.70south.com/news/976459732/index_html

Antarctica New Zealand, www.antarcticanz.govt.nz

Antarctica Online (USA), www.antarcticaonline.com

The Antarctican (online news service), www.antarctican.com

Argentine Antarctic Institute, www.dna.gov.ar

Australian Antarctic Division, www.antdiv.gov.au

Belgian Scientific Research Programme on the Antarctic, www.belspo.be/antar

British Antarctic Survey, www.antarctica.ac.uk

CCAMLR, www.ccamlr.org

Chilean Antarctic Institute, www.inach.cl

CIA's Antarctica page, www.odci.gov/cia/publications/factbook/

Council of Managers of National Antarctic Programs, www.comnap.aq

French Polar Institute, www.ifremer.fr/ifrtp/wwwenglish/index.html

Friends of the Earth, www.foei.org

Gateway Antarctica, www.anta.canterbury.ac.nz

Greenpeace, www.greenpeace.org

Heritage Antarctica, www.heritage-antarctica.org

The Ice (USA), www.theice.org

International Whaling Commission, www.iwcoffice.org

Italian Antarctic Program, www.pnra.it/index_inglese.html

McMurdo Station (virtual tour), astro.uchicago.edu/cara/vtour/mcmurdo/

NASA, Antarctic history, http://quest.arc.nasa.gov/antarctica

National Institute of Polar Research (Japan), www.nipr.ac.jp

National Science Foundation (USA), www.nsf.gov

New South Polar Times (online newspaper), 205.174.118.254/nspt/home.htm

Norwegian Polar Institute, www.npolar.no

Philately, www.south-pole.com

Russian Arctic and Antarctic Research Institute, www.aari.nw.ru

SCAR, www.scar.org

South Pole Research Institute (Britain), www.spri.cam.ac.uk

Swedish Polar Research Secretariat, www.polar.kva.se

United Nations, www.un.org

Index